Reforming Water Law and Governance

Cameron Holley · Darren Sinclair
Editors

Reforming Water Law and Governance

From Stagnation to Innovation in Australia

Springer

Editors
Cameron Holley
UNSW Sydney
Sydney, NSW
Australia

Darren Sinclair
University of Canberra
Bruce, Canberra, ACT
Australia

ISBN 978-981-10-8976-3   ISBN 978-981-10-8977-0 (eBook)
https://doi.org/10.1007/978-981-10-8977-0

Library of Congress Control Number: 2018937342

© Springer Nature Singapore Pte Ltd. 2018

This work is subject to copyright. All rights are reserved by the Publisher, whether the whole or part of the material is concerned, specifically the rights of translation, reprinting, reuse of illustrations, recitation, broadcasting, reproduction on microfilms or in any other physical way, and transmission or information storage and retrieval, electronic adaptation, computer software, or by similar or dissimilar methodology now known or hereafter developed.

The use of general descriptive names, registered names, trademarks, service marks, etc. in this publication does not imply, even in the absence of a specific statement, that such names are exempt from the relevant protective laws and regulations and therefore free for general use.

The publisher, the authors and the editors are safe to assume that the advice and information in this book are believed to be true and accurate at the date of publication. Neither the publisher nor the authors or the editors give a warranty, express or implied, with respect to the material contained herein or for any errors or omissions that may have been made. The publisher remains neutral with regard to jurisdictional claims in published maps and institutional affiliations.

Printed on acid-free paper

This Springer imprint is published by the registered company Springer Nature Singapore Pte Ltd. part of Springer Nature
The registered company address is: 152 Beach Road, #21-01/04 Gateway East, Singapore 189721, Singapore

Preface

On 21 September 2016, the United Nations High Level Panel on Water made a joint statement calling for action to deliver a 'water secure world'. The urgency of this call is underpinned by the centrality of water to our very existence, which is being threatened in an unprecedented fashion by increasing water disasters and scarcity arising from economic and population growth, as well as the compounding negative impacts of climate change. Fittingly, water security assumes a prominent position in the United Nations' Sustainable Development Goals. While it clearly has global and international dimensions, achieving water security is first and foremost a challenge for national and regional governments. It is incumbent, then, on advocates of evidenced-based public policy to put their 'shoulder to the wheel' in seeking, articulating and providing better approaches to water law and governance.

The aspiration of this book is, in our own small way, a response to this call for action. In particular, it aims to complement and expand existing research and policy thinking on water law and governance by distilling contemporary and practical legal and governance insights from Australia's now mature water reform journey, by providing comparisons with the experiences of other nations, particularly the United States of America, and by projecting how best to steer water governance onto a more sustainable path.

The book's focus on Australia arose from recent suggestions that Australia's water governance reforms have stagnated. Indeed, around the same time as the High Level Panel on Water was forming, Australia, arguably a global leader in taking innovative and comprehensive action on water governance, was becoming increasingly equivocal and apathetic in responding to existing and future water security challenges. Australia's 20-year law and governance journey of developing market systems, collaborative planning and enforcement had achieved much (and received much praise) in responding to major droughts. However, by 2014 there was a palpable mood that water law and governance was no longer a political or policy priority despite many issues remaining unresolved. The risks of reform complacency are substantial. As Karlene Maywald, former chair of Australia's National Water Commission, warned on 20 October 2014: 'Now is not the time to walk away from water reform and leadership … Instead, we urge governments to lock in their hard-won gains … This is the only way to ensure our communities—and Australia's next generations—against future water security risks'.

It was in response to this context that we, through the Faculty of Law and the Connected Waters Initiative Research Centre at UNSW Sydney, organised a workshop on 7–8 December 2015 on the broad theme of Rethinking Water Governance. Many of the chapters of this book emerged from this workshop, where detailed and in-depth reviews and discussions were held amongst participants as they looked into the past and projected forward to highlight key successes and limits of Australia's hybrid water governance system. The new lines of vision and collective voices this workshop brought to the analysis of Australian water law and governance led us to invite participants to revise their papers for submission in a series of collective outputs, including as chapters in this book. A number of other contributions were also received from those who had otherwise been unable to attend the workshop.

Our work has benefited significantly from two Australian Research Council Discovery Early Career Researcher Awards (DE140101216 and DE170101536) and an Australian Research Council Linkage Grant (LP130100967). Some material within the book was first developed from and appeared in Emma Carmody, Barbara Cosens, Alex Gardner, Lee Godden, Janice Gray, Cameron Holley, Louise Lee, Bruce Lindsay, Liz Macpherson, Rebecca Nelson, Erin O'Donnell, Lily O'Neill, Kate Owens, Darren Sinclair (2016). 'The future of water reform in Australia—starting a conversation' *Australian Environment Review,* 31(4), 132–137 and contributions to a Special Issue ('Rethinking Water Law and Governance', Volume 33, Part 4, 2016, p. 275) of the *Environmental and Planning Law Journal* edited by Cameron Holley. The special issue was first published by Thomson Reuters in the Environmental and Planning Law Journal and should be cited as Cameron Holley (ed), Special Issue, Rethinking Water Law and Governance (2016) 33(4) *EPLJ* 275. For all subscription inquiries please phone, from Australia: 1300 304 195, from Overseas: +61 2 8587 7980 or online at legal.thomsonreuters.com.au/search. The official PDF version of the special issue and its contributions can also be purchased separately from Thomson Reuters at http://sites.thomsonreuters.com.au/journals/subscribe-or-purchase.

This book was only made possible by the dedication and hard work of the authors, and we are extremely grateful to all of them. We also thank our colleagues Wendy Timms, Fleur Johns, Andy Baker, Martin Anderson, Bryce Kelly, Bronwen Morgan, Rosemary Rayfuse, Alex Gardner, Paul Martin, Neil Gunningham, James Patterson, Lauren Butterly, Gabriela Cuadrado Quesada, and two anonymous reviewers who reviewed and commented on the proposal, the work contained within this book, or who extended in other ways our understanding of water, and its law and governance issues. Many thanks must also go to Springer for their assistance, understanding and tireless efforts, and to Amelia Brown, Trent Wilson, Genevieve Wilks, Bonnie Perris, Antonia Ross and Ganur Maynard for the unstinting research assistance and editorial work.

Sydney, Australia Cameron Holley
Bruce, Canberra, Australia Darren Sinclair
November 2017

Contents

Replenishing Australia's Water Future: From Stagnation to Innovation 1
Cameron Holley and Darren Sinclair

Part I The Murray-Darling Basin—Progress and Challenges in Multi-jurisdictional Water Governance

The Unwinding of Water Reform in the Murray-Darling Basin: A Cautionary Tale for Transboundary River Systems 35
Emma Carmody

Multi-jurisdictional Water Governance in Australia: Muddle or Model? 57
Bradley C. Karkkainen

Environmental Water Transactions and Innovation in Australia 79
Katherine Owens

Part II Water Markets—Property, Regulation and Implementation

Water Entitlements as Property: A Work in Progress or Watertight Now? 101
Janice Gray and Louise Lee

Regulatory and Economic Instruments: A Useful Partnership to Achieve Collective Objectives? 123
Adam Loch, C Dionisio Perez-Blanco, Dolores Rey, Erin O'Donnell and David Adamson

Water Markets and Regulation: Implementation, Successes and Limitations 141
Cameron Holley and Darren Sinclair

vii

Part III Collaboration and Participation—Litigation, Coordination and Water Rights

Public Participation in Water Resources Management in Australia: Procedure and Possibilities 171
Bruce Lindsay

A Governance Solution to Australian Freshwater Law and Policy 193
Jennifer McKay

Lessons from Australian Water Reforms: Indigenous and Environmental Values in Market-Based Water Regulation 213
Elizabeth Macpherson, Erin O'Donnell, Lee Godden and Lily O'Neill

Part IV Future Governance Challenges—Cumulative Impacts, Resource Industries and Climate Change

Regulating Cumulative Impacts in Groundwater Systems: Global Lessons from the Australian Experience 237
Rebecca Louise Nelson

Compromising Confidence? Water, Coal Seam Gas and Mining Governance Reform in Queensland and Wyoming 257
Poh-Ling Tan and Jacqui Robertson

Governing the Freshwater Commons: Lessons from Application of the Trilogy of Governance Tools in Australia and the Western United States 281
Barbara Cosens

Replenishing Australia's Water Future: From Stagnation to Innovation

Cameron Holley and Darren Sinclair

Abstract This chapter charts Australia's leading-edge water law and governance reforms. It discusses progress on implementation and the challenges this has posed. Connections are drawn between Australia's experience and the water law and governance literature. After outlining the book's chapters, four fundamental questions are analysed and answered, namely how successful is Australia's approach to designing and implementing water governance? What conditions have enabled or blocked its success, including environment, social, political and legal? How does Australia's water governance system compare and contrast with different international water governance practices? And what are the broader insights for future water governance practice and theory?

We are grateful for the research assistance and editing of Amelia Brown and Trent Wilson. Parts of this chapter first appeared in Carmody et al 2016; Holley and Sinclair 2016 and Cameron Holley and Darren Sinclair, 'Governing Water Markets; Achievements, Limitations and the need for regulatory reform' (2016) 33(4) *Environmental and Planning Law Journal* 301–324. The latter two articles were first published by Thomson Reuters in the Environmental and Planning Law Journal and should be cited as Cameron Holley and Darren Sinclair, 'Rethinking Australian water governance: successes, challenges and future directions' (2016) 33(4) *EPLJ* 275–283; Cameron Holley and Darren Sinclair, 'Governing Water Markets; Achievements, Limitations and the need for regulatory reform' (2016) 33(4) *EPLJ* 301–324. For all subscription inquiries please phone, from Australia: 1300 304 195, from Overseas: +61 2 8587 7980 or online at legal.thomsonreuters.com.au/search. The official PDF version of this article can also be purchased separately from Thomson Reuters at http://sites.thomsonreuters.com.au/journals/subscribe-or-purchase.

C. Holley (✉)
UNSW Law School, Connected Waters Initiative Research Centre (UNSW Sydney), the National Centre for Groundwater Research and Training, and the Global Risk Governance Programme (University of Cape Town), Sydney, Australia
e-mail: c.holley@unsw.edu.au

D. Sinclair
University of Canberra, Visiting Fellow at Connected Waters Initiative Research Centre (UNSW Sydney) and Member of the National Centre for Groundwater Research and Training, Bruce, ACT, Australia
e-mail: darren.sinclair@canberra.edu.au

© Springer Nature Singapore Pte Ltd. 2018
C. Holley and D. Sinclair (eds.), *Reforming Water Law and Governance*,
https://doi.org/10.1007/978-981-10-8977-0_1

1 Introduction

In the age of the Anthropocene, stresses on water are local, transboundary and global (High Level Panel on Water 2016, p. 3; Crutzen and Stoermer 2000; Crutzen 2002). Within and across these multiple scales, the impacts of climate change, population growth, and accompanying energy and food demands, are all placing increasing pressure on water systems and supplies (Pahl-Wostl et al. 2010; WWAP 2015). Agriculture alone consumes the large majority of water, and almost one-fifth of the world's population already live in areas with physical water scarcities (FAO 2007). Projections suggest that if the world continues on its current trajectory, we may face a 40% shortfall in water availability by 2030, threatening our environment, our security and our economy (High Level Panel on Water 2016, p. 3).

It is not surprising then that since 2012, water crises have been consistently listed as a top global risk (WEF 2017). As the United Nations Sustainable Development Goals make clear, we must urgently change how we manage the availability and use of water if we, and our fellow species, are to survive (see e.g. Sustainable Development Goal 6; High Level Panel on Water 2016, p. 3).

The traditional response to our shared water challenge has been to pursue technological and infrastructure fixes (Jury and Vaux 2005). In contrast, law, governance and regulation, the subjects of this book, have been far from front and centre (Kidd et al. 2014; Pahl Wostl 2015; Gray et al. 2016). According to the UN High Level Panel on Water (2016, p. 6), water is 'often left out of budgeting, legislation, and human resources mobilization decisions and discussions'. This lack of legal and governance leadership has marginalised water law and governance systems, and arguably left them poorly equipped to cope with increasing pressure on water resources (Gupta and Pahl-Wostl 2013). As the OECD (2011, p. 17) succinctly puts it, today's '"water crisis" is largely a governance crisis'.

Over the latter part of last century, and early this century, water laws, policies and evolving governance apparatuses have been gradually tested, explored and refined at international, regional and national levels. Internationally, there has been increasing attention on global responses, through multi-lateral agreements and frameworks, as well as trans-boundary water sources involving integrated water resource management (Biswas 2004; Salman 2007; Tarlock 2010; Gupta and Pahl-Wostl 2013; Moynihan 2014). However, state sovereignty remains a governance hurdle for numerous water resources, and there have been few sustainable successes in the complexity of managing transboundary water problems (Tujchneider et al. 2013; Hearns et al. 2014; Gray et al. 2016). At the regional level, the European Union has developed experimental governance and basin-focused mechanisms that encourage cooperation between nation states and civil society under its Water Framework Directive (van Rijswick 2016). Nevertheless, widespread and demonstrable achievements have been slow to arrive, with a number of studies highlighting the challenge of engaging non-government stakeholders and moving beyond planning to implementation (Pahl-Wostl et al. 2010; Papas 2014; Kochskämper et al. 2018). Similarly, within nation states, many national and sub-national systems have evolved to confront daunting water governance challenges. Along with recent innovations, such as California's Sustainable Groundwater Management Act (Kiparsky et al. 2017), countries have often subscribed to sustainable development, integrated water resource management, water rights and markets (see also the general principles in, for example, the Helsinki Rules 1966; Berlin Rules 2004; Convention on the Law of Non-navigational Uses of International Watercourses 1997; ILC Articles on Transboundary Aquifers 2008; and other related approaches under the Sustainable Development Goals). Again, many of these approaches have struggled with poor implementation, a continuing emphasis on facilitating expert-led technology driven solutions, often to the exclusion of stakeholders and indigenous peoples, and barriers to integrating energy and other policies (Biswas and Tortajada 2010; Mukhtarov and Gerlak 2014; Araral and Wang 2013; Cosens et al. 2014; Pittock et al. 2015). Many of the above problems have been recognised and what has emerged is something of an 'experimental' process, where criticism and revision is a feature of evolving water governance developments (De Burca et al. 2014; Holley 2016). In effect, this has resulted in 'living laboratories' of water law and governance (Pahl-Wostl 2015, p. 267).

One such laboratory, held up as a leader in water governance reform, has been Australia (Godden and Foerster 2011). Historically, water governance in Australia has been a responsibility of State governments, with rights closely linked with individual land ownership. By the end of the 1980s, it was clear a radically different approach was needed. A nationally agreed water framework (COAG 1994; Cwth of Aus. 2004) was developed, spurred in part by the growing recognition of 'sustainability' and persistent droughts (World Commission on Environment and Development 1992; NSESD 1992), the popularity of market based instruments (Hayek 1945; WMO 1992) and a view that joined-up Commonwealth and State government programs were the most efficient and effective way to govern Australia's environment (IGAE 1992; Holley and Sinclair 2016, p. 276).

Australia's national water framework commenced in 1994, with the support of the Commonwealth and State governments, under the guidance of the intergovernmental forum known as the Council of Australian Governments (COAG). Subsequently, Australia's water reforms have matured into a hybrid governance system involving three key governance pillars: top-down regulation (hierarchy), catchment water planning involving stakeholder consultation (place based collaboration) and water trading (involving the separation of land and water rights and establishing a cap and trade scheme for delivering public outcomes) (Hussey and Dovers 2007).

Other nations, like the United States, and many in Europe, have followed these developments with interest as they confront their own present and future water crises (FAWLMI 2010; Jasper 2015; Australian Water Partnership 2017). Similarly, academics have analysed, interrogated and lauded Australian water law and governance (Godden and Foerster 2011), leading to a rich vein of scholarship examining this system and distilling its lessons using various 'theoretical lenses', including multilevel governance (Evans and Dare 2013), federalism (Kildea and Williams 2011), adaptive management (McDonald and Styles 2014), water management (Hussey and Dovers 2007; Baldwin and Hamstead 2015) and most prominently, economics (Grafton et al. 2016; Holley and Sinclair 2016, p. 276).

While this work has documented a range of early successes and challenges in the initial design and roll out of Australia's reforms, its governance system has evolved, matured and in some cases even begun to stagnate (NWC 2014a). Two distinct and recent weaknesses have emerged in Australia's national water reforms. First, there has been a lack of critical attention on the utility of law and governance in Australia's water policy, particularly a lack of multi-disciplinary integration of the legal dimension, and an absence of legal, policy and regulatory insights into addressing the water governance crisis (for other contexts, see Gupta and Pahl-Wostl 2013). Arguably, this is most acute in terms of ground-level policy implementation (Moore et al. 2014). Secondly, as foreshadowed above, water reform in Australia has fallen rapidly down the list of national priorities and progress has stalled. Despite once being heralded as a world leader in water reform, as of 2017, assessments of Australia's leadership are now far more cautious, if not downright critical. As leading Australian scientist John Williams (2017) puts it: 'There has been a policy silence on water reform from Federal and state governments. Absolutely nothing has happened to take matters forward. In fact, there is mounting evidence of not just policy stagnation but rather policy retreat'. But the threats to sustainable water management are not receding, and, indeed with growing populations and climate change, are only increasing.

At such a critical juncture, it is significant and timely to interrogate the legal and governance experiences, challenges, lessons and future direction of Australia's water reforms, so as to identify how best to steer water law and governance towards a more sustainable future path. We put these issues to a group of leading water law specialists and this edited collection presents their insights and solutions. Reflecting Australia's natural water scarcity and its heavy use of water for agriculture (totalling around 60% or more of all water use), their focus is on governing water quantity rather than quality, with many (although not all) of our chapters addressing rural water governance issues (rather than urban). In this introductory chapter, we briefly summarise Australia's water reform journey before outlining several fundamental questions posed by Australia's mature water governance system, and how our authors tackled them. This is followed by an examination and reflection on the book's chapters and what they suggest are the core implications for more successful water governance systems that are capable of responding to persistent and new challenges, and to develop a grounded vision of water law and governance reform for the 21st century (Moore et al. 2014).

2 Australia's Water Reform Journey—from Innovation to Stagnation

The early history of Australia's water management has been discussed in detail elsewhere, but in simple terms, Australia's initial Indigenous water management traditions were marginalised following British colonisation (see Marshall 2017). This was followed by a transition, during the late 1800s and early 1900s, from common law riparian and rule of capture rights, to state legislative regimes (Gardner et al. 2009). As Barbara Cosens points out in her chapter (discussed below), this mirrored to some extent what occurred in the western United States. Over time, Australia's states progressively vested control over water in the Crown and abolished or displaced existing common law rights, creating a system of water licencing (Gardner et al. 2009). It was during this period that attention focused on a growing water crisis in Australia and globally (WWAP 2015).

As noted above, Australia's modern water reform journey commenced in the early 1990s with the Commonwealth Government's national competition reform packages. Founded on ideas of joined up governance, intergovernmental action was taken to arrest widespread water resource degradation (COAG 1994, 3). The 1994 national water framework set out aspirations for market based reforms, supported by new institutional and administrative arrangements and decision-making processes. This entailed state and territory governments implementing nationally agreed goals regarding integrated catchment management (e.g. 'an integrated catchment management approach to water resource management') and collaboration (e.g. 'consult with the representatives of local government and the wider community in individual catchments') (COAG 1994, 6–7). Perhaps the most iconic aspect of Australia's approach was the creation of property rights to extract water, within extraction limits set using scientific methods. This included consumption-based pricing and full-cost recovery, the separation of water property rights from land title, and the implementation of water allocations and entitlements (COAG 1994).

Water entitlements provide an on-going share of a consumptive pool of water within a given catchment. Water allocations are the volume of water assigned to a water entitlement in a specific water season (e.g. annually) under a specified water resource plan (Grafton and Horne 2014). Users are not required to utilise the allocated water to maintain an entitlement, but limits and rules can apply to the physical carry over of unused water allocation from one year to the next (Grafton and Horne 2014). Allocations for the environment were also created to maintain the health and viability of river systems and groundwater basins (COAG 1994).

Trading arrangements for allocations and entitlements were instituted, with the intention that water would be used in the most efficient (and productive) way, primarily within state boundaries and individual catchments (COAG 1994). This system of trade was to be underpinned by new administrative arrangements and decision-making processes to ensure an integrated approach to water management, involving consultation with local government and the wider community within individual catchments (COAG 1994).

In addition to collaborative approaches to catchment planning, and markets, the reforms also encouraged a system of regulatory enforcement (COAG 1994, 6c) and periodic monitoring and peer review at the state government level (COAG 1994, 11b). Commonwealth, state and territory ministerial management councils were to report annually to COAG on 'progress in implementing the various initiatives and reforms' (COAG 1994, 11b).

Despite such reporting to COAG, the constitutionally entrenched position of the states (where they enjoy concurrent formal powers with the Commonwealth government) enabled them to act as a potential veto player in the coordination process (Kay 2013). Additional accountability drivers were thus provided by the Commonwealth through what was known as the National Competition Policy and Related Reforms. These reforms incentivised implementation though establishing a system of payments to the states upon successfully meeting implementation goals (COAG 1995). These accountability mechanisms were overseen by an independent advisory body known as the National Competition Council, who conducted progressive assessment on state progress over a five to seven-year period (National Competition Council 2005).

While progress was made implementing the water reforms (e.g. separating land and water rights), it was substantially slower than expected. Indeed, by the time the competition payments had ended, 'substantial work' still remained to be done (National Competition Council 2005, p. xvi). There were many possible reasons for this slow implementation, including its national scale, the complexity of the reform (e.g. identifying place based catchment boundaries, setting numerous catchment based caps), and the agricultural sector's political pressure against water allocation reductions at a time when rainfall was limited by the Millennium Drought.

Confronting continuing drought conditions (intensified during the Millennium Drought), and with significant work remaining on the 1994 framework, a subsequent 2004 Intergovernmental Agreement between the states, territories and Commonwealth Government on water was developed. The agreement, known as the National Water Initiative (NWI), consolidated the 1994 reforms and extended their aspirations for joined-up, place-based approaches, market reforms, collaboration, enforcement and continuous monitoring and improvement. It aimed to embed a nationally-compatible water market, progressively remove barriers to water-trading, facilitate efficient water use and address structural adjustment issues (Cwth of Aus. 2004, paras 23 and 58). These second wave reforms also aspired to return over-allocated or overused systems to environmentally-sustainable levels of extraction by encouraging the development and finalisation of placed-based statutory water allocation plans, and making statutory provision for environmental and other public benefit outcomes (Cwth of Aus. 2004, para 23).

Under the NWI, state and territory governments maintained significant discretion in pursuing goals. For example, with respect to water planning, states were to 'determine whether a plan is prepared, what area it should cover, the level of detail required, its duration or frequency of review, and the amount of resources devoted to its preparation based on an assessment of the level of development of water systems, projected future consumptive demand and the risks of not having a detailed plan' (Cwth of Aus. 2004, para 38).

Collaboration was also reinforced under the NWI, with 'community partnerships' and 'engaging water users and other stakeholders in achieving [NWI] objectives' identified as key reform outcomes (Cwth of Aus. 2004, para 93). This was to be achieved through transparency and open and timely consultation with all stakeholders throughout plan development and review (Cwth of Aus. 2004, para 95 and Schedule E; NWC 2011). These engagement aspirations sought to encourage trust and buy-in (Tyler 2005), and arguably reflected a pragmatist understanding of the importance of public participation to develop local governance responses (Karkkainen and Fung 2000). In this case, the aim was to identify appropriate responses to address the impacts of water cutbacks (Tan et al. 2012; Cwth of Aus. 2004, para 97). While 'governments and the community' were to determine water management and allocation decisions, the government often remained the final arbiter and was to 'consult' and make 'judgements informed by best available science, socio-economic analysis and community input' (Cwth of Aus. 2004, para 36).

Perhaps most significantly, the NWI extended previous COAG commitments on monitoring and continuous improvement. This involved several reforms (e.g. benchmarking pricing and service quality of water delivery agencies) (Cwth of Aus. 2004, para 2iii-75 Schedule D), including three fundamental changes. First, recognising that continuous improvement and innovation requires good monitoring, the NWI committed to 'water resource accounting' and ensuring 'adequate measurement, monitoring and reporting systems are in place in all jurisdictions'. This included establishing a national framework for comparison of water accounting systems to 'encourage continuous improvement leading to adoption of best practice' (Cwth of Aus. 2004, para 81; McKay and Gardner 2013).

Second, at the local level, plans and water access entitlements were to implement continuous improvement systems in the form of 'adaptive management' (Cwth of Aus. 2004, para 25). While the specific term 'adaptive management' is not defined in the NWI, its objectives included monitoring the performance of water plan objectives, outcomes and water management arrangements; factoring in knowledge improvements as provided for in the plans; and providing regular public reports. This reporting was designed to help water users and governments to manage risk, and to give early indications of possible changes to water management decisions (Cwth of Aus. 2004, para 40).

Third, the NWI led to the establishment of the National Water Commission (NWC). The NWC was an independent body whose commissioners were nominated by the states, territories and the Commonwealth government to provide a skills-based national perspective that assisted with the effective implementation of the NWI (Cwth of Aus. 2004, para 10; NWC 2014a). In the absence of the national competition payments and National Competition Council, which ended their roles in water in 2005, the NWC was given responsibility for key monitoring, improvement and innovation roles, including: (1) undertaking a baseline assessment of the water resource and governance arrangements; (2) accrediting and assessing State implementation plans; (3) developing and revising performance indicators for the NWI (along with the Natural Resource Management Ministerial Council; Cwth of Aus. 2004, para 12, 104); and (4) conducting publicly available assessments on progress with the NWI Agreement, including advising on actions required to better realise the objectives and outcomes (Cwth of Aus. 2004, paras 10, 11, 12; *National Water Commission Act 2004* (Cth), s7). Taken together, these NWC roles reflect what some pragmatists have called a 'new central body' that can reduce the costs of information flows, foster benchmarking processes and facilitate horizontal diffusion of best practice and continual improvements (Scheuerman 2004; Karkkainen and Fung 2000, p. 691).

From 2004, States progressively began to implement the NWI reform in legislation and on the ground. While substantial progress was made, it was often slower than expected, which led to a new major national initiative announced in 2007. The National Plan for Water Security (later known as Water for Our Future) aimed to ensure efficient water use and improve environmental outcomes, with a particular focus on Australia's largest river catchment, the Murray-Darling Basin (MDB). The $10 billion investment was directed towards various goals (e.g. improving monitoring and metering), however the flagship of this announcement related to the MDB. Often referred to as Australia's food bowl, its management has been subject to numerous cooperative arrangements between the states of Queensland, New South Wales, Victoria and South Australia and the Commonwealth Government (see Connell and Grafton 2011). However, through the new National Plan, the Commonwealth Government proposed 'once and for all' to address water over-allocation in the MDB, introducing a new set of governance arrangements involving the creation of a sustainable cap on surface and groundwater use and major engineering works at key sites. The *Water Act 2007* (Cth) was subsequently introduced and premised on a complete federal takeover of management of the MDB. However, this Act was never to receive the necessary support from state governments to enable the takeover to fully occur. Rather, the Act came to rest on various Commonwealth constitutional powers (e.g. external affairs) and new cooperative arrangements, including the *2008 Intergovernmental Agreement on Murray–Darling Basin Reform* and the *2013 Intergovernmental Agreement on Implementing Water Reform in the Murray-Darling Basin* (Commonwealth of Australia 2014, *Water Act 2007* (Cth), s 9).

While these new agreements would continue to be based on NWI principles, the Commonwealth and a new independent MDB Authority (MDBA) were to be responsible for MDB-wide planning and management (*Water Act 2007* (Cth), part 9; Commonwealth of Australia 2014). States within the MDB remained responsible for managing water resources within their own jurisdictions, but they agreed to align their water management with a new MDB Plan and its new cap by 1 July 2019 (*Basin Plan* 2012 (Cth), cl 6.04). Other core elements of the reforms included removing restrictions on trade, especially across state boundaries through a new common trade framework; the empowerment of the Australian Competition and Consumer Commission (ACCC) to enforce water market rules and water charge rules with rafts of water brokers, exchanges and irrigation infrastructure operators (Australian Government 2009; ACCC 2017; *Competition and Consumer Act 2010*;

Water Market Rules 2009 (Cth); *Water Charge (Termination Fees) Rules 2009* (Cth)); and the creation of the Commonwealth Environmental Water Holder (CEWH) to support government purchase of water entitlements and substantial investment in more efficient water infrastructure for the purpose of protecting or restoring the environmental assets of the MDB (DEE 2016; ANAO 2011).

Although these MDB reforms would need to overcome a range of challenges, not least community anger at the MDB planning and consultation process, a Basin Plan was agreed to in 2012 (Gross 2011; Tan et al. 2012; Bowmer 2014). This set new MDB and catchment caps to recover 2,750 gigalitres of water through a combination of investment in infrastructure efficiency and water buybacks (Murray-Darling Basin Authority 2015).

Simultaneously with the passing of the *Basin Plan* 2012 (Cth), a new challenge for the NWI reforms emerged, namely the intensification and expansion of coal, coal seam gas and other unconventional sources of extraction (Whitehead 2014; Parsons et al. 2014). Historically such extractive industries had been excluded from water rights and water allocation plans under the NWI (Cwth of Aus. 2004, para 34), but this led to growing social concern over their adverse impacts on local water resources. Various state and Commonwealth reforms followed, including integrating unconventional gas into NWI water planning, bioregional assessments and new forms of environmental assessment for unconventional gas and coal mines likely to significantly impact on water resources (*Environment Protection and Biodiversity Conservation Act 1999* (Cth) s 24E; NWC 2014b, p. 2; COAG 2013).

After 20 years of this lengthy reform story, the most recent chapter was the Commonwealth announcing the development of northern Australian water resources, while simultaneously abolishing the NWC (Commonwealth of Australia 2015). The Commonwealth views Australia's northern region as having the untapped potential of being the 'food bowl' of Asia. Although it receives 60% of national rainfall, this is highly variable and has limited irrigation utility (Commonwealth of Australia 2015, p. 40). In order to increase agricultural production, it suggests enhancing 'water resource development by providing access to secure water rights to encourage investment' (Commonwealth of Australia 2015, p. 41). While acknowledging the dangers of over-allocation that have occurred in the MDB, it proposes water markets, as well as new resource assessments and investment in water infrastructure, such as new dams (Commonwealth of Australia 2015, pp. 118, 125).

In terms of the NWC, it was formally dismantled on the basis that progress in implementing the NWI was such that monitoring of national reforms was no longer needed, and statutory functions could be transferred to other Commonwealth agencies and Australia's Productivity Commission (*National Water Commission (Abolition) Act 2015* (Cth); Hannam 2015).

Following the abolition of the NWC, there has been a growing belief that Australian water policy is beginning to stagnate (NWC 2014a; NWC 2014b). Arguably gripped by the hydro-illogical cycle (Wilhite 2012), the absence of a Millennium-scale drought has seen weaknesses overlooked and law and policy leadership wane. As the Wentworth Group of Concerned Scientists (2014, p. 0) affirm, 'it appears Australian governments are walking away from strategic water reform at the very time when we should be preparing for the next inevitable drought'. Indeed, it is increasingly unclear how resilient the NWI blueprint will be in the face of shifting political agendas, growing complexity, reform fatigue, shrinking public resources at state levels and the absence of an independent oversight body like the NWC (NWC 2014a, pp. 3–4).

It was of some relief to commentators that 2017 saw a response to these concerns, with an inquiry into Australian water reforms from the Productivity Commission, and recent announcements of policy frameworks to guide further implementation of the NWI (e.g. *National Groundwater Strategic Framework 2017*). However, whether and where these reforms will lead Australia's national water governance system, and whether they will be sufficient to meet water scarcity challenges, remains to be seen.

3 The Water Scarcity Governance Challenge

With any local, national or indeed global context, it is paramount to maintain 'good' water law and governance given that the next few decades will see major increases in population and food production (both dependent on water), as well as likely water scarcity due to droughts and climate change (Steffen 2015). Our starting point for this book was to examine this issue through the Australian experience of water governance. In particular, Australia is suffering from two deficits: the first is a lack of commitment to tackling existing and emerging legal and governance challenges, and the second is a stalling and even unwinding of crucial water reforms.

The book seeks to respond to these deficits and extract further value out of Australian water governance. The intention is to provide insights for both Australian and international audiences. In particular, it asks:

1. How successful is Australia's approach to designing and implementing water governance?
2. What conditions have enabled or blocked its success, including environment, social, political and legal?
3. How does Australia's water governance system compare and contrast with different international water governance practices, particularly in Northern America?
4. What are the broader insights for future water governance practice and theory?

The chapters in the book seek to address these questions by examining, contrasting and interrogating different issues, facets and thinking within water law and governance, offering insights from practice, doctrine, theory and comparison with other nations. Below is a brief overview of each of their relative contributions.

The MDB is Australia's most well know site of water law and governance in the world. As Bradley C. Karkkainen notes, it is for many outside observers 'something of a marvel' (see Chap. 3). It is fitting then to commence Part One of our book with three chapters that interrogate the MDB, its transboundary complexities and the recent innovations, successes and failures in governing this prominent water resource.

In Chap. 2, Emma Carmody offers a cautionary tale on water reform in the MDB and its recent unravelling. Her chapter commences with the passing of the *Water Act 2007*, something of a historic moment in the management of this transboundary river system. Carmody outlines the core objectives of the new Act and its subsequent Basin Plan that aimed to introduce new limits on water extraction across the MDB. However, her analysis raises serious questions over the sustainability of the limits imposed, highlighting the challenge of implementing statutory and policy changes without marginalizing environmental outcomes. She concludes by setting out a series of recommendations intended to restore faith in water reform processes in the MDB, as well as assist regulators and stakeholders to identify and manage possible barriers to proper implementation of domestic and international water laws in other transboundary basins.

Following Carmody's analysis, Chap. 3 offers a comparative perspective on the Australian experience. Posing the question of whether Australia's approach to its multijurisdictional water challenges is a 'muddle' or a 'model', Karkkainen finds the truth probably lies somewhere between these extremes. His analysis highlights both strengths and weaknesses of multijurisdictional governance in the MDB, and does so through comparing it with governance arrangements in two critically important multijurisdictional North American basins, namely the Colorado River and the Laurentian Great Lakes. Karkkainen proceeds to compare these systems, concluding that while not without its flaws, the MDB is arguably the most advanced along several key dimensions of multijurisdictional water resources governance.

Chapter 4 rounds out the analysis of the MDB reforms by considering a modern innovation that has arisen from the Basin Plan and its recent attempts to address water-stresses and over-allocation. Here, Katherine Owens hones-in on the use of large-scale government buybacks of water entitlements, a strategy that has proven to be unpopular politically. In response, Owens finds innovation in the use of markets, impact investment and the MDB Balanced Water Fund. The chapter analyses the benefits and risks of this approach, as well as the legal supports and safeguards required to advance these initiatives in meeting their environmental objectives. Owens argues that when accompanied by appropriate legislative safeguards, private initiatives of the kind represented by the Fund have the potential to function as an effective complementary measure, and improve system resiliency for the benefit of a range of uses, including the environment.

Such innovations in the use of MDB water markets provides a useful transition to Part Two of the book, which focuses more squarely on the theory and practice of economic and market instruments. In particular, this part evaluates the existing laws and policies applying to what many see as the flagship of Australia's water reforms, tradable water rights and water markets. This presents a stock-take of current successes and remaining challenges in Australia's water market reform journey, with a particular focus on three crucial issues, namely property rights, allocating water to the environment and legally enforcing cap and trade market approaches.

Central to Australia's system of markets has been a threshold concept of property in water. In Chap. 5, Janice Gray and Louise Lee consider the extent to which Australian water markets sufficiently address the treatment of water as a traditional property right. To understand the extent and impacts of propertisation, they interrogate what appears at first blush a decidedly straightforward question: are water entitlements property? Their answer is that property is, in fact, quite omnifarious in nature. Referencing Australian states and other international jurisdictions, Gray and Lee explore whether water entitlements, in fact, need to be characterised as property to support trading, and whether keeping water outside the proprietary frame may open up opportunities for a wider range of governance tools. Drawing on the Australian context, their insights raise concerns of relevance for other jurisdictions where water trading is either being relied on, or being considered as a water management tool.

Continuing the detailed discussion of market approaches, Chap. 6 sees Adam Loch, Dioni Perez-Blanco, Dolores Rey, Erin O'Donnell and David Adamson explore how water governance and demand management arrangements can be linked to economic instruments like water markets to address the broad range of water reallocation problems that exist in Australia and other global contexts. They usefully examine common types of economic instruments and detail their pros and cons, arguing that policy-makers can learn from growing evidence of successful partnerships between regulatory and economic instruments.

Chapter 7 adds further flesh to the often-skeletal economic framing of water markets. Directing critical attention to the legal and governance issues of Australia's water markets in practice, Cameron Holley and Darren Sinclair argue that markets can and have promoted flexible responses to past and future droughts; efficiencies that contribute to economic and environmental benefits; and are steadily increasing in trade functionality. However, they find Australia's multiple water cap and trade schemes are falling well short of their potential in a range of areas, including regulatory enforcement, water accounting, managing universality, environmental benefits and social impacts. They conclude that markets themselves can be a barrier to areas of future water law and policy innovation and argue that Australia's lessons suggest a need to look beyond markets for new complementary regulatory tools (see also Cosens in Chap. 13).

Part Three of the book focuses on the vexing issues of collaboration and participation in water governance. Alongside the development of water markets, water planning has been a key reform goal and characteristic of trends in integrated water resource management. This part reflects on this planning, examining the successes, failures and future of consultation and collaboration. The chapters take a broad view of participation, and examine the prominence and relevance of a range of non-government and civil society actors in water planning, markets and regulatory systems. At the heart of all three chapters is an argument for a re-conception of the governance landscape, particularly to make space for and engage with a wider range of Indigenous interests and other non-government actors in the water sector.

Bruce Lindsay begins this part in Chap. 8 with a focus on procedure and possibilities for public participation in water resources management. Arguing that there has been government and commercial/industrial hegemony over water resources, Lindsay critiques the two main procedural vehicles for public participation in Australian water governance, namely consultation and litigation. He finds limitations in both areas: consultation lacking in policy direction and sophistication and the infrequent use of civil litigation. He argues these two procedural frameworks present possible pathways to widen community engagement in other international water contexts and draws on Australian insights to map the way forward for both these consultative and adjudicative modes.

In Chap. 9, Jennifer McKay turns the focus to the collective goal of sustainable water use of rivers, aquifers and manufactured water in urban and rural Australia. Acknowledging that water governance comprises public and private interactions, McKay examines the coordination and lack thereof between civil society (viewed through epistemic communities and non-government organisations), law and organisations. The chapter offers a comparative aspect with China, and points to ways to increase coordination by suggesting Australia's *Water Act 2007* be applied to the entire nation. Finding that civil society is no longer playing the crucial role it once did in Australia, McKay argues for much stronger epistemic communities and non-government organisations in water governance arrangements.

Chapter 10 sees Elizabeth Macpherson, Erin O'Donnell, Lee Godden and Lily O'Neill conclude Part Three with a detailed look at Indigenous and environmental values in Australia's market-based approach to water regulation. They argue that despite the successive reforms, the implementation of Australia's model of water governance has yet to completely redress the historical exclusion of Indigenous peoples from water law frameworks, and has struggled to account for the needs of a healthy and sustainable aquatic environment. Offering very useful reflections on two of the most recent developments in Australia's water reform trajectory, the northern Australian *White Paper* and the collaborative planning approach set in the *Water for Victoria* policy, they highlight the difficulty of ensuring fairness in the operation of hybrid governance systems for water regulation. The chapter reveals significant lessons for international policy-makers, including the need for water law reforms to be founded on meaningful engagement with Indigenous peoples, and embedded Indigenous and environmental values and rights in water planning and governance.

Part Four concludes the book by looking more exclusively and explicitly towards some of the unresolved and future challenges of water law and governance. The chapters consider issues that have neither been fully resolved nor given adequate consideration in the governance landscape, namely cumulative impacts, extractive industries and future adaptive capacities. Given their focus on these new challenges, it is perhaps unsurprising that a common thread across these final three chapters is the habitually overlooked issue of groundwater, which remains a vital source of global water supplies.

In Chap. 11, Rebecca Nelson draws sharp focus to the seismic shifts in groundwater extraction in Australia and the need to deal with the cumulative impacts of extraction. Reviewing how groundwater reforms in Australia have approached cumulative impacts, Nelson finds that despite a much broader view of the impacting activities that are relevant to an assessment of cumulative impacts, both regulatory and non-regulatory measures have not focused significantly on the implications of connections between natural systems. Nelson identifies remaining weaknesses in approaches to managing cumulative impacts, and helpfully charts key challenges and a reform agenda for Australian and international policy makers.

The theme of impacting activities on groundwater is picked up by Poh-Ling Tan and Jacqui Robertson in Chap. 12, where they examine water, coal seam gas and mining governance reform in Queensland and Wyoming. Their comparison finds that in both Queensland and Wyoming, it has been difficult to develop and implement a water governance model that promotes sustainability. Their chapter explores the reasons for this, arguing Queensland's bifurcated legislative regime for petroleum and gas activities, where issues of quality and quantity are separately assessed, provides a barrier to substantive consideration of water impacts. They also draw attention to risks of reactionary responses, amidst a regulatory intent of adaptive management of the water resources.

Following Tan and Robertson, Barbara Cosens concludes Part Four of this book with her analysis of water governance and the need for greater adaptive capacity. Examining lessons from Australia and the Western United States, Cosens explores a trilogy of approaches to governance of common pool water resources: (1) government regulation; (2) the division of the resource into private property; and (3) self-organisation involving both governmental and nongovernmental actors. Offering insights on the combination of governance approaches that lead to greater adaptive capacity, Cosens' chapter concludes that the distinctive nature of water necessitates that all three approaches are used in combination if we are to collectively rise to the dire impacts on earth systems from burgeoning populations and climate change.

In the remainder of this introductory chapter, we synthesise broad lessons and insights from these leading water law academics and practitioners, and respond to the questions posed above. As we will see, the diversity of governance instruments, contexts and responses explored in each chapter meant that there were few categorical answers to the questions raised. However, some shared themes emerged, providing important and novel understandings and proposals for water law and governance.

First, we interrogate the success of Australia's approach to designing and implementing water governance. Second, we examine the conditions that have enabled or blocked its success, including environmental, social, political and legal. Third, we explore how Australia's hybrid governance system of collaborative planning, markets and regulation compares or contrasts with different international water governance practices. Fourth, and finally, we ask what are the broader insights (be they on policy, law or theory) for future water law and governance?

4 Water Law and Governance: Insights from Australia

4.1 How Successful Is Australia's Approach to Designing and Implementing Water Governance?

As outlined in the chapters, Australia's legal and policy architecture for water governance has been on a significant journey since 1994. Guided by intergovernmental agreements, and national oversight, Australian states and territories continue to implement the National Water Initiative, the Basin Plan (2012) and related reforms. Not all states have implemented the NWI reforms to the same degree, and there has been substantial experimentation with various terminologies and legislative frameworks, statutory and non-statutory planning instruments for setting market caps and water regulatory institutions (Tan et al. 2012).

Notwithstanding this diversity, these laws and policies have collectively sought to balance Australia's water allocation amongst consumptive and environmental uses to efficiently achieve sustainable management. This has inevitably led to conflicts and social struggles between different interests and pressures on water use (Tan et al. 2012; Syme and Nancarrow 2008). The scale and pace of national water reform has also given rise to practical challenges, delays, implementation failures and reform fatigue (Hussey and Dovers 2007; NWC 2014a; Commonwealth of Australia 2014), including court challenges to 'caps' established as entitlement reductions to secure sustainability (Millar 2005; *ICM Agriculture Pty Ltd & Ors v The Commonwealth of Australia & Ors* [2009] HCA 51), social protests against extractive industries and water allocation decisions affecting regional communities (Gross 2011), allegations of corruption and collusion (see Matthews 2017) and critiques of water recovery efforts (Wentworth Group of Concerned Scientists 2017). Even so, Australia's reforms have spurred change in societal values and attitudes relating to water use and management (Jackson et al. 2015), with farmers, government agencies, environmental water holders and service providers becoming more sophisticated as managers and architects of water governance schemes (Jackson et al. 2015).

Such a complex, fraught and continuing reform process means finding definitive answers on the success of Australia's approach to designing and implementing water governance is almost impossible. Indeed, the chapters highlighted numerous and often contrasting views on these matters. For instance, Gray and Lee, Macpherson et al. and Cosens recognise the innovation of Australia's hybridised regulatory design of trade, planning and enforcement reforms. Cosens in particular sees this design as central to adaptive capacity, suggesting this combination 'substantially increased Australia's ability to cope with long-term drought'. However, such praise was typically more qualified, and often limited to certain instruments, certain locations and even certain times. McKay, for instance, suggests that despite some success, there is a persistent lack of efficiency arising from 'too many jurisdictional differences', with disparate implementation undermining equitability and effectiveness. Others, like Lindsay, argue that both effectiveness and equity have suffered because of the absence of a coherent policy framework or legal support structures for deliberation, and in Macpherson et al's words, a continuing failure 'to completely redress the historical exclusion of Aboriginal peoples'. Despite the 'seismic shifts' identified by Nelson in her chapter on groundwater management, a range of issues also remain unresolved. These include limited integrated management of surface and groundwater (Cosens), the impact of development on groundwater (Tan and Robertson) as well as regulatory systems for addressing 'the cumulative impacts of groundwater extraction on water entitlement holders and dependent ecosystems' (Nelson).

Mixed views on effectiveness were also evident in consideration of the MDB, where authors such as Karkkainen and Carmody identified significant success in the MDB reforms and the Water Act, which in Carmody's words 'constituted a significant step forward'. However, as both acknowledge, over time such success has succumbed to various implementation challenges. As Karkkainen pragmatically puts it, the MDB represents 'reasonably successful multijurisdictional governance of a critically important freshwater system'.

Findings of mixed success in water markets were also discussed across the chapters. Given Australia's reputation as a 'global leader in the use of water markets as a policy' (Karkkainen), it is perhaps unsurprising that water markets were a significant focus of analysis. Cosens, for instance, shows how the experience of both Australia and the US (albeit to a lesser extent) illustrates the substantial adaptive capacity of water markets as a response to a changing climate. This is echoed by Holley and Sinclair, who cite empirical evidence and examples of markets achieving 'flexible responses to past and future droughts' and 'efficiencies that contribute to economic and environmental benefits'. Loch et al. also note how Australia and other nations have used tradable rights as a way to help to both clarify expectations and reduce conflicts over water.

A number of the chapters drill-down into the details of how markets operate as regulatory systems to find that strong governmental roles are a core foundation of water market success in Australia (for further see Gunningham et al. 1998). Loch et al. neatly capture this sentiment when they note that markets (and economic instruments generally) 'rely on the capacity of the state to control access, or create and enforce property rights'. Cosens provides a useful illustration of this governmental role and its contribution to success in her discussion of markets contributing to resilience and adaptive capacity. As she put it, it was 'the authority of government as owner of the water' that enabled it to set and adjust water allocations and market caps, so as to ensure the market instrument adapted quickly and responded to variability in water supplies over time. Owens' explores a further dimension of this governmental roles, noting that the government's (and indeed other actors') ability to buy and hold water entitlements for environmental purposes has enabled the use of markets (albeit controversially) to recover water for the environment and protect ecosystem values. Holley and Sinclair also explore the importance of government (often in response to identified shortcomings) in the monitoring and enforcement of water use practices.

Despite this recognised importance of governmental roles in market success, a number of chapters identify ongoing and significant implementation challenges that have limited the effectiveness and efficiency of markets in Australia. This includes failures to pay sufficient attention to environmental water, particularly in Northern Australia (Macpherson et al.), slow progress by some states in developing preconditions of tradable property for market operation (see Gray and Lee's review of the NWI progress), as well as 'negotiations and political skirmishes' that have stymied the efficiency of markets as a tool to ensuring compliance with sustainable diversion limits (Carmody). As Karkkainen explains, tensions are 'rife' in the MDB, where a singular focus on maintaining minimum in-stream flows has arguably 'straightjacketed' broader considerations of environmental quality and ecological integrity. Owens confirms such concerns and notes in her chapter that a lack of responsiveness to economic, social and cultural values in the MDB has led to a general resistance from irrigation communities to water buybacks, leading to a shift to less efficient, and potentially less assured, recovery measures, like infrastructure upgrades. Echoing Owens' concerns about the potential for detrimental impacts of water transactions, Cosens queries whether Australia has gone too far in reducing transaction costs at the expense of local economies and ecosystems, potentially harming environmental and cultural values as well as local economies. As she notes, third parties and communities harmed by a transfer that is nevertheless consistent with the water plan, have no recourse to redress such impacts.

In short, while the chapters suggest the NWI and its market-based approach have made substantial progress as a tool for adaptive and drought resilient water management, there are still many limits and gaps to their overall efficiency, effectiveness and equity. As Loch et al. concludes, 'markets are far from a panacea for the reallocation of scarce water resources and progress toward sustainable water-use outcomes'.

4.2 What Conditions Have Enabled or Blocked Its Success, Including Environmental, Social, Political and Legal?

The contributors identify a mix of conditions that have both enabled or blocked the success of Australia's governance approaches, which are viewed in the context of specific instruments and issues in each chapter. Here we do not aim to revisit these, but rather identify three general conditions common to many of the chapters and which appear to be broader enablers and/or barriers to success.

The first condition is the presence or absence of a broad water scarcity crisis that appears to act as an enabler or barrier to success in water governance, respectively. As those familiar with the hydro-illogical cycle (Wilhite 2012) will recognise, droughts are a common driver of water law and governance action. And this was certainly the case in Australia, where, in McKay's words 'The Millennium Drought focused the minds of all governments Federal, State and Local on water issues'. Of course, there are many ways to respond to such a crisis, and in Australia 'value and emphasis' was placed on water propertisation, as well as 'the introduction of water-use restrictions, water-use targets, increased dam building plans, the construction of desalination plants, and the growth of sewer mining and recycled water operations' (Gray and Lee). As is often the case with a crisis, particularly regarding common goods (Hardin 1968; c.f. Ostrom 1990), direct and immediate action was placed in the hands of government, who tried to address public concern and ward off further crises. As Macpherson et al. note, this a familiar story for Australian water reform, where long before the Millennium, severe droughts as far back as the 1880s saw Australians 'demand public "responsibility" for water governance'. However, such a top down response is not without risks, and as a number of the chapters detail, the recent absence of an ongoing crisis (with the breaking of the Millennium Drought) has seen the previous government-dominated strategy begin to solidify as a barrier to the continuing implementation and success of water reform.

Cosens illustrates this point by showing how Australia's Hobbesian motivated response to water insecurity lacked a base in existing local community and civil society capacity. This, Cosens argues, led to a 'loss of legitimacy' and community anger as the crisis itself ended. Such anger and backlash to regulation since the ending of the Millennium Drought is clearly evident in Owens, Carmody and Holley and Sinclair's chapters, where details of resistance regarding water buybacks are explored in more detail. In the face of climate change, the re-emergence of drought may very well tip the scales back to the need for urgent action, but it appears that governance approaches will suffer if they are not carefully calibrated between top down and bottom up during shifts from crisis to stability.

Interlinked with conditions of water scarcity and drought is a second condition, namely the politics of environment versus economic development, and the brake that economic narratives can apply to successful water reform. A common theme across several chapters is the dominance of economic and development narratives within politics and their negative impact on successful water governance. While this is far from a new revelation, it is clear that conflicts between extractive use and environmental protection remain a thorn in the side of effective water reforms and implementation. Tan and Robertson clearly capture this point in their analysis of mining and coal seam gas, where economic and political reasons favour employment, development and royalties over environmental concerns. This, they argue, has catalysed evident failures in Queensland's water governance approach, and its ability to comply with national water priorities and frameworks.

In addition to mining and gas companies, a number of chapters also point to irrigation lobbies and related 'partisan local interests' (McKay), such as state governments, forming a key barrier to successful water governance. Carmody, for instance, explores the political skirmishes within the MDB, noting how economic interest 'threatens the survival of the Basin's water resources' as it undermines visions for environmental sustainability. Macpherson et al. reinforce this point, noting that, along with the continuing exclusion of Indigenous people from water frameworks, accounting for the needs of sustainable aquatic environments within a paradigm of development has been a persistent challenge. Owens similarly casts the Basin Plan as being 'undermined by politics during its development process, which compromised the recovery targets'. However, Owens also acknowledges that 'irrigation communities' resisted water buyback programs (in Loch et al's terms, 'Payment for Ecosystem services') because these efforts were seen to take advantage of farmers already stressed from the ongoing Millennium Drought and undermined rural communities.

Of course, such politics are likely to remain an ongoing challenge in Australia (as in other nations), and as Karkkainen reminds us, will not always see economic concerns win out. As he puts it: 'Both environmentalists and irrigators frequently complain that their side is getting short shrift, and this has led to policy reversals, seeking to rebalance the equation in favour of one set of interests or the other. Such tensions are perhaps inevitable under the conditions of water scarcity that pervade in Australia's interior'. Lindsay perhaps offers the most complete potential remedy to such pervasive political concerns. He suggests Australia needs to both 'front' and 'back load' the governance system to ensure more genuine engagement to enhance and reduce policy backsliding, while also empowering countervailing interests to bring court challenges to confront potential weaknesses of politics writ large on water policy. Absent such reforms, Carmody recommends 'constant vigilance - from civil society' to ensure that laws are properly implemented and enforced over time. Moreover, in the face of ongoing political contestation, it is important to acknowledge that tension can also be a source of innovation in law and governance practice. A range of chapters identify pioneering examples, such as experiments with Indigenous engagement (Macpherson et al.) and 'reimagined legal institutions, actors and combinations of instruments' for mobilising environmental water transactions (Owens).

A final overlapping issue common across many of the chapters is that of implementation. A number of contributors explicitly or implicitly recognise that even where policy settings have been put in place, outcomes will not be realised without close attention to implementation. Implementing laws, particularly water focused ones, are of growing interest to academics (see e.g. Martin and Kennedy 2015). Attention here can be varied, but often has concerns with empirical or observable phenomena that focuses on organisational structures, resourcing, responsiveness and learning to respond to real world developments. In the words of Martin and Kennedy (2015, p. 4), these types of studies are based around a shared understanding of effectiveness judged as 'moving law "off the page"'.

Failures in facilitating law into action were repeatedly identified as a major barrier to success across the book's chapters. Some pointed to negative impacts arising from a failure to fully or adequately redress gaps that emerged in the implementation of Australia's policies, including impacts on groundwater from development (Tan and Robertson; Nelson) and Indigenous engagement (Lindsay; Macpherson et al.). Others, such as Holley and Sinclair, point to failures of states adequately resourcing, training and building policies around compliance and enforcement, metering, monitoring, and trade (see also Gray and Lee on property rights and progress taken in different states). Their findings are echoed by Loch et al. who discuss institutional, physical and social barriers (i.e. transaction costs) that can impede the implementation of markets. As they wisely note 'water markets are the most challenging to design, implement and sustain over time ... [requiring] ongoing institutional capacity-building and adaptive governance arrangements for the best probability of sustainable implementation'. McKay similarly raises concerns regarding funding and capacity building in the context of urban water supply projects. As she explains 'the focus needs to be on implementing institutional change through reform approaches'. Karkkainen's comparative analysis of Australia and international cases also recognise failures (albeit ones that run deeper even than implementation itself) arising from parties' inability to hold themselves accountable for implementing objectives.

In many respects, these implementation concerns in Australia are a function of the federal and cooperative structure, where the states have been left to implement national objectives and policy settings. In doing so, they have faced obvious challenges, including limited resources, as well as susceptibility to political influence (see the above discussion of politics). Such challenges are, of course, familiar to federal structures. It is true that significant, conscious and accountable experimentation, as has occurred within the broader field of environmental governance, can be a success and produce many innovations (Sabel and Simon 2011, p. 83; Nourse and Shaffer 2014; Holley et al. 2012). What is evident from the chapters, however, is that there must be consistent and close attention paid to issues of implementation to ensure that ideals are met in practice and good water outcomes are delivered. As Carmody nicely sums up in her analysis of the Basin Plan: 'newly minted, best-practice water laws are more of an armistice than an outright victory'.

What conditions, then, are needed for fostering successful implementation? While this will inevitably vary with context (Ostrom 1990) some insights are evident from the chapters, including ongoing processes of reflection, responsiveness, learning and adaption (Macpherson et al.), adequately resourcing government agencies (Holley and Sinclair), and facilitating an active civil society capable of holding governments to account over the long-term (Carmody; Lindsay). These and other issues are taken up in more detail below in response to our fourth and final question.

4.3 How Does Australia's Water Governance System Compare and Contrast with Different International Water Governance Practices, Particularly in Northern America?

Although Australian water law and governance is our central focus, many of the chapters identify useful points of comparison and insights regarding international practices. A range of national and sub-national contexts are explored, contrasted and interrogated to various degrees, such as the United Kingdom, Europe, China,

Canada and the United States. This included identifying differences between and insights from water rights and property in Australia and trends of renationalisation in the United Kingdom (Gray and Lee); Australia's consultative approach to public participation in water governance and distinct public participation approaches in Europe and the United States (Lindsay); and Australia's top down approach to water recovery activities that have crowded out contrasting activities conducted by water trusts and NGOs in places such as the United States (Owens). Other chapters shed light on potential points of similarity between Australia and other contexts, such as the use of payment of ecosystem services in Australia, the United States, Chile, Mexico, South Africa and China (Loch et al.); and the transboundary basins challenges confronting the MDB, the Colorado, Danube and Jordan Rivers (Carmody—see also Karkkainen, discussed below).

Across the book, four chapters in particular stand out for adopting a more in depth and directly comparative approach. First, McKay's analysis of urban water supplies sees her compare Australian water governance with China's traditional, but changing, public private partnership model. Second, Tan and Robertson interrogate water, gas and mining through a comparative study of Wyoming and Queensland. They find that despite very different historical roots in their approach to water governance and the issuing of water permits and rights, there is congruence in both regulatory frameworks for the unconventional gas industry when it comes to water. Third, Cosens conducts a comparison of water governance in the western United States and Australia through the lens of adaptive capacity. For Cosens, water planning in Australia is more extensive than in the western United States, where coordinated action has tended to rest at the state level (such as the California State Water Action Plan) rather than a watershed-based process. While there is evidence of higher level state process in Australia, such as Northern Victoria's Sustainable Water strategies (see Lindsay), the more programmatic approach of Australia towards catchment focused water allocation planning sits in contrast to any watershed focused efforts in Western United States, which Cosens notes 'develop on an ad hoc basis when conflict arises'. Cosens comparative study also addresses markets, and echoes Gray's and Lee's analysis of property rights in her findings that the western United States context has placed a 'higher value' on the 'property' nature of the right to water. As Cosens sees it, this played a significant role in limiting the development of markets, because 'water transfers are heavily regulated, leading to much less robust water markets than developed in Australia's Murray-Darling Basin'.

Fourth, Karkkainen dissects the Great Lakes and Colorado River to draw comparisons between these sites of water governance practice and the MDB. Karkkainen's comparison leads him to argue that horizontal coordination among states (and provinces in the Great Lakes Basin) is not a strong suit of any of the three management regimes, perhaps due in part to the heavily top-down nature of the governance arrangements currently in place. Notably, he finds that compared to the MDB, 'environmental protection and human economic uses of water are not nearly as well integrated on the Colorado' as environmental protection remained an afterthought for this major United States river basin. However, as he goes on to argue, environmental protection efforts in the MDB (particularly around flows) could learn much from environmental protection and ecological restoration efforts in the Great Lakes system and even on the Colorado, where recent efforts to return a seasonal pulse to the Colorado Delta may contain valuable lessons.

4.4 What Are the Broader Insights for Future Water Governance Practice and Theory?

Throughout the book, our authors have provided numerous lessons and raised varied questions that are central to the next wave of research and on ground practice in water law and governance. In this final section we identify two general questions that arose from their collective contributions.

4.4.1 The Future of Australia's Reforms

One significant question is what future reforms are needed in water law and governance in Australia? As noted above, Australia's system has matured and begun to stagnate as the recent drought has waned. Australia accordingly stands at a critical juncture: while there are signs that attention is beginning to shift to the next steps in reform, it is clear from the book's contributions that, although Australia has come a long way in water management, the design and implementation of the reforms to date do not appear sufficient to meet future water challenges. Each of the chapters has pinpointed specific issues that require attention, and offer suggestions for how these may be addressed. Broadly speaking, Australia will need to capitalise on its successes and redress past weaknesses if future water challenges are to be successfully confronted. As the former Commission Chair of the National Water Commission (2014c, p. 1) has stated: 'We can avoid costly mistakes in the future by learning the lessons of the past. Future generations should never have to endure the social, economic and environmental costs of another Murray Darling Basin'. Below, we highlight seven priorities that emerged from the collective work (Carmody et al. 2016, pp.132–134) of many of our contributors and stand out as vital issues that need to be addressed.

Regulate the Water Market to Ensure Equity, Enhance Efficiency and Protect the Environment Numerous contributions revealed insufficient attention and resources directed towards the implementation of Australia's water market approach. A common theme was the need to regulate markets to enhance compliance, mitigate secondary impacts on other water users and on ecosystems, address the adverse impacts on vulnerable communities from water markets (and broader water policy), fix the compatibility of markets with managing groundwater, and provide for improved telemetered metering, monitoring and data collection (which can also assist with adaptation and learning, as well as market efficiency). Such regulation will be vital to ensuring a level playing field, building confidence in market systems and improving outcomes for the community and the environment alike (Holley and Sinclair 2012).

Reform Water Buybacks Politics and local concerns were repeatedly raised as a major challenge for setting and implementing sustainable diversion limits (e.g. under the Basin Plan), including the imposition of legislative caps on water buybacks (see e.g. Carmody's chapter). This arguably necessitates a reimagining of environmental water transactions in Australia, which may include reconsidering the cap on buybacks, strengthening rules around environmental flows to ensure water for the environment and opening up collaboration between government and non-governmental actors in water transactions (e.g. water trusts and non-profit investment in environmental water transactions) (see e.g. Owens chapter). It will be critically important to design regulatory environments that allow for both collaboration between government and non-governmental actors, in order to bolster water recovery efforts, as well as institutional checks and balances to ensure sustainable water management (Holley et al. 2012; Owens 2017).

New Systems for Dealing with Cumulative Impacts As some of our contributors detailed (e.g Tan and Robertson; Nelson), future water governance reform needs to deal adequately with the cumulative impacts of water extraction on groundwater and groundwater-dependent ecosystems, particularly from mining and unconventional gas (such as Coal Seam Gas and shale gas). To date, state and national efforts have been evolving and are subject to ongoing reviews, however, laws and policies in most jurisdictions do not adequately address this issue (NWC 2014b). It is true that some progress is being made (for example, bioregional assessment being conducted by the Australian Government). However, innovative and rigorous laws and policies will be needed, particularly ones that incorporate strategic planning. This could also include an obligation to prohibit development where there is a risk of irreversible damage to water resources.

Protect Environmental Water Despite billions of dollars being invested in acquiring the rights to water licences to rectify over-allocation and improve environmental outcomes, environmental water remained a clear challenge across several chapters. Adding to this, recent reports in New South Wales suggest that, at times, environmental water may have been extracted for consumptive use (see Matthews 2017), thereby propping up the reliability of other licence holders (particularly during drier periods). Protecting the public's investment in sustainable water outcomes will require the imposition of rules to protect environmental water as it moves through the system, as well as environmental water for groundwater-dependent ecosystems (O'Donnell and Garrick 2017).

Implement Strategic Planning While significant progress has been made across Australia in water planning, the evolution has been slow and complicated, in part because the uncertainties confronted are high and the water management challenges complex. Acknowledging and addressing these challenges demands significantly improved strategic planning in future water reforms. Key issues raised across the chapters include:

- Greater comprehension of surface/ground water connectivity.
- Building in capacity for learning and adaptive management (see also Holley and Sinclair 2011; Cosens 2015).
- Accommodating the impact of climate change as a driver of future policy reform, water infrastructure development and limits on water consumption (c.f. Neave et al. 2015); and
- The impacts of characterising water entitlements as property rights and whether there should be greater consistency across State/Territory legislation.

Improve Models and Tools for Participation in Water Governance Australia's national water reforms embraced a 'top down' policy approach where public participation has been either via water markets or through non-binding processes of consultation undertaken by decision-makers. As a number of our contributors make clear (e.g. Lindsay), broad-based participation and the inclusion of local and traditional knowledge is crucial to the future management of water resources, not least to enhance decision making and to build trust and satisfaction with laws and plans (EJA 2015). The latter is particularly important to guard against further unwinding of Australia's water policy goals, ensure more successful implementation and enhance the legitimacy of the water. The suite of participatory models and tools will need to further mature and evolve, beyond mere opportunities to be 'consulted', and in addition, embrace third party rights to participate (Kallies and Godden 2008), more 'deliberative' democratic procedures (Tan et al. 2012), greater use of public inquiries and hearings, new 'co-management' arrangements (Son 2012; Holley et al. 2016), and continued evolution of environmental 'water trusts'. These various models and tools also need to be backed up by both policy support and legal mechanisms (e.g. improving access to water data) to better respond to power dynamics, resources and capacity among relevant interests and groups (Karkkainen 2002), with a view to better integrating distributive justice into planning and management.

Ensure Full Recognition of Indigenous Interests Despite some progress, the engagement and integration of Indigenous communities' needs and concerns, particularly commercial and native title rights, has not been a priority of Australian water governance (Marshall 2017). As Macpherson et al. and others make clear, improvements and fixes in this area are vital for equity and effectiveness of future water governance, particularly in relation to the development of northern Australia where Indigenous land tenure (land rights or native title) is highly significant (O'Neill 2014, p. 39). Indeed, any development of Australia's northern water resources will require this, and all of the above priorities to be addressed, in order to avoid repeating past mistakes (Martin 2016; Carmody et al. 2016, pp.132–134).

4.4.2 Hybrid Water Governance

A second set of questions raised by our contributors was the implications for theory and practice of hybrid systems of water law and governance. Understandings of hybrid policy approaches and their compatibility have roots in legal and regulatory pluralism (Gunningham, Grabosky and Sinclair 1998). Recent theories flowing from this include Smart Regulation, whose central normative argument is that, in the majority of circumstances, the use of multiple (and complementary) rather than single policy instruments, and a broader range of regulatory actors, can and should be used to produce better regulation outcomes (Gunningham and Holley 2016). This concept is (albeit implicitly) inherent to Australia's approach to water governance, involving collaborative water planning, cap and trade markets and the regulation of extraction limits. Although the chapters throughout the book draw on a broad range of thinking, including economics, deliberative democracy, adaptive governance, regulatory and multijurisdictional governance, economic and property theory, epistemic communities and institutional change, collectively their reflections on Australia's approach to water suggest that there is a need for further research on hybrid water governance approaches, their limits and when and how they can be compatible together or with other approaches.

As Cosens explains, there are strengths, such as enhanced adaptive capacity, that can arise from hybrid approaches of markets, government and self-organised capacity being chosen and used effectively. However, many of the diagnosed weaknesses within the book's chapters arise from a lack of complementarity or an inability to find new instruments and governance actors that can harmoniously address palpable weaknesses. McKay, for instance, refers to the need for better coordination between 'the three governance players' namely the law, the organisations and civil society's epistemic communities and NGOs. As she puts it 'all three players seem to be holding different shaped balls and rules and hence are poorly coordinated'. Nelson and Tan and Robertson also identify weaknesses and suggest the need for great attention to the interaction between Environmental Impact Studies, water planning and licensing regulation. Others raised challenges with water markets, but point out that remedies can be difficult due to market exclusivity. Indeed, the nature of economic instruments, is that they tend to 'crowd out' other regulatory approaches (for example, process based standards) that might otherwise be employed to deliver flexibility to regulated entities (briefly, this is because any constraints on the market or pricing that other regulatory policies might induce have the potential to undermine the efficiency of the market) (for a more detailed discussion see Gunningham et al. 1998). Holley and Sinclair acknowledge this challenge in querying the suitability of markets to groundwater, and call for exploration of new forms of complementary self-organisation. Gray and Lee also note this problem with markets, suggesting propertisation for markets may cause as many disputes as it solves, particularly because of the robust nature of the remedies available in cases where a property right is breached (such as compensatory measures when policy shifts). A range of chapters also point to the need for additional policy mixes, such as Lindsay's calls for additional participation approaches (see also Carmody) and Owens' analysis of collaborative approaches for NGOs in managing environmental water.

Addressing these challenges raises a further fundamental issue, namely who facilities and steers the existing or new regulatory mix. Across the range of chapters, it was clear that government has dominated the Australian policy space. As contributors acknowledged, the government designed and driven agenda has had major benefits, with plans in place, markets beginning to operate and drought being managed (e.g. Loch et al; Holley and Sinclair). It has also enabled distinctive efforts, such as the 'robust central role played by the Murray-Darling Basin Authority' in transboundary contexts like those detailed by Karkkainen in his comparison with United States and Canadian practice. Even so, many authors lamented the difficulty of shifting policy to optimise social, economic and environmental outcomes within this largely government driven system (e.g. Owens) where private environmental water is crowded out, public participation in governmental activities and conduct is limited, and self-organisation and Indigenous interests are marginalised (Cosens; Macpherson et al.).

These insights raise critical issues for further examination in water governance thinking, including whether hybrid systems of collaborative planning, markets and regulation are sufficient to steer water governance to a sustainable future? When and how can other credible water governance and policy alternatives operate in combination with such approaches? Which of the combinations will be most complementary and effectively fit within the broad parameters of a market-based water trading system? How can water governance bridge the gap between high-level policy intent and effective implementation on the ground? What institutional structures, responsibilities and relationships are best able to support policy implementation? How can robust regulatory and governance cultures, policies and practices be adopted and fostered within those agencies responsible for effective water governance on the ground (and act as a ballast against potential corruption, collusion and political interference)? And finally, how can water governance generate greater engagement, ownership, steering and ultimately commitment on the part of water users and other stakeholders?

These and other concerns will be particularly important given the ongoing challenges of water scarcity in the age of the Anthropocene, as well as the challenges it poses for achieving innovation in the future of water governance and law.

References

Araral, E., & Wang, Y. (2013). Water governance 2.0: A review and second generation research agenda. *Water Resources Management, 27*(11), 3945–3957.
Australian Water Partnership. (2017). Overview. http://waterpartnership.org.au/about/#page-12018. Accessed 10 Oct 2017.
Australian Competition and Consumer Commission. (2017). ACCC role in water. https://www.accc.gov.au/regulated-infrastructure/water/accc-role-in-water. Accessed 10 Oct 2017.
Australian Government (Senate Standing Committee on Rural and Regional Affairs and Transport). (2009). *Implications for long-term sustainable management of the Murray-Darling Basin system*. Canberra: Australian Government.
Australian National Audit Office (ANAO). (2011). *Restoring the balance in the Murray-Darling Basin: Audit Report No.27 2010–11*. Canberra: Commonwealth of Australia.
Baldwin, C., & Hamstead, M. (2015). *Integrated water resource planning*. Abingdon: Earthscan.

Basin Plan. (2012). (Cth).

Berlin Rules on Water Resources (Berlin Rules) (adopted 21 August 2004). Seventy-First Report of the International Law Association, 2004.

Biswas, A. (2004). Integrated water resources management: A reassessment. *Water International, 29*(1), 248–256.

Biswas, A., & Tortajada, C. (2010). Future water governance: Problems and perspectives. *International Journal of Water Resources Development, 26*(2), 129–139.

Bowmer, K. (2014). Water resources in Australia: Deliberation on options for protection and management. *Australasian Journal of Environmental Management, 21,* 228–240.

Carmody, E., Cosens, B., Gardner, A., Godden, L., Gray, J., Holley, C., et al. (2016). The future of water reform in Australia—starting a conversation. *Australian Environment Review, 31*(4), 132–137.

Commonwealth of Australia (Cwth of Aus). (2004). *Intergovernmental Agreement on a National Water Initiative.* Canberra: Commonwealth of Australia.

Commonwealth of Australia. (2014). *Report of the Independent Review of the Water Act 2007.* Canberra: Commonwealth of Australia.

Commonwealth of Australia. (2015). *Our north, our future: white paper on developing northern Australia.* Canberra: Commonwealth of Australia.

Competition and Consumer Act. (2010). (Cth).

Connell, D., & Quentin Grafton, R. (Eds.). (2011). *Basin futures: Water reform in the Murray-Darling Basin.* Canberra: ANU Press.

Convention on the Law of Non-Navigational Uses of International Watercourses, 21 May 1997, into force 19 May 2014, 36 International Legal Materials 700 (1997).

Cosens, B. (2015). Application of the adaptive water governance project to management of the Lake Eyre Basin and its connections to the Great Artesian Basin. *Australian Environment Review, 30*(6–7), 146–152.

Cosens, B., Gunderson, L., & Chaffin, B. (2014). The adaptive water governance project: assessing law, resilience and governance in regional socio-ecological water systems facing a changing climate. *Idaho Law Review, 1,* 1–27.

Council of Australian Governments (COAG). (1994). Council of Australian Governments' Communiqué, 25 February 1994, Hobart. http://ncp.ncc.gov.au/docs/Council%20of%20Australian%20Governments%27%20Communique%20-%2025%20February%201994.pdf. Accessed 10 Oct 2017.

Council of Australian Governments (COAG). (1995). *Agreement to Implement the National Competition Policy and Related Reforms.* Canberra: Australian Government.

Council of Australian Governments (COAG). (2013). National Harmonised Regulatory Framework for Natural Gas from Coal Seams. http://www.coagenergycouncil.gov.au/sites/prod.energycouncil/files/publications/documents/National-Harmonised-Regulatory-Framework-for-Natural-Gas-from-Coal-Seams_1.pdf. Accessed 10 Oct 2017.

Crutzen, P. J. (2002). Geology of mankind. *Nature, 415*(68/67), 3–23.

Crutzen, P. J., & Stoermer, E. F. (2000). The 'Anthropocene'. *IGBP Newsletter, 41,* 17–18.

De Burca, G., Keohane, R., & Sabel, C. (2014). Global experimentalist governance. *British Journal of Political Science, 44,* 477–486.

Department of the Environment and Energy. (2016). About commonwealth environmental water. https://www.environment.gov.au/water/cewo/about-commonwealth-environmental-water. Accessed 10 Oct 2017.

Environment Protection and Biodiversity Conservation Act. (1999). (Cth).

Environmental Justice Australia (EJA). (2015). Water citizenship: Advancing community involvement in water governance in Victoria. https://envirojustice.org.au/major-reports/water-citizenship-advancing-community-involvement-in-water-governance-in-victoria. Accessed 10 Oct 2017.

Evans, M., & Dare, L. (2013). *Multi-level governance and institutional layering: The case of national water governance.* Paper presented at the International Conference on Public Policy, Grenoble, France, 28 June 2013.

French-Australian Water and Land Management Initiative (FAWLMI). (2010). Canberra: ANU and French Embassy. http://politicsir.cass.anu.edu.au/events/french-australian-water-and-land-management-initiative-publications-launch. Accessed 10 Oct 2017.

FAO. (2007). Coping with water scarcity challenge of the twenty–first century. http://www.fao.org/3/a-aq444e.pdf. Accessed 10 Oct 2017.

Gardner, A., Bartlett, R., & Gray, J. (2009). *Water resources law*. Sydney: LexisNexis Butterworths.

Godden, L., & Foerster, A. (2011). Introduction: Institutional transitions and water law governance. *The Journal of Water Law, 22*(2/3), 53–57.

Gray, J., Holley, C., & Rayfuse, R. (Eds.). (2016). *Trans-jurisdictional water law and governance. Studies in Water Resource Management Series*. Abingdon: Earthscan.

Grafton, R. Q., Horne, J., & Wheeler, S. (2016). On the marketization of water: Theory and evidence from the Murray-Darling Basin, Australia. *Water Resource Management, 30*(3), 913–926.

Grafton, R. Q., & Horne, J. (2014). Water markets in the Murray-Darling Basin. In R. Q. Grafton, P. Wyrwoll, C. White, & D. Allendes (Eds.), *Global water: issues and insights* (pp. 37–44). Canberra: ANU Press.

Gross, C. (2011). Why justice is important. In D. Connell, & R. Quentin Grafton (Eds.), *Basin Futures Water reform in the Murray-Darling Basin* (pp. 149–162). Canberra: ANU Press.

Gunningham, N., & Holley, C. (2016). Next generation environmental regulation. *Annual Review of Law and Social Science, 12*, 273–293.

Gunningham, N., Grabosky, P., & Sinclair, D. (1998). *Smart regulation; designing environmental policy*. Oxford: Oxford University Press.

Gupta, J., & Pahl-Wostl, C. (2013). Editorial on global water governance. *Ecology and Society, 18*(4), 54.

Hannam, P. (2015). Parched NSW seeks help as National Water Commission axed. Sydney Morning Herald Online. http://www.smh.com.au/environment/parched-nsw-seeks-help-as-national-water-commission-axed-20150513-gh0ork.html. Accessed 18 June 2017.

Hardin, G. (1968). The tragedy of the commons. *Science, 13*, 1243–1248.

Hayek, F. (1945). The use of knowledge in society. *The American Economic Review, 35*(4), 519–530.

Hearns, G., Henshaw, T., & Paisley, R. (2014). Getting what you need: Designing institutional architecture for effective governance of international waters. *Environmental Development, 11*, 98–111.

Helsinki Rules on the Uses of Waters of International Rivers, 20 August 1966, Fifty-Second report of the International Law Association (1976), 484.

High Level Panel on Water. (2016). Action plan. https://sustainabledevelopment.un.org/content/documents/11280HLPW_Action_Plan_DEF_11-1.pdf. Accessed 10 Oct 2017.

Holley, C. (2016). Linking law and new governance: Examining gaps, hybrids and integration in water policy. *Law & Policy, 38*(1), 24–53.

Holley, C., & Sinclair, D. (2011). Collaborative governance and adaptive management: (Mis) Applications to groundwater, salinity and run-off. *The Australasian Journal of Natural Resources Law and Policy, 14*(1), 37–69.

Holley, C., & Sinclair, D. (2012). Compliance and enforcement of water licences in NSW: Limitations in law, policy and institutions. *Australasian Journal of Natural Resources Law and Policy, 15*(2), 149–189.

Holley, C., & Sinclair, D. (2016). Rethinking Australian water governance: successes, challenges and future directions. *Environmental and Planning Law Journal, 33*(4), 275–283.

Holley, C., Gunningham, N., & Shearing, C. (2012). *The new environmental governance*. Abingdon: Earthscan.

Holley, C., Sinclair, D., Lopez-Gunn, E., & Schlarger, E. (2016). Collective management of groundwater. In Tony Jakeman, Olivier Barreteau, Randall Hunt, Jean-Daniel Rinaudo, & Andrew Ross (Eds.), *Integrated groundwater management: Concepts approaches and challenges* (pp. 229–252). Open Access: Springer.

Hussey, K., & Dovers, S. (Eds.). (2007). *Managing water for australia*. Collingwood: CSIRO.

Intergovernmental Agreement on the Environment. (1992). Canberra: Australian Government.

International Law Commission (ILC). (2008). *Articles on trans-boundary aquifers*. Geneva, Switzerland: United Nations.

Jackson, Sue, Pollino, Carmel, Maclean, Kirsten, Bark, Rosalind, & Moggridge, Bradley. (2015). Meeting Indigenous peoples' objectives in environmental flow assessments: Case studies from an Australian multi-jurisdictional water sharing initiative. *Journal of Hydrology, 522,* 141–151.

Jasper, C. (2015). As California enters its fifth year of drought, state lawmakers undertake study mission to Australia. ABC. 20 Oct. http://www.abc.net.au/news/2015-10-20/californian-lawmakers-study-australian-drought-response/6869076. Accessed 10 Oct 2017.

Jury, W., & Vaux, H. (2005). The role of science in solving the world's emerging water problems. *PNAS 102*(44), 15715–15720.

Karkkainen, B. (2002). Collaborative ecosystem governance: scale, complexity and dynamism. *Virginia Environmental Law Journal, 21*(2), 189–243.

Karkkainen, B. A., & Fung, C. Sabel. (2000). After backyard environmentalism. *American Behavioural Scientist, 44*(4), 690–709.

Kallies, A., & Godden, L. (2008). What price democracy? Blue Wedges and the hurdles to public interest environmental litigation. *Alternative Law Journal, 33*(4), 194–199.

Kay, A. (2013). Multi-level governance in Australian federalism: The open method of coordination in open economy policy-making. 26–28 June 2013. In: International Conference on Public Policy, Grenoble. http://archives.ippapublicpolicy.org/IMG/pdf/panel_45_s3_kay.pdf. Accessed 10 Oct 2017.

Kidd, M., Feris, L., Murombo, T., & Iza, A. (Eds.). (2014). *Water and the law: Towards sustainability*. Cheltenham: Edward Elgar.

Kildea, P., & Williams, G. (2011). The water act and the murray-darling basin plan. *Public Law Review, 22,* 9–14.

Kiparsky, M., Milman, A., Owen, D., & Fisher, A. T. (2017). The importance of institutional design for distributed local-level governance of groundwater: The case of california's sustainable groundwater management act. *Water, 9*(10), 755.

Kochskämper, Elisa, Challies, Edward, Jager, Nicolas W., & Newig, Jens. (2018). *Participation for effective environmental governance: Evidence from european water framework directive implementation*. New York: Routledge.

Marshall, V. (2017). *Overturning aqua nullius: Securing aboriginal water rights*. Canberra: Aboriginal Studies Press.

Martin, P., & Kennedy, A. (2015). Introduction: A jurisprudence of environmental governance. In P. Martin & A. Kennedy (Eds.), *Imlementing environmental law* (pp. 1–26). Cheltenham: Edward Elgar.

Martin, P. (2016). Creating the next generation of water governance. *Environmental and Planning Law Journal, 33*(4), 388–401.

Matthews, K. (2017). *Independent investigation into NSW water management and compliance: Interim Report*. Sydney: NSW Government.

Millar, I. (2005). *Testing the waters: Legal challenges to water sharing plans in NSW*. Paper presented at the Water Law in Western Australia Conference, 8 July. http://www.edonsw.org.au/testing_the_waters_legal_challenges_to_water_sharing_plans_in_nsw. Accessed 10 Oct 2017.

McKay, C., & Gardner, A. (2013). Water accounting information and confidentiality in Australia. *Federal Law Review, 41,* 127–220.

McDonald, J., & Styles, M. (2014). Legal strategies for adaptive management under climate change. *Journal of Environmental Law, 26,* 25–53.

Moore, M., von der Porten, S., Plummer, R., Brandes, O., & Baird, J. (2014). Water policy reform and innovation: A systematic review. *Environmental Science & Policy, 38,* 263–271.

Moynihan, R. (2014). The Rising role of regional approaches in international water law: Lessons from the UNECE water regime and himalayan asia for strengthening transboundary water cooperation. *Review of European Community and International Environmental Law, 23*(1), 43–58.

Mukhtarov, F., & Gerlak, A. (2014). Epistemic forms of integrated water resource management: towards knowledge versatility. *Policy Sciences, 47*(2), 101–120.

Murray-Darling Basin Authority. (2015). Sustainable diversion limit adjustment mechanism. http://www.mdba.gov.au/basin-plan-roll-out/sustainable-diversion-limits. Accessed 10 Oct 2017.

National Competition Council (NCC). (2005). *Assessment of Governments' Progress in Implementing the National Competition Policy and Related Reforms*. Melboune: Commonwealth of Australia.

National Strategy for Ecologically Sustainable Development. (1992). Canberra: Australian Government.

National Water Commission (NWC). (2014a). *Australia's water blueprint: National reform assessment 2014*. Canberra: Australian Government.

National Water Commission (NWC). (2014b). *Water for mining and unconventional gas under the national water initiative*. Canberra: Australian Government.

National Water Commission (NWC). (2014c). 'Don't drop the ball on water' urges national water commission. Canberra: Australian Government. http://webarchive.nla.gov.au/gov/20160615063005/http://www.nwc.gov.au/media-centre/media/dont-drop-the-ball-on-water-urges-national-water-commission. Accessed 10 Oct 2017.

Neave, I., McLeod, A., Raisin, G., & Swirepik, J. (2015). Managing water in the Murray-Darling Basin under a variable and changing climate. *AWA Water Journal, 42*(2), 102–107.

Nourse, V., & Shaffer, G. (2014). Empiricism, experimentalism, and conditional theory. *Southern Methodist University Law Review, 67*, 101–142.

OECD. (2011). *Water governance in OECD countries: A multi-level approach*. London: OECD Publishing.

O'Donnell, E., & Garrick, D. (2017). Defining success: a multi-criteria approach to guide evaluation and investment. In A. Horne, A. Webb, M. Stewardson, B. Richter, & M. Acreman (Eds.), *Water for the environment* (pp. 625–648). London: Elsevier.

O'Neill, L. (2014). The role of state governments in native title negotiations: A tale of two agreements. *Australian Indigenous Law Review, 18*(2), 29–42.

Ostrom, E. (1990). *Governing the commons*. Cambridge: Cambridge University Press.

Owens, K. (2017). *Environmental water markets and regulation: A comparative legal approach*. United Kingdom: Routledge.

Pahl-Wostl, Claudia, Holtz, Georg, Kastens, Britta, & Knieper, Christian. (2010). Analyzing complex water governance regimes. *Environmental Science & Policy, 13*(7), 571–581.

Pahl-Wostl, C. (2015). *Water governance in the face of global change*. Switzerland: Springer.

Papas, M. (2014). River basin management between european union member states: What can australia learn from another multilevel governance system? *Australasian Journal of Natural Resources Law and Policy, 17*(2), 153–188.

Parsons, R., Lacey, J., & Moffat, K. (2014). Maintaining legitimacy of a contested practice: How the minerals industry understands its 'social licence to operate'. *Resource Policy, 41*, 83–90.

Pittock, J., Hussey, K., & Dovers, S. (Eds.). (2015). *Climate, energy and water*. New York: Cambridge University Press.

Sabel, C., & Simon, W. (2011). Minimalism and experimentalism in the administrative state. *Georgetown Law Journal, 100*(1), 53–93.

Salman, S. (2007). The united nations watercourse convention ten years later: Why has its entry into force proven difficult? *Water International, 32*, 1–15.

Steffen, W. (Climate Council of Australia). (2015). Thirsty Country: climate change and drought in Australia. www.climatecouncil.org.au/droughtreport2015. Accessed 10 Oct 2017.

Scheuerman, W. (2004). Democratic experimentalism or capitalist synchronization? Critical reflections on directly-deliberative polyarchy. *Canadian Journal of Law and Jurisprudence, 17*, 101–127.

Son, C. (2012). Water reform and the right for Indigenous Australians to be engaged. *Journal of Indigenous Policy, 13,* 3–26.
Sustainable Development Goals. (2015). http://www.un.org/sustainabledevelopment/water-and-sanitation. Accessed 10 Oct 2017.
Syme, G., & Nancarrow, B. (2008). The social and cultural aspects of sustainable water use. In Lin Crase (Ed.), *Water policy in Australia* (pp. 230–247). Washington DC: RFF Press.
Tan, P. L., Bowmer, K. H., & Baldwin, C. (2012). Continued challenges in the policy and legal framework for collaborative water planning. *Journal of Hydrology, 474*(12), 84–91.
Tarlock, D. (2010). Four challenges for international water law. *Tulane Environmental Law Journal, 23*(2), 369–408.
Tujchneider, O., Christelis, G., & Van der Gun, J. (2013). Towards scientific and methodological innovation in transboundary aquifer resource management. *Environmental Development, 7,* 6–16.
Tyler, T. (2005). *Readings in procedural justice*. Burlington: Ashgate.
van Rijswick, M. (2016). Trans-jurisdictional water governance in the European Union. In Janice Gray, Cameron Holley, & Rosemary Rayfuse (Eds.), *Trans-jurisdictional Water Law and Governance* (pp. 62–78). Abingdon: Earthscan.
Water Act. (2007). (Cth).
Water Market Rules. (2009). (Cth).
Water Charge (termination Fees) Rules. (2009). (Cth).
Whitehead, I. (2014). Better protection or pure politics? Evaluating the 'water trigger' amendment to the EPBC Act. *Australian Environmental Law Digest, 1,* 23–36.
Williams, J. (2017). Turning the tide of water reform: Time to reignite Australia's stagnant policy debate. APPS Policy Forum, 8 Feb. https://www.policyforum.net/turning-tide-water-reform/. Accessed 10 Oct 2017.
WEF (World Economic Forum). (2017). The global risks report 2017, 12th edn. http://wef.ch/risks2017. Accessed 10 Oct 2017.
Wentworth Group of Concerned Scientists (B Tucker). (2014). Statement on the future of Australia's water reform.
Wentworth Group of Concerned Scientists. (2017). Five actions necessary to deliver the Murray-Darling Basin Plan 'in full and on time'. 5 June. http://wentworthgroup.org/wp-content/uploads/2017/06/Five-actions-to-deliver-Murray-Darling-Basin-Plan-Wentworth-Group-June-2017.pdf. Accessed 10 Oct 2017.
Wilhite, Donald A. (2012). Breaking the hydro-illogical cycle: Changing the paradigm for drought management. *EARTH Magazine, 57*(7), 71–72.
World Commission on Environment and Development. (1992). *Our Common Future*. Oxford University Press 1987.
World Meteorological Organization. (1992). International conference on water and the environment: Development issues for the 21st Century. Dublin, 26–31 January 1992: Dublin Statement and Report of the Conference (World Meteorological Organization, 1992).
WWAP (United Nations World Water Assessment Programme). (2015). *The united nations world water development report 2015: Water for a sustainable world*. Paris: UNESCO.

Part I
The Murray-Darling Basin—Progress and Challenges in Multi-jurisdictional Water Governance

The Unwinding of Water Reform in the Murray-Darling Basin: A Cautionary Tale for Transboundary River Systems

Emma Carmody

Abstract The passage of the *Water Act 2007* by the Australian Government was an historic moment in the management of the Murray-Darling Basin, a transboundary river system home to 16 Ramsar wetlands and responsible for generating approximately 50% of the country's irrigated produce (Murray-Darling Basin Authority 2017). The Act sought to end decades of unsustainable water use—principally for irrigated agriculture—by introducing new limits on water extraction across the Basin. The vehicle for achieving this goal, the Basin Plan, was in turn passed in late 2012. However—and contrary to the requirements of the Act—the limits imposed under the Plan were not sustainable, not least of all because they failed to take into account likely, future climate change. Implementation of the Plan's various sub-instruments and strategies has also been beset by statutory and policy changes that privilege consideration of socio-economic factors over environmental outcomes. This chapter examines the aforementioned reorientation and sets out a series of recommendations intended to restore faith in water reform processes in the Murray-Darling Basin. It is also hoped that this analysis will assist regulators and stakeholders to identify and manage possible barriers to proper implementation of domestic and international water laws in other, transboundary basins.

1 Introduction

The National Plan for Water Security was outlined by conservative Prime Minister John Howard on 25 January 2007. In his landmark speech, Prime Minister Howard (Howard 2007) described water as 'Australia's greatest conservation challenge. No single substance has a greater impact on the human experience or on our

---

E. Carmody (✉)
Visiting Fellow, Faculty of Law, University of New South Wales,
Environmental Defenders Office NSW, Sydney, Australia
e-mail: emma.carmody@edonsw.org.au

© Springer Nature Singapore Pte Ltd. 2018
C. Holley and D. Sinclair (eds.), *Reforming Water Law and Governance*,
https://doi.org/10.1007/978-981-10-8977-0_2

environment.' He went on to announce a $10 billion plan to 'improve water efficiency and to address the over-allocation of water in rural Australia, particularly in the Murray-Darling Basin [MDB]' (National Plan for Water Security). The legal cornerstone of this program—the *Water Act 2007*—was enacted by the Australian Parliament shortly thereafter. An ambitious and complex piece of legislation, its overriding objective is distilled in the requirement to reinstate an 'environmentally sustainable level of take' (ESLT) (*Water Act* 2007, s. 23) in Australia's largest and most productive river basin, the MDB. The Act sets out the two main mechanisms by which this is to be achieved: the development of a legislative instrument known as the 'Basin Plan' imposing—*inter alia*—limits on water extraction across the Basin and the establishment of a statutory authority known as the 'Commonwealth Environmental Water Holder' (CEWH) tasked with purchasing water entitlements from willing sellers in order to 'return' water to the environment and ensure that these limits are met.

While the imposition of a statutory limit on water extraction predates the Water Act, the Act nonetheless constituted a significant step forward in the management of water resources in the MDB, not least of all because—if properly implemented—it would result in the long-term protection of the river system and its water-dependent ecosystems, which includes 16 Ramsar-listed wetlands. However, in the ten years following the Act's passage through Parliament, pressure from the irrigation lobby and certain Basin States has significantly undermined Prime Minister Howard's vision for the MDB. In short, the focus has shifted from reinstating an ESLT and restoring the health of the Basin to minimising the socio-economic impacts that *may* be associated with water recovery for the environment. This reorientation, coupled with the absence of any meaningful reference to climate change in the Basin Plan, not only threatens the survival of the Basin's water resources, it ignores the link between sustainable water extraction and the long-term viability of Basin communities. It also risks embedding—for decades to come—unsustainable extraction limits and rules in legislative instruments that are being developed under the Basin Plan. Given the virtually impenetrable nature of these rules and Australia's reputation as a global leader in water management, there is a real possibility that the Basin Plan will nonetheless provide a credible façade of world's best practice. As this is arguably more detrimental than unsustainable practices and policies that are clearly identified as such, there is an urgent need to catalogue these shortcomings and to propose solutions.

This chapter will therefore examine the various legislative amendments, policy decisions and processes underpinning the aforementioned reorientation and analyse whether the Basin Plan—which is in the process of being implemented—is consistent with the *Water Act* and, to that extent, constitutionally valid. It will then advance a set of legal, policy and governance strategies designed to restore faith in water management practices in the MDB and, crucially, to reinstate an ESLT and protect environmental water. Finally, it is hoped that this analysis will in turn assist other jurisdictions in the process of revising existing—or developing new—water management laws and policies. To that end, while the MDB falls within a single nation state, it is nonetheless a transboundary basin and accordingly beset by a range of social, political and environmental challenges. As such, it is possible to draw certain parallels between the regulation of the MDB and that of other, well-known transboundary basins such as the Colorado, Danube and Jordan Rivers, respectively.

1.1 Environmental Background

The MDB covers approximately 1 million square km and traverses parts of four separate state jurisdictions and one entire territory (Basin states). It contains Australia's three largest rivers: the Darling, the Murray and the Murrumbidgee. Its climate has been described as 'relatively dry compared to other regions of Australia', with approximately 94% of all rainfall lost through evapotranspiration (Australian Bureau of Statistics 2010). Irrigated agriculture and horticulture cover two percent of its surface area, but account for approximately 90% of Basin-wide water extractions (Grafton et al. 2014). Decades of over-extraction by these industries—particularly from the 1960s onwards—has resulted in significant decline in the health of water resources and water-dependent ecosystems across the Basin (Jones et al. 2002). This culminated with the revelation in 2010 that 20 out 23 river valleys in the Basin were in 'poor' or 'very poor' ecological condition (Davies et al. 2008). While 'much of the data was collected during drought conditions' (Murray-Darling Basin Authority 2011b), the findings were consistent with studies conducted in the pre-drought period (Norris et al. 2001). Significantly, data from a range of sources indicated that the Basin's 16 Ramsar-listed wetlands—a metaphorical flock of canaries in the coal mine—were in serious decline, with many having undergone radical changes in both extent and character in the preceding decades (Kingsford et al. 2011).

The relationship between extraction and drought is significant, with the former exacerbating the impacts of the latter (Grafton et al. 2014). Indeed, certain species, populations and communities have been unable to recover from periods of scarcity or adapt to the conditions imposed by this dramatically altered environment, while others have proven more resilient. However, many ecosystems that have managed to adapt appear to be functioning at the 'low end of historically much higher ranges' (Colloff et al. 2015). This prognosis extends to many of the aforementioned Ramsar-listed wetlands (Cunningham et al. 2011). For example, in 2009, the Australian Government lodged an Article 3.2 notice with the Ramsar Secretariat, indicating that the Macquarie Marshes were likely to experience a change in ecological character (Australian Government 2009).

2 The Refocussing of Basin Law and Policy Between 2007 and 2017

2.1 Water Act

It was against the backdrop of this slowly evolving ecological catastrophe that the *Water Act* was passed in mid-2007. Sitting at the pointy intersection between federalism and natural resource management, it was the most significant—and controversial—piece of water legislation that had ever been passed by an Australian Parliament. The Act itself brought the allocation and protection of Basin water resources within the purview of the Australian Government, thereby ending 100 years of near-exclusive control of Basin water resources by state and territory governments. However, execution of the Act still depended on inter-governmental co-operation. As such, its passage ushered in a new era of negotiations and political skirmishes regarding the precise nature and implications of the updated regulatory framework. These skirmishes have come to define the legal and policy evolution—or devolution—that has occurred in the ten years following its introduction. All this despite—or possibly because of—the fact that the Act is one of the most rigorous pieces of environmental legislation in Australia.

This rigour is largely due to the constitutional underpinnings of the statute. Specifically, the Act derives the majority of its constitutional validity from a suite of environmental treaties to which Australia is signatory (*Water Act 2007* (Cth) s 9). Known as the 'Relevant International Agreements' (*Water Act 2007* (Cth) s 4), they include the Convention on Biological Diversity (Biodiversity Convention), the Convention on Wetlands of International Importance especially as Waterfowl Habitat (Ramsar Convention), the Convention on the Conservation of Migratory Species of Wild Animals (Bonn Convention) and a number of bilateral migratory bird conventions. As a consequence, the Act and any of its subordinate legislative instruments must satisfy the High Court's test for 'proper implementation' of any treaty to which Australia is a party. Specifically, the legislative framework must be 'appropriate and adapted' to the task of implementation (*State of Victoria v Commonwealth 1996*).

Relevantly, the two most constitutionally significant of the Relevant International Agreements are the Biodiversity Convention and Ramsar Convention. This is due to the breadth of the former and the fact that 16 wetlands in the MDB are listed under the latter. According to the Australian Government Solicitor (AGS), these two Conventions 'establish a framework in which environmental objectives have primacy but the implementation of environmental objectives allows consideration of social and economic factors' (AGS 2010). This hierarchy is reiterated by two of Australia's most eminent constitutional lawyers, Professor George Williams and Dr Paul Kildea, who note that while 'some social and economic factors can be taken into account in the meeting of the core environmental objectives', proper implementation of these Agreements requires environmental obligations to be given greater weight than socio-economic factors (Williams and Kildea 2011). This provides a certain inviolable environmental 'safety net'; failure to maintain this net would render the Act and its instruments unconstitutional. Accordingly, the objects (ss. 3(b), (c)) and other key provisions of the *Water Act* (ss. 20(a), 21(1), 36(1)(a), 60, 86AA and 105(3)) provide for implementation of the Relevant International Agreements, in particular 'to address the threats to the Basin water resources' (s.3 (b)). As noted in the introductory paragraph, the two principal means by which this is to be achieved are the creation of a legislative instrument known as the 'Basin Plan' and the voluntary sale of water entitlements to the CEWH, known colloquially as 'buybacks'. These will be discussed in turn.

The Basin Plan is a framework agreement insofar as it provides for the creation and implementation of subordinate instruments and strategies over the twelve years following its enactment. Arguably the most important of these are 'water resource plans' (WRPs) which must be made for each 'water resource area' (WRA) in the Basin (Water Act, ss. 22(1) (Item 11); 22(3)). Each WRP is required to specify the Sustainable Diversion Limit for the relevant WRA (Water Act, ss. 22(3)(b), (6A), (6B)), which will in turn come into effect by mid-2019 (Basin Plan, cl. 6.04(1)). In other words, they must identify how water is to be allocated within the area, including to the environment (Water Act, s. 22(3)). The *Water Act* also stipulates that they must include, *inter alia*, requirements regarding: the regulation of interception activities with a significant impact (individually or cumulatively) on the water resources within the WRA; environmental water; water quality and salinity objectives; trading rules; and metering and monitoring (Water Act *2007* (Cth) s 22 (3)(d)(e)(f)(g)(i)). The WRP must also be prepared having regard to the 'social, spiritual and cultural matters relevant to Indigenous people' in relation to the water resources of the WRA (Water Act, s. 22(3)(ca)).

Relevantly, the Act requires the Basin Plan—via WRPs—to impose SDLs that reflect an ESLT for all Basin water resources (*Water Act 2007* (Cth) s 20(b)). The concept of an 'ESLT' is unique to the Act and is defined as the level of take from a water resource which, if exceeded, would compromise any one of the following elements of that resource: its key environmental assets, its ecosystem functions, its productive base or key environmental outcomes (*Water Act*, s. 4). In order to meet this requirement, the Basin Plan specifies SDLs for each of the relevant surface water and groundwater resource areas in the Basin. Indeed, the Act states that the SDLs 'must' reflect an ESLT (Water Act, ss. 22(1)(Item 6); 23(1)). It is arguable that the reinstatement of an ESLT is the principal means by which the Act and Basin Plan give effect to the Relevant International Agreements. Specifically, the impacts of overconsumption of water on biodiversity—including Ramsar wetlands—can only be adequately addressed by first, reallocating more water to the environment (Pittock and Finlayson 2011b) and second, by ensuring that this water is protected from extraction for irrigation (Murray-Darling Basin Authority 2011a).

The second key means by which the Water Act seeks to address the over allocation of water resources is via the purchase of water entitlements. This water is known as 'Commonwealth Environmental Water Holdings'. As the Act prohibits the compulsory acquisition of entitlements (Water Act, s. 255), they are acquired by the Commonwealth through a public tender process (Australian Government, Cth

Water Purchasing 2016). To date, the Commonwealth has purchased approximately 2,106 GL of permanent entitlements (measured as an annual long-term average yield) (Department of Agriculture and Water Resources 2018) and has used its water to improve environmental outcomes in a number of catchments (CEWO 2013). Indeed, the purchase of entitlements is arguably the single most important means of reducing the pool of consumptive water and ensuring compliance with the SDLs set out in the Basin Plan. It is also the most economical: the Productivity Commission estimated that 'the Australian Government may pay up to four times as much for recovering water through infrastructure upgrades than through water purchases' (Productivity Commission 2010). Notwithstanding these advantages, amendments to the Water Act (ss.23A, 23B; Part 2AA) and the final version of the Basin Plan (Chap. 7) have resulted in a limit being imposed on the outright purchase of entitlements (*Water Act*, Division 5, Part 2). 'Buybacks' have been supplanted by heavily subsidised infrastructure upgrades intended to improve on-farm efficiency (with the resulting water savings being purchased by the CEWH). As these developments support the central thesis of this chapter, they will be discussed in more detail in the following sections. Before doing so, it is necessary to consider the science that was originally developed to underpin the creation of the Basin Plan, and in particular the setting of SDLs.

2.2 Guide to the Basin Plan

The complex suite of environmental, social and economic obligations set out in the *Water Act*, in particular the requirement to reinstate an ESLT, necessitated the development of a body of scientific and socio-economic literature to inform the Basin Plan. This preparatory work was undertaken by the statutory authority empowered to develop and deliver the Plan, namely the Murray-Darling Basin Authority (MDBA or Authority) (*Water Act*, s. 41, Part 9). The MDBA's analysis was eventually published in August 2010 in a document entitled 'The Guide to the proposed Basin Plan' (Guide) (MDBA 2010). The Guide stipulated a water recovery range that reflected the MDBA's best estimate of what might constitute an ESLT. This range was based on detailed modelling which, in simple terms, calculated the ecological outcomes that could be achieved for 18 'hydrologic indicator sites' across the Basin using different volumes of water and watering regimes. Information regarding the watering requirements for these sites—which included 10 Ramsar wetlands—was then used to determine the overall reduction in SDLs needed to reinstate an ESLT.

The MDBA noted that the *Water Act* identified two 'broad requirements' regarding the development of SDLs: to 'establish SDLs that reflect an [ESLT]' and 'in doing so [ensure] the economic, social and environmental outcomes are optimised and the net economic returns maximised' (MDBA 2010). On this basis, it tested three water recovery scenarios: 3,000 GL/year, 3,500 GL/year and 4,000 GL/year and concluded that a minimum of 3,000 GL and a maximum of 7,600 GL/year would need to be recovered, with a preference for a recovery figure of between 3,000 and 4,000 GL/year (which represents 27–37% of current consumption) (MDBA 2010). The Authority's conclusions were highly contentious, to say the least. On the one hand, the proposal to abstract even 3,000 GL/year from the consumptive pool was met with outright hostility by certain communities, culminating in copies of the Guide being burnt in irrigation-dependent towns such as Griffith (Rogers 2010). On the other, a number of scientists and conservation groups insisted that a sustainable level of take would require Basin-wide extractions to be reduced by a figure at the upper end of the bracket (Pittock 2011).

The Authority's preference for a figure located at the lower end of the proposed recovery spectrum was based on concerns regarding socio-economic impacts beyond this range (MDBA 2010). However, its own analysis seemed to indicate that the lower end of the bracket was unlikely to yield optimal environmental outcomes across the entire Basin, noting that it 'believes the environmental water requirements for key environmental assets and key ecosystem functions can be achieved with a high level of uncertainty with a Basin-wide reduction in diversions of 3,000 GL/y…' (MDBA 2010). The inability to meet these requirements with anything other than a 'high level of uncertainty' suggested that 3,000GL/year did not, in fact, reflect an ESLT and to that extent would fall foul of Australia's obligations under the Relevant International Agreements, in particular the Ramsar Convention. While 7,600 GL/year could 'address current degradation of key ecological assets' (Pittock 2013) in the MDB at a low level of uncertainty, the MDBA considered that this figure 'would not optimise economic, social and environmental outcomes' (MDBA 2010). As such, speculation remained (and debate continues) regarding a legally acceptable reduction figure. This issue will be addressed in Part 3.

2.3 Draft Basin Plan to Final Basin Plan

The 3,000–4,000 GL/year bracket recommended in the Guide would come to foreshadow the progressive erosion of the volume of environmental water provided for under the Basin Plan, as well as the refocusing of MDB law and policy more generally. The first clear indication of this trajectory coincided with the release of the Draft Basin Plan (Draft Plan) in late 2011. The much-anticipated Draft Plan proposed a reduction figure of 2,750 GL/year, 250 GL/year less than the absolute minimum amount considered viable by the Authority. Unsurprisingly, a scientific review of the MDBA's modelling for a 2,800 GL/year reduction scenario conducted by the Commonwealth Scientific and Industrial Research Organisation (CSIRO) indicated that this figure would be insufficient to meet 'several of the specified hydrologic and ecological targets' set by the MDBA for Basin 'indicator sites' (which include ten Ramsar wetlands) (Young et al. 2011). The Draft Plan was further criticised by the Wentworth Group of Concerned Scientists as lacking detail (in particular details regarding ecological targets reached under this scenario), and for increasing groundwater extractions by 2,600 GL/year (Wentworth Group 2012).

The Draft Plan was followed in May 2012 by a revised Proposed Basin Plan (Revised Proposed Plan) which maintained the 2,750 GL/year scenario. However, the accompanying letter of transmission from the Chair of the MBDA to the Chair of the Murray-Darling Basin Ministerial Council introduced the notion of an 'adjustment mechanism' which would allow this figure to be increased or decreased:

> Starting from the proposed 2750GL reduction, such a mechanism could operate on the basis of:
> - allowing for a decrease in SDLs only if the social and economic outcomes are at least equivalent (or better than) those proposed in the final Plan; or
> - allowing for an increase in SDLs only if the environmental outcomes are at least equivalent (or better than) those proposed in the final Plan (MDBA Transmittal Letter 2012).

Though this description is schematic in nature, the mechanism was clearly and unapologetically intended to appease concerns in certain Basin States about the possible socio-economic impacts associated with a reduction in water allocations for irrigation. The adjustment mechanism itself was provided for in the Amended Proposed Basin Plan (Amended Proposed Plan), released in August 2012. Briefly, the Amended Proposed Plan provided that the 2,750 GL/year baseline figure could move upward if 'efficiency measures' (that is, on-farm irrigation upgrades) were implemented, thereby enabling more water to be delivered to the environment without reducing irrigators' productivity. Conversely, the baseline figure could move downward if 'supply measures' (such as regulators delivering water to wetlands) were given effect, allowing less water to be delivered to the environment without compromising environmental outcomes. Net adjustments were limited to 5% of the total surface water SDL of the Basin. These provisions, while moved to another chapter in the final version of the Basin Plan (passed in late 2012), essentially remained unchanged. That is, the Final Basin Plan legislated the 2,750 GL/year baseline figure that could be reduced by up to 650GL/year via supply measures and increased by up to 450GL/year if efficiency measures were implemented (Basin Plan, Chap. 7). As these 'works and measures' have become an important and ultimately contentious component of Basin law and policy, they will be examined in greater detail, below.

2.4 Amendments to the Water Act

Before doing so, it is necessary to touch on two amendments to the *Water Act* which were passed through Parliament in late 2012. The first amendment, the *Water Amendment (Long-term Average Sustainable Diversion Limit Adjustment) Bill 2012* adding sections 23A and 23B of the *Water Act*, was practical in nature, essentially ensuring that there was a solid legislative basis for incorporating the aforementioned adjustment mechanism into the final Basin Plan. The second amendment, *Water Amendment (Water for the Environment Special Account) Act 2013*, Sects. 86AA—86AJ of the *Water Act* created an 'Environmental Water Holdings Special Account' (Special Account) designed to fund 'efficiency works'. Specifically, these amendments provided for up to 450 GL/year of water 'saved' through on and off-farm efficiency works and other measures. Ostensibly, the amendment was to facilitate the purchase and delivery of this water to the end of the river system, located near the South Australian capital, Adelaide. In particular, it reflected a political deal struck between the South Australian and Commonwealth Governments to use this extra water to deliver specific environmental outcomes in the Coorong, Lower Lakes and Murray Mouth (CLLMM) region. However, the wording of the amendment was sufficiently flexible and non-binding so as to avoid a formal legislative breach in the event that the 450 GL/year could not be purchased by the requisite date.[1] In the event that this volume of water could be saved and the corresponding licenses purchased by the Commonwealth, it would bring the recovery figure up to approximately 3,200 GL/year. Modelling completed by the MDBA indicated that a 3,200 GL/year scenario which took into account the removal of several 'operational constraints'[2] (such as roads and bridges which prevent the delivery of water to floodplains) would still only achieve 66% of the Authority's own environmental targets for the Basin (Wentworth Group of Concerned Scientists 2012). It remains unclear what percentage of targets will be met if 450 GL/year of water is recovered via efficiency works. However, for the reasons outlined below, it is unlikely to be more than 66%.

2.5 Supply Measures and Efficiency Works

As previously noted, the inclusion of an adjustment mechanism was underpinned by a strong desire to mitigate real or perceived socio-economic impacts associated with reductions in SDLs across the Basin. From the outset, state governments (in particular NSW and Victoria) expressed a strong preference for the implementation of supply measure projects which would increase SDLs by 650 GL/year, thereby reducing the aggregate reduction figure to approximately 2,100 GL/year (before

---

[1]The provisions regarding environmental outcomes in the Basin generally and the CLLMM region specifically are provided for in the objects of the amending Part 2AA, rather than substantive provisions. Furthermore, the language employed in this Part is discretionary, which means that the outcomes sought in the CLLMM region are not strictly binding (except to the extent that failure to achieve them would result in a breach of the Ramsar Convention, which could in turn undermine the constitutional validity of the *Water Act* and *Basin Plan*). The requisite date is mid-2024: *Basin Plan*, cl. 7.11.

[2]The final *Basin Plan* requires the MBDA to produce a Constraints Management Strategy. Constraints are physical and policy constraints which inhibit the delivery of environmental water onto floodplains. The Constraints Management Strategy was published by the MDBA in 2013.

factoring in efficiency works). While the Basin Plan requires supply measures to procure 'equivalent environmental outcomes' within the prescribed 'limits of change', questions remain regarding the methodology underpinning these outcomes (Wentworth Group of Concerned Scientists 2012). Specifically, these measures are as yet unproven and will nonetheless result in the river system as a whole receiving less water. As the success of supply measures is in part based on a spatial rationalisation of environmental outcomes, they will also result in trade-offs. That is, good environmental outcomes in one area will 'compensate' for less favourable outcomes in another.

As the new SDLs mandated under the *Basin Plan* commence in 2019, supply measures and efficiency works must be finalised by June 2017 (*Basin Plan*, cl. 7.10) with a reconciliation date in 2024 (*Basin Plan*, cl. 7.11) to account for any discrepancies between the proposed and final adjustment figure. Accordingly, business cases have been prepared for both supply and efficiency projects by the NSW, Victorian and South Australian Governments. A stocktake and audit of the proposed projects was undertaken by independent consultants at the request of the Murray-Darling Basin Ministerial Council (Stocktake Report) (Martin et al. 2015). The Stocktake Report indicated that there was a 'high level of confidence' that the proposed supply projects would deliver approximately 500 GL/year in water 'savings', and 'moderate to high' confidence that the efficiency work projects would deliver the 'Commonwealth's program objective of 106 GL/year by mid-2019' (Martin et al. 2015). However, the Stocktake Report went on to note that 'there is considerable risk that the program aim of 450 GL by 2023/4 will not be met' (Martin et al. 2015).

The aforementioned findings regarding the efficiency works program compound those concerns outlined above about the flexible nature of the legal obligations set out in the Special Account amendment. The absence of any statutory obligation to purchase any particular class of licence, or licences that are physically capable of resulting in water being delivered to a particular area, also raises questions about the likely impact of the program on the ailing CLLMM region. Finally, the 1,500GL/year statutory limit on 'buybacks' that was introduced in 2015 (discussed below), assumes that a sufficient quantity of water will be 'saved' via supply works to ensure compliance with SDLs by mid-2019. It is entirely plausible that technical difficulties or budgetary constraints will mean that actual water savings will be less than projected savings. Thus, in the event that supply works do not 'save' the requisite quantity of water, the Commonwealth will be unable to 'bridge the gap' due to the aforementioned legislative embargo.

2.6 *1,500 G/L Cap on the Purchase of Entitlements*

The *Water Amendment Bill 2015* was introduced to Federal Parliament in May 2015 and referred to the Senate Environment and Communications Legislation Committee two weeks later. It was subsequently passed by both Houses on 14

September 2015 (as recommended in the Committee's report) and assented to in mid-October.[3] The Bill's central feature was an amendment to the Water Act which placed a 1,500 GL/year cap on the purchase of water entitlements by the Commonwealth Government (exempting the purchase of entitlements associated with on-farm water efficiency works). The Replacement Explanatory Memorandum for the Bill succinctly outlined the basis for such a cap:

> The Bill will address the concerns of rural and irrigation communities regarding the potential socio-economic impacts of Commonwealth environmental water purchases by placing a 1500 gigalitre limit on the amount of surface water that can be purchased ('water purchase contracts') by the Commonwealth in 'bridging the gap' to the sustainable diversion limits (SDLs) set out in the Basin Plan (Water Amendment Bill 2015, Replacement Explanatory Memorandum).

Concerns regarding the CEWH's impact on rural communities are longstanding, and have generated considerable animosity across the Basin. While the notion that the CEWH's immense purchasing power distorts water markets is deeply ingrained, there is little objective evidence to support this claim. For example, the 2014 'Report of the Independent Review of the Water Act 2007' stated that 'the NWC's (National Water Commission's) assessments have found no evidence of environmental water holders distorting market outcomes or of allocation trades by environmental water holders distorting market prices' (Commonwealth of Australia 2014).

Resistance to the purchase of entitlements by the Commonwealth Government is likely driven by three additional factors. In the first instance, irrigation corporations do not look favourably on individual farmers who sell their entitlements to the CEWH as this increases the per capita maintenance cost borne by the corporation's remaining 'shareholders' (see, e.g., the comments made by Mr. Les Gordon of the National Farmers' Federation during the Senate Committee inquiry into the *Water Amendment Bill 2015*; see Environment and Communications Legislative Committee, *Water Amendment Bill (Provisions)*, September 2015, p. 11). Second, Commonwealth funding for on-farm efficiency works and the subsequent purchase of excess entitlements is clearly a boon for eligible farmers. Indeed, documents obtained under the *Freedom of Information Act 1982* (Cth) (FOI Act) reveal that an irrigator lobby group sought to be paid more than four times the market value of particular entitlements, stipulating that they should be purchased as part an efficiency upgrade program.[4] Third and foremost, 'buybacks' have become the proverbial patsy for all manner of rural mischance, from unemployment to reduced

---

[3]Though at the time of writing, the *Water Act* had not been amended to include the newly assented to provisions.

[4]Documents obtained under the FOI Act reveal that this group sought to strike a deal with the MDBA to be paid $5,500 for entitlements purchased through an efficiency works scheme.

school enrolments (e.g. Jeffery et al. 2017; NSW Government 2012). However, modelling indicates that the socio-economic impacts associated with buybacks are relatively small at a Southern-Basin scale, and far less significant than those associated with drought (Wittwer and Dixon 2011).

While the legislated cap may have appeased NSW and Victoria and their relevant constituencies, it may (as previously noted) prevent the Commonwealth from 'bridging the SDL gap' by the requisite date. To reiterate, the SDLs outlined in clause 10.11 of the *Basin Plan*—and which currently require the Commonwealth to recover 2,750 GL—apply from 1 July 2019. These SDLs, in accordance with clauses 7.10, 7.11, 7.20 and 7.21 of the *Basin Plan*, may change pursuant to supply measure projects put foward by Basin states by 30 June 2017 and operational by 30 June 2024. An amendment to the Basin Plan to reduce SDLs by 605 GL (as a consequence of 37 proposed supply measure projects) is currently before the Australian Parliament: *Basin Plan Amendment (SDL Adjustments) Instrument 2017* (Cth). If the amendment is disallowed, the Commonwealth will have to recover approximately 644 GL of additional water by 01 July 2019 (in order to ensure that it has 2,750 GL worth of entitlements by this date). This logic is reflected in clause 3, 'Bridging the Gap', of the 'Intergovernmental Agreement on Implementing Water Reform in the Murray-Darling Basin.' Specifically, subclause 3.2 notes the following:

> ...the Parties note that if the volume of SDL offsets is less than 650 GL, any shortfall in a jurisdiction's apportioned share of the 2750 GL water recovery target (after taking full account of their committed water recovery projects and their share of SDL offsets) can be purchased in that jurisdiction by the Commonwealth between 2016 and 2019 (Intergovernmental Agreement on Implementing Water Reform in the Murray Darling Basin, June 2013.)

However, in accordance with the *Water Amendment Bill* section 85C (2) and the *Water* Act section 50(5), the cap may prevent the Commonwealth from purchasing enough water to reduce Basin-wide diversions by the required amount by mid-2019. This would result in an inconsistency between the Act and the Plan, ultimately frustrating the purpose of the latter. Furthermore, clause 6.12(4)(b) of the *Basin Plan* provides that Basin States may claim that they have a 'reasonable excuse' for exceeding the mandated SDLs due to circumstances beyond their control (with these circumstances including the Commonwealth failing to achieve the SDLs for a water resource unit). As the 1,500 GL cap may constitute a 'reasonable excuse', states may be justified in exceeding SDLs until the Commonwealth is able to provide additional water from supply measures. Furthermore, states may legitimately avoid taking proactive measures (such as reducing allocations) to make up for any deficit in Commonwealth water.

2.7 Northern Basin Review

Following the introduction of a 1,500 GL/year limit on 'buybacks', the MDBA embarked upon a scientific and socio-economic review, under clause 6.06 of the *Basin Plan*, of the Northern Basin with a view to determining whether the volume of water to be returned to the environment under the *Basin Plan* could be reduced from 390 GL/year without further compromising environmental outcomes (NB Review). While the pretext for the NB Review was a need to identify and fill data gaps (Murray-Darling Basin Authority 2015), the revised figure recommended by the Authority—namely 320 GL/year plus a suite of non-enforceable 'toolkit' measures—has been widely criticised as being motivated by politics rather than a strong evidence base. Specifically, the recommended 320 GL/year scenario recommended by the Authority was not based on any of the model scenarios discussed in the supporting hydrological modelling report that was made available during the consultation period (Murray-Darling Basin Authority 2016), an astonishing admission given the purpose of the exercise was to see whether new evidence justified amending the volume of water being returned to the environment under the *Basin Plan*.[5] This revelation was compounded by the fact that the socio-economic analysis undertaken by the Authority excluded a number of downstream communities (such as Louth and Wilcannia), and may have misreported employment figures for certain towns.[6] Furthermore, documents obtained under the *FOI Act* indicated that socio-economic modelling was provided to certain industry groups for comment and amendment prior to being finalised (EDOs of Australia 2017). This, combined with the fact that other stakeholder groups were not afforded such access, inevitably raised questions about the objectivity of the evidence itself, as well as the transparency of the underlying process. Significantly, the aforementioned 'toolkit' measures—which essentially amount to Natural Resource Management offsets for water and ill-defined actions to protect environmental flows—have no statutory basis and in some instances are highly unlikely to work for either policy or operations reasons (EDOs of Australia 2017).

It is therefore unsurprising that submissions made during the public consultation process for the NB Review by individual members of the Northern Basin Advisory Committee (NBAC)—a statutory committee tasked with advising the MDBA throughout the Review as per section 203 of the *Water Act 2007* (Cth)—were highly critical of the process, purported evidence base and the final, 320 GL/year

---

[5]See also the submission by the Wentworth Group of Concerned Scientists (2017), which critiqued the Authority's scientific methodology.

[6]Documents obtained under the FOI Act indicated that total jobs in Warren actually increased after the Millennium Drought. This information has not been objectively reported in the publicly available report entitled 'Northern Basin Review—technical overview of the socio-economic analysis.' Rather, this report focuses on job loss during the Millennium Drought; it also imputes job losses to water recovery rather than water scarcity during the drought (Murray-Darling Basin Authority 2016, pp. 43–44; EDOs of Australia 2017, p. 2).

recommendation.[7] While the amendments to the *Basin Plan* proposed under the NB Review were evetually disallowed by Parliament, an amending instrument could be re-tabled at a later date. However, if substantively similar amendments are indeed passed there are likely to be strong grounds for a legal challenge, in particular for breaches of Australia's international obligations, for failing to ensure that the *Basin Plan* has been developed on the basis of best available scientific knowledge and socio-economic analysis and for authorising SDLs that do not reflect an ESLT (*Water Act* s. 21(4)(b), s. 23; for further, see EDOs of Australia 2017, p.5).

2.8 Protection of Environmental Water

Efforts to protect and restore the MDB have for the most part focused on water recovery. While this is fundamental to the reinstatement of an ESLT and the health of water-dependent ecosystems, it is also crucial that Basin laws and policies provide for the protection of environmental water (Murray Darling Basin Authority 2011a, pp. 6, 8, 9, 10, 15; EDO NSW 2012). In simple terms, this requires the imposition of cease-to-pump rules prohibiting the extraction of the CEWH's water at ecologically significant times, including during periods of low flow. MDBA policy statements concerning the implementation of the *Basin Plan* reveal a narrow and legally questionable interpretation of key provisions in the Plan (See in particular: Murray-Darling Basin Authority, *Basin Plan Water Resource Plan Requirements, Position Statement 1H, Potential Reliability Changes,* 6 September 2016), which if unchallenged may dissuade States from including new rules to protect environmental water in WRPs. While the 'toolkit' put forward under the Northern Basin Review included measures to protect the CEWH's water, as noted above they are ultimately unenforceable and to that extent less likely to be effective.

Furthermore, the success of certain supply measures, and by way of corollary, the final supply measure contribution to SDLs, depends upon the implementation of 'prerequisite policy measures' (PPMs), also known as 'unimplemented policy measures' (*Basin Plan* cl. 7.15). PPMs were assumed in the original hydrological modelling undertaken by the MDBA to inform the development of the *Basin Plan* (Murray-Darling Basin Authority 2012). Accordingly, they are provided for in Chap. 7 of the Plan, and in practical terms consist of two core strategies: releasing held environmental water on top of unregulated headwater flows, and environmental flow reuse designed to facilitate multi-site watering. These strategies have the potential to generate good environmental outcomes, on the condition that the water is adequately protected as it moves through the system.

---

[7]See submissions by the following NBAC member: Mal Peters (Chair); Sarah Moles; Geoff Wise; Ed Fessey. At the time of writing, they were available on the MDBA website: https://getinvolved.mdba.gov.au/bp-amendments-submissions.

Under the *Basin Plan*, adjustments to SDLs based on supply measures must result in two outcomes. First, they must give rise to 'equivalent environmental outcomes', with equivalent outcomes measured against the 'benchmark environmental outcomes'. Second, supply measures must avoid 'detrimental impacts on reliability of supply of water to the holders of water access rights that are not offset or negated (Detrimental Impacts Clause)' (*Basin Plan* cl. 7.15). In its Draft PPM Policy,[8] the NSW Government chose a particularly narrow—and arguably specious—interpretation of the Detrimental Impacts Clause, assuming that the imposition of cease-to-pump rules in WRPs to protect this environmental water would constitute a 'detrimental third party impact' and was therefore unauthorised (EDO NSW 2015). At the time of writing—almost two years after the Draft PPM Policy was released for comment—a detailed, final document is still not publicly available. Furthermore, a 'limited assurance review' conducted by the Australian National Audit Office in late 2017 found that 'there were risks that NSW was not delivering environmental water consistent with the *Basin Plan* (ANAO 2017–2018).' In response to this - and other inquiries prompted by serious allegations of water theft and mismanagement of water resources in NSW (Matthews 2017, Ombudsman NSW 2017)—the NSW Government released a consultation paper in March 2018 outlining five possible options for protecting environmental water (NSW Government 2018). While encouraging, the ultimate success of any reform program will depend on the mix of protection measures chosen, how they are implemented and whether they are legally enforceable.

2.9 Complementary Measures

'Complementary measures'—natural resource management (NRM) actions that act as a substitute or an offset for water—are the most recent attempt to further reduce the volume of water available to the environment under the *Basin Plan*. Examples include managing cold water pollution and releasing a herpes virus to kill carp. While NRM actions such as pest eradication are required to optimise environmental watering, they are not a legitimate offset for water in a river system that will arguably remain over-allocated under a fully implemented *Basin Plan*. Nor are they legal under the regulatory framework or, in the case of the herpes virus, currently classified as fit-for-release.[9]

Complementary measures are heavily supported by the NSW Government (NSW Department of Primary Industries—Water 2016), Commonwealth ministers including the former Deputy Prime Minister (who was also the Minister for

---

[8]The 'Draft NSW Prerequisite Policy Measures - Implementation Plan' was released for targeted consultation in May 2015.

[9]The carp herpes virus is still being tested, and as yet is not considered safe for release in the MDB (CSIRO 2017).

Agriculture and Water Resources) (Joyce, Hunt, Ruston 2016) and the irrigation lobby (see e.g. Cotton Australia 2016). Relevantly, the former CEO of the National Irrigators' Council, Tom Chesson, accepted an appointment within the National Carp Control Plan, a unit within the Australian Department of Agriculture and Water Resources that has been charged with developing a plan for the 'potential release' of the herpes virus by 2018 (Department of Agriculture and Water Resources 2016). Pre-empting the results of scientific trials that were still underway, Mr Chesson was quoted in an Australian newspaper as saying that '[w]hilst I have no doubt that the virus will work, it will not be as effective as it can be if there are not "complimentary [sic] measures" undertaken' (Bettles 2016). It is unclear why the virus—if it is indeed deemed safe for release in Australia's waterways—would be any less effective if it were not used to offset water. It is particularly unclear why this is the case when considered against the backdrop of scientific and legal evidence, which indicate a need to return at least 4,000 GL/year to the environment.

3 Legal Validity and the Path Forward

Analysis undertaken by this author of the legislative framework,[10] data contained in the Guide (Murray Darling Basin Authority 2010, p. 115), and other sources (Pittock and Finlayson 2011a, p. 39; Grafton 2011) indicates that a minimum reduction figure of 4,000GL/year (Pittock 2011) is likely to be necessary to satisfy the requirements of the Act and to discharge Australia's international obligations. While the primary objection to returning this volume of water to the environment is socio-economic impacts—real or perceived—independent research suggests otherwise:

> A review of the results of socioeconomic models of the Basin, and also empirical evidence over the past decade-long drought, provides no support for the recommendation by the MDBA that increases in environmental flows should be limited to 4000 GL/yr due to the high socioeconomic impacts of reduced water diversions (Grafton 2011).

It is also arguable that these requirements and obligations can only be met if measures are put in place to protect water recovered for the environment from extraction by irrigators, particularly in relation to the 16 Ramsar wetlands located in the Basin (Murray-Darling Basin Authority 2011a, pp. 6, 8, 9, 10, 15; EDO NSW 2012). Again, objections grounded in a particularly narrow conception of 'detrimental third party impacts' are legally questionable and if used to inform the ongoing implementation of the *Basin Plan*, may give rise to judicial review proceedings.

As litigation is a tool of last resort, it is timely to consider what other means could be employed to arrest and reverse the erosion in Basin laws and policies. First

---

[10]This includes consideration of the AGS's (2010) advice.

and foremost, governance arrangements need to be rethought in order to reduce irrigator bias. At the national level, the water and agricultural portfolios should be separated and responsibility for the former returned to the Minister for the Environment. Similarly, the *Water Act* Division 2, Part 9 should be amended to strengthen the conflict-of-interest provisions that apply to MDBA board members. For example, a new provision could be added prohibiting the appointment of more than two out of the six members with financial links—past or present—to the irrigation industry. The largest and at times most recalcitrant Basin state, NSW, should take steps to improve integration between the Office of Environment and Heritage (which is responsible for environmental watering) and the Department of Primary Industries—Water (which is responsible for rural water allocation and implementation of the *Basin Plan*).

At a regulatory level, *Basin Plan* implementation needs to be re-conceived with a view to including rules in WRPs that protect environmental water (including a range of flow events) before they are deemed fit for accreditation in mid-2019. Indeed, the WRP development process requires greater scrutiny and involvement by the MDBA and CEWH in order to avoid a 'technical dump' of information by Basin States just prior to the deadline for accreditation in mid-2019. Furthermore, as the proposed SDLs are highly unlikely to satisfy the requirement to reinstate an ESLT, a statutory review under clause 6.06 of the *Basin Plan* is necessary in order to reconsider the volume of water being returned to the environment. As the SDLs that are currently mandated under the *Basin Plan* do not take into account climate change, the likely, future impacts of water scarcity must be built into any such review (Pittock et al. 2015). Where an amendment to the *Basin Plan* and SDLs requires a reduction in allocations, compensation should be paid to eligible entitlement holders as per the requirements of the *Water Act* Division 4, Part 2.

The 1,500GL/year cap on 'buybacks' should also be repealed, as should the requirement that monies in the Special Account be used to purchase entitlements associated with on-farm efficiency works. This would enable the outright purchase of entitlements with funds from this Account. Further, the *Water Act* and *Basin Plan* should be amended to ensure a range of conditions are satisfied (including in relation to future water scarcity, risk mitigation and long-term governance arrangements) before supply measure projects are funded and implemented. Finally, Part 2AA of the Act should be amended to not only remove any ambiguity surrounding the requirement to deliver the additional 450 GL/year to the CLLMM, but to enable this water to be purchased outright (as opposed to via efficiency works).

That such recommendations are necessary only five years after the passage of the *Basin Plan* highlights the complex nature of transboundary water management. It also reminds us—yet again—that newly minted, best-practice water laws are more of an armistice than an outright victory. The latter requires well-resourced government agencies, political champions committed to the success of the implementation process and an active civil society capable of holding governments to account over the long-term.

4 Conclusions

The refocussing of Basin laws and policies has proceeded at breakneck pace over the last five years in particular. This has given rise to speculation regarding the extent to which hard-won reforms will be eroded and the consequences of this regression for water resources and water-dependent biodiversity in the MDB. It is hoped that by documenting this process and suggesting a path forward, policy-makers and the community will work together to ensure that one of Australia's most important pieces of NRM legislation—the *Water Act*—is properly implemented before it is too late. It is also hoped that the reorientation of water laws and policies in the Basin will remind governments and stakeholders in other jurisdictions that the passage of new, innovative laws is only the beginning of the reform journey. Constant vigilance—from civil society in particular—is required to ensure that these laws are properly implemented and enforced over time.

References

Australian Bureau of Statistics. (2010). 1301.0: Year book 2009–10. Feature article: Murray-Darling Basin. Retrieved May 2, 2017, from http://www.abs.gov.au/AUSSTATS/abs@.nsf/Lookup/1301.0Chapter3042009-10.

Australian Government. (2009). Letter to Ramsar Secretariat. Retrieved February 2, 2017, from http://www.environment.gov.au/water/topics/wetlands/database/pubs/28-art-3-2-notification-20090717.pdf.

Australian Government. (2016). Commonwealth water purchasing in Murray-Darling Basin. Retrieved March 3, 2017, from http://www.agriculture.gov.au/water/markets/commonwealth-water-mdb/.

Australian Government Solicitor (AGS). (2010).The role of social and economic factors in the Basin Plan. Retrieved May 2, 2017, from http://www.aph.gov.au/DocumentStore.ashx?id=dd6cb9d1-a591-48de-97aa-ec31cf91e259.

Australian National Audit Office (ANAO). ANAO Report No. 17, Assurance Review. (2017–2018). *Department of Agriculture and Water Resources' Assessment of New South Wales' Protection and use of Environmental Water under the National Partnership Agreement on Implementing Water Reform in the Murray-Darling Basin.*

Bettles, C. (2016). Irrigators fishing for new CEO. In *Queensland country life*. 2 November 2016. Retrieved March 5, 2017, from http://www.queenslandcountrylife.com.au/story/4266461/irrigators-fishing-for-new-ceo/?cs=4704.

Colloff, M. J., Caley, P., Saintilan, N., Pollino, C. A., & Crossman, N. D. (2015). Long-term ecological trends of flow-dependent ecosystems in a major regulated river basin. *Marine & Freshwater Research, 66*(11), 957–969.

Commonwealth Environmental Water Office. (2013). Environmental Outcomes Report 2012–13. Retrieved May 2, 2017, from https://www.environment.gov.au/system/files/resources/cd8b1f80-3ca7-4275-9100-30524335ef75/files/environmental-outcomes-report-12-13.pdf.

Cotton Australia. (2016). Growers call for basin authority to implement complementary measures in northern basin. 13 September 2016. Retrieved May 2, 2017, from http://cottonaustralia.com.au/news/article/growers-call-for-basin-authority-to-implement-complementary-measures-in-nor.

Cunningham, S. C., Griffioen, P., White, M., & Mac Nally, R. (2011). *Mapping the condition of river red gum (Eucalyptus camaldulensis Dehnh.) and Black Box (Eucalyptus largiflorens F. Muell.) stands in the living murray icon sites. stand condition report 2010*. Canberra: Murray-Darling Basin Authority.

CSIRO. (2017). Reducing Australia's carp invasion. Retrieved March 05, 2017, from https://www.csiro.au/en/Research/BF/Areas/Invasive-species-and-diseases/Biological-control/Biocontrol-of-carp.

Davies, P. E., Harris, J. H., Hillman, T. J., & Walker, K. F. (2008). *A report on the ecological health of rivers in the Murray–Darling Basin 2004–2007, Report by the Independent Sustainable Rivers Audit Group for the Murray–Darling Commission*. Canberra: Murray–Darling Basin Commission.

Department of Agriculture and Water Resources. (2016). National carp control plan. Retrieved March 05, 2017, from http://www.agriculture.gov.au/pests-diseases-weeds/pest-animals-and-weeds/national-carp-control-plan.

Department of Agriculture and Water Resources. (2018). *Progress of Water Recovery towards Bridging the Gap to the SDLs As at 31 December 2017*. Retrieved 24 March 2018 from http://www.agriculture.gov.au/SiteCollectionDocuments/water/progress-towards-bridging-gap.pdf.

EDO NSW. (2012). Submission in response to Proposed arrangements for shepherding environmental water in NSW-Draft for consultation, 6 July 2012. Retrieved March 04, 2017, from https://d3n8a8pro7vhmx.cloudfront.net/edonsw/pages/351/attachments/original/1473997289/EDO_NSW_Sub_Shepherding_July_2012.pdf?1473997289.

EDO NSW. (2015). Submission responding to the Draft NSW Pre-requisite Policy Measures-Implementation Plan, 29 May 2015. Retrieved March 5, 2017, from https://d3n8a8pro7vhmx.cloudfront.net/edonsw/pages/2096/attachments/original/1436144311/EDO_Letter_re__PPMs_290515.pdf?1436144311.

EDOs of Australia. (2017). *Submission responding to the Northern Basin Review*, 24 February 2017. Retrieved March 04, 2017, from https://d3n8a8pro7vhmx.cloudfront.net/edonsw/pages/3625/attachments/original/1488240454/Northern_Basin_Review_EDOs_of_Australia_Submission_Feb_2017.pdf?1488240454.

Grafton, R. Q. (2011). Economic costs and benefits of the proposed basin plan. In D. Connell, & R. Q. Grafton (Eds.), *Basin futures: water reform in the Murray-Darling Basin* (pp. 245–262). Canberra: ANU E Press.

Grafton, R., Pittock, J., Jiang, Q., Possingham, H., & Quiggin, J. (2014). Water planning and hydro-climatic change in the Murray-Darling Basin, Australia. *Ambio, 43*(8), 1082–1092.

Howard, J. (2007). National Plan for Water Security, speech delivered to the National Press Club on 25 January 2007. Retrieved February 2, 2017, from http://parlinfo.aph.gov.au/parlInfo/download/media/pressrel/K81M6/upload_binary/k81m68.pdf;fileType=application%2Fpdf#search=%22media/pressrel/K81M6%22.

Jeffery, C., Delaney, A., Gorman, V., & Shuhvta, B. (2017). Murray-Darling Basin Plan southern review not until 2026. ABC News. 24 January 2017. Retrieved March 4, 2017, from http://www.abc.net.au/news/2017-01-25/10-year-wait-for-murray-darling-basin-plan-southern-review/8211332.

Joint media release with the Deputy Prime Minister, the Hon Barnaby Joyce MP, the Minister for the Environment, the Hon Greg Hunt MP, and the Assistant Minister for Agriculture and Water Resources, Senator the Hon Anne Ruston (2016). *Cleaning up Australia's waterways*. Retrieved March 24 2018, from http://minister.industry.gov.au/ministers/pyne/media-releases/cleaning-australias-waterways.

Jones, G., et al. (2002). *Independent report of the expert reference panel on environmental flows and water quality requirements for the river Murray system*. Canberra: Cooperative Research Centre for Freshwater Ecology.

Kingsford, R., et al. (2011). A Ramsar wetland in crisis–the Coorong, Lower Lakes and Murray Mouth, Australia. *Marine and Freshwater Research, 62*, 255–265.

Martin, W., & Turner, G. (2015). *SDL Adjustment Stocktake Report*.

Matthews, K. (2017). *Independent investigation into NSW water management and compliance—Interim report.*
Murray-Darling Basin Authority. (2010). *Guide to the proposed Basin Plan: Technical background.* Murray-Darling Basin Authority: Canberra.
Murray-Darling Basin Authority. (2011a). *River management–challenges and opportunities.* Murray-Darling Basin Authority: Canberra.
Murray-Darling Basin Authority. (2011b). The Proposed 'environmentally sustainable level of take' for surface water of the Murray-Darling Basin: Method and outcomes. Retrieved May 2, 2017, from https://www.mdba.gov.au/sites/default/files/pubs/eslt-mdba-report.pdf.
Murray-Darling Basin Authority. (2012). *Hydrologic modelling to inform the proposed Basin Plan: Methods and results.* Murray-Darling Basin Authority: Canberra.
Murray-Darling Basin Authority. (2015). The Northern Basin review. Retrieved May 2, 2017, from http://www.mdba.gov.au/sites/default/files/pubs/northern-basin-review.pdf.
Murray-Darling Basin Authority. (2016). Northern Basin review-Technical overview of the socioeconomic analysis. Retrieved May 2, 2017, from https://www.mdba.gov.au/publications/mdba-reports/northern-basin-review-technical-overview-socio-economic-analysis.
Murray-Darling Basin Authority. (2017). Economy of the Basin. Retrieved May 2, 2017, from https://www.mdba.gov.au/discover-basin/people/economy-basin.
Norris R. H., et al. (2001). *The Assessment of River Condition (ARC). An audit of the ecological condition of Australian Rivers.* Final report submitted to the National Land and Water Resources Audit Office, Canberra, Australia.
NSW Government. (2012). *NSW Government submission on the proposed Murray Darling Basin Plan.* DPI Water. April 2012. Retrieved March 4, 2017, from http://www.water.nsw.gov.au/__data/assets/pdf_file/0006/548898/law_reform_mdb_murray_darling_basin_plan_nsw_govt_response.pdf.
NSW Government. (2018). *Better management of environmental water—Consultation paper.*
NSW Department of Primary Industries. (2016). Water, Northern Basin review, NSW Synopsis, November 2016. Retrieved March 5, 2017, from http://www.water.nsw.gov.au/__data/assets/pdf_file/0011/688790/Northern-Basin-Review-NSW-Synopsis-fact-sheet_161130.pdf.
Ombudsman NSW. (2017). *Investigation into water compliance and enforcement 2007-17: A special report to Parliament under section 31 of the Ombudsman Act 1974.*
Pittock, J. (2011). *Six reasons why the 2011 draft Murray-Darling Basin Plan fails*, Global Water Forum, UNESCO. 21 December 2011. Retrieved March 03, 2017, from http://www.globalwaterforum.org/2011/12/21/six-reasons-why-the-2011-draft-murray-darling-basin-plan-fails/.
Pittock, J. (2013). Lessons from adaptation to sustain freshwater environments in the Murray-Darling Basin, Australia, *WIREs Clim Change* (unpaginated).
Pittock, J., & Finlayson, M. (2011a). Freshwater ecosystem conservation: Principles versus policy. In D. Connell, & R. Q. Grafton (Eds.), *Basin Futures* (pp. 39–58). Acton: ANU E Press.
Pittock, J., & Finlayson, M. (2011b). Australia's Murray-Darling Basin: freshwater ecosystem conservation options in an era of climate change. *Marine & Freshwater Research, 62*(3), 232–243.
Pittock, J., Williams, J., & Grafton, R. Q. (2015). The Murray-Darling Basin Plan fails to deal adequately with climate change. *Water, 42*(6): 28–32.
Rogers, E. (2010). Inquiry called as basin plan anger grows. ABC News, 13 October 2010. Retrieved November 10, 2017, from http://www.abc.net.au/news/2010-10-14/inquiry-called-as-basin-plan-anger-grows/2298200.
Wentworth Group of Concerned Scientists. (2012). *Statement on the 2011 Draft Murray-Darling Basin Plan*, 18 January 2012. Retrieved November 17, 2017, from http://wentworthgroup.org/wp-content/uploads/2013/10/Statement-on-the-2011-Draft-Basin-Plan.pdf.
Wentworth Group of Concerned Scientists. (2017). *Submission on the Proposed amendments to the Murray-Darling Basin Plan*, 24 February 2017. Retrieved November 17, 2017, from http://wentworthgroup.org/wp-content/uploads/2017/06/Submission-on-Proposed-Amendments-to-Basin-Plan-24Feb.pdf.

Williams, G., & Kildea, P. (2011). The Water Act and Murray-Darling Basin Plan. *Public Law Review, 22*(1), 9–14.

Wittwer, G., & Dixon, P. (2011). *Water trading, buybacks and drought in the Murray-Darling Basin: Lessons from Economic Modelling.* Centre of Policy Studies Monash University, General Paper No. G-222 September 2011.

Young, W. J., Bond, N., Brookes, J., Gawne, B., & Jones, G. J. (2011). Science review of the estimation of an environmentally sustainable level of take for the Murray–Darling Basin. A report to the Murray–Darling Basin Authority from the CSIRO Water for a Healthy Country Flagship.

Legislation

Water Sharing Plan for the Barwon-Darling Unregulated and Alluvial Water Sources 2012 (NSW).
Water Act 2007 (Cth).
Water Amendment Bill. (2015) Replacement Explanatory Memorandum.

Cases

Basin Plan Amendment (SDL Adjustments) Instrument 2017 (Cth).
State of Victoria v Commonwealth (1996)186 CLR 416; 138 ALR 129.

Multi-jurisdictional Water Governance in Australia: Muddle or Model?

Bradley C. Karkkainen

Abstract From a certain angle of vision, management of the Murray-Darling Basin (MDB) may appear a hopeless muddle, riven by seemingly intractable conflicts between competing interests, punctuated by a series of compromises that seem to leave all parties dissatisfied. From another perspective, the MDB regime stands out as an attractive and innovative model of integrated multi-jurisdictional water resources governance. Like the MDB, North America's Colorado River and Laurentian Great Lakes are critically important freshwater systems in multi-jurisdictional settings. The Colorado River is governed by 'The Law of the River,' a multilayered set of legal arrangements that have proven durable, albeit inflexible and in important respects dysfunctional. Straddling an international boundary between two federal systems, the Great Lakes are governed primarily through bilateral agreements between the United States and Canada that express laudably ambitious goals but frequently fall short of aspirations, in part because they fail to effectively integrate subnational actors (particularly the U.S. states) into the governance regime. This chapter critically examines the governance regimes of the North American examples and then draws comparisons among all three systems, concluding that while not without its flaws, the MDB is arguably the most advanced along several key dimensions of multi-jurisdictional water resources governance.

1 Introduction

To many Australians, management of water resources in the Murray-Darling Basin —Australia's premier interstate freshwater system—might appear a hopeless muddle, part top-down regulation, part bottom-up collaboration, part fitful efforts to let markets and price signals set the tone and direction, all set against a backdrop of

---

B. C. Karkkainen (✉)
University of Minnesota Law School, Minneapolis, USA
e-mail: bradk@umn.edu

seemingly intractable conflict, inconstant policy, and lack of consensus over even the most basic goals.

To observers from other quarters of the globe, management of water resources in the Murray-Darling Basin might appear as something of a marvel—a genuine effort at integrated water resources management in a complex multi-jurisdictional environment, in which multiple perspectives and interests, national, state, and local, have both a stake and a voice, and in which novel policy tools are regularly tested. From this perspective, the cacophony of clashing interests, ideas, and policy proposals is nothing less than democracy at work.

As is often the case, the truth probably lies somewhere between these extremes. This chapter will attempt to highlight both strengths and weaknesses of multi-jurisdictional governance in the Murray-Darling Basin by comparing it with governance arrangements in two critically important multi-jurisdictional North American basins, the Colorado River and the Laurentian Great Lakes.

Like its North American counterparts, the Murray-Darling system poses special management challenges because its vast expanse crosses multiple jurisdictional boundaries within a federal system, requiring both horizontal (inter-state) and vertical (state-federal) policy integration. Specifically, the chapter will compare these systems along several dimensions: integration of environmental policies into water resources management, multi-level governance, horizontal policy coordination, regional institutions scaled to the resource, in-basin versus out-of-basin water uses, and the use of water markets as a policy tool.

Because the Murray-Darling governance regime is amply described elsewhere in this volume (see e.g. Carmody; Owens), this chapter will forego an extensive description of it and focus instead on the governance regimes for the Colorado River and the Great Lakes before drawing comparisons among the three systems.

2 The Colorado River: The Limitations of a Rights-Based Approach

It would be inaccurate to say the Colorado River—the most important transboundary river in the arid American Southwest—lacks a governance regime. Governance on the Colorado is defined by a dense thicket of judicial rulings, federal statutes, inter-state compacts, water delivery contracts, agency rules and guidelines, treaty law, and miscellaneous other institutional arrangements known collectively as 'The Law of the River' (Adler 2008). It is not hyperbole, however, to say that the Colorado River is governed by no *coherent* governance regime.

The 1,450 mile (2,334 kilometre) Colorado River rises on the Western Slope of the Rocky Mountains in the state of Colorado and flows southwesterly to its outlet at the Gulf of California (Sea of Cortez), an arm of the Pacific Ocean. Including its tributaries, the Colorado River Basin encompasses parts of seven U.S. and two Mexican states. Along the way, the Colorado and its tributaries generate hydroelectric power and provide vital urban and agricultural water supplies through a comprehensive series of dams, reservoirs, diversions, and aqueducts serving some 40 million people both within the basin (including major urban centers like Phoenix and Las Vegas) and outside it (Los Angeles, San Diego, Denver, Salt Lake City). The Colorado's spectacular canyons and whitewater rapids are also a vital aesthetic, recreational, and economic resource, with the Grand Canyon National Park alone drawing nearly 5 million visitors annually and generating an estimated half-billion dollars in tourism-related spending. Demands on the Colorado's precious water are so great, however, that at least 90% of its flow is consumed before it leaves the United States, and in most years the river never actually reaches the sea (Glennon and Culp 2002). Some 45% of the basin's water is used outside the basin itself, supporting urban and industrial uses throughout the American Southwest as well as highly productive irrigated agricultural regions like California's Imperial and Coachella Valleys and Colorado's Arkansas River Valley, where inter-basin diversions from the Colorado's tributaries augment natural flows from the Eastern Slope of the Rocky Mountains.

Interjurisdictional governance on the Colorado began with one foot in amicable cooperation and, in characteristically American fashion, with the other foot in litigation. Deciding that they had a common interest in the economic development of their then lightly populated region, the seven Basin states joined together to form the League of the Southwest in 1917 to promote economic development, and specifically irrigated agriculture, looking to the Colorado River as a major source of water (Hundley 2009). In 1921, the U.S. Congress sought to further these ambitions by authorising the Basin states to enter into a legally binding interstate compact to allocate shares of the Colorado's water resources (Pub. Law 67-56, 42 Stat. 171 (1921)), although some Basin states were initially skeptical about binding themselves by a compact.

Early in 1922, however, the U.S. Supreme Court issued a landmark ruling in *Wyoming v. Colorado* 259 U.S. 419 (1922), resolving a long-festering dispute over uses of the Laramie River, which rises on the Eastern Slope of the Colorado Rockies and flows north into Wyoming, outside the Colorado River Basin. When Colorado began to develop its portion of the Laramie by building upstream dams and diversions, Wyoming sought to enjoin these developments, arguing that Colorado's proposed diversions would interfere with already-established downstream uses of water in Wyoming. Wyoming argued this violated principles of the prior appropriation doctrine that governed water law in both states (Hundley 2009). Under prior appropriation law, the first user of previously unappropriated water acquires priority rights to its continued use, to which later appropriators must give way. Colorado countered that while each state embraced prior appropriation law to allocate water within its own territorial jurisdiction, the Court should resolve the inter-state dispute by applying the principle of equitable allocation, which the Court had articulated in an earlier case, *Kansas v. Colorado* 206 U.S. 46 (1907) involving one prior appropriation state (Colorado) and one riparian law, non-prior appropriation state (Kansas) (Hundley 2009). But the Supreme Court held for Wyoming, setting off a panic among the Colorado Basin states which envisioned both a cascade of interstate litigation over water allocations and a race among the states to develop the Colorado's water resources and thereby gain protection as the 'senior' rights-holder under prior appropriation law (Lochhead 2001).

In particular, both the Upper Basin states (principally Colorado, Wyoming, Utah, and New Mexico) and the relatively undeveloped Lower Basin state of Arizona feared that rapidly growing California would appropriate and thus 'lock up' the lion's share of Colorado River water before they had a chance to develop (Adler 2008). The burgeoning population of Los Angeles had already shown a seemingly limitless thirst for water, and that city had demonstrated the technical, fiscal, and administrative capacity to stretch its tentacles to the farthest reaches of California in pursuit of water. In fact, Los Angeles had long eyed the Colorado as a potential source of hydroelectric power and a major addition to its water supply system that would enable its continued rapid growth (Hundley 2009). For its part, California feared the ability of upstream states to cut off its (future) water supply by virtue of superior geographic advantage, especially through out-of-basin transfers that would produce no return flow to the Colorado (Hundley 2009). To preempt these looming threats, the states negotiated and adopted in 1922 what amounted to a mutual non-aggression pact known as the Colorado River Compact—the cornerstone document of the Law of the River, still in force today.

The Colorado River Compact, 45 Stat. 1057 (1928) divides the basin into two parts, the Upper Basin (Colorado, Utah, Wyoming, New Mexico, and a small part of Arizona) and the Lower Basin (Arizona, California, and Nevada). The Upper Basin states are obligated to deliver 7.5 million acre-feet (maf) of water to the Lower Basin at Lee's Ferry, Arizona, a few miles south of the Arizona-Utah border, designated by the Compact as the boundary between the Upper and Lower Basins. Based on calculated annual flow of 16.5 million acre-feet (maf), this was expected to leave 7.5 maf available for use by the (generally less developed) Upper Basin states, thus splitting the water between the Upper and Lower Basins in equal fractional shares while leaving an additional 1.5 maf for use in Mexico (Adler 2008). The Compact left it to the states in each sub-basin to negotiate state-by-state allocations. This allayed the fears of both the Upper Basin states and California. Still warily eyeing its bigger and rapidly growing neighbour California, however, Arizona initially declined to ratify the Compact (Hundley 2009).

When the states failed to agree on state-by-state allocations, Congress again intervened, the states litigated some more, and the U.S. Supreme Court issued a series of far-reaching decrees that further shaped the Law of the River. Congress' first contribution was the 1928 *Boulder Canyon Project Act* 45 Stat. 1057, codified at 43 U.S.C. Chapter 12A, authorising the federal Bureau of Reclamation to build a large hydroelectric, flood control, and storage dam on the Colorado River at Black Canyon along the Nevada-Arizona border, now known as the Hoover Dam (earlier proposals had called for such a dam at Boulder Canyon, but when that site proved geologically infeasible the relocated project retained its original name). The Act also specified numerical allocations for the Lower Basin states—2.8 maf for Arizona, 4.4 maf for California, and 0.3 maf for then sparsely populated Nevada, which at the time had seemingly abundant groundwater resources for its modest population of only 89,000.

Arguing it was inequitable to award California the lion's share of the water just because it was further along in its development, Arizona sued California (and sometimes the other basin states) repeatedly, and usually unsuccessfully, in a lengthy series of cases decided by the U.S. Supreme Court sitting in original jurisdiction, all captioned *Arizona v. California*.[1] California countered that by diverting the waters of the Gila River, a major tributary of the Colorado, Arizona had already claimed its allocated share of the Colorado River Basin's waters. The Court held, *inter alia*, that:

- The Colorado River Compact is not unconstitutional, as Arizona had claimed (283 U.S. at 457–458).
- Allocations among the Lower Basin states are subject neither to the law of prior appropriation nor to the principle of equitable allocation, but instead are definitively governed by the *Boulder Canyon Project Act* (373 U.S. at 564–567).
- The Act gives exclusive and plenary authority to the U.S. Secretary of the Interior to contract for distribution of water from the Colorado River mainstream in a manner consistent with the Act, and this includes power to decide which water users in each state would receive water (373 U.S. at 580–581).
- Water diverted from tributaries does not count toward any state's allocated share of water from the mainstream (a major win for Arizona and Colorado, which were farthest along in damming and diverting tributaries) (373 U.S. at 572–573).
- If 7.5 maf is available in the Colorado River mainstream below Lee's Ferry, California is entitled to its full 4.4 maf share. If there is a surplus above 7.5 maf, California is entitled to 50% of the surplus, Arizona to 46%, and Nevada to 4% (376 U.S. at 343). If there is a shortage below 7.5 maf, the Secretary of the Interior has discretion to allocate the shortfall in a manner consistent with the purposes of the Act, the best interests of the basin states, and the welfare of the nation (373 U.S. at 592–594).
- The Secretary also has authority, and a legal duty, to allocate Colorado River water to certain Indian tribes, consistent with reserved water rights implicitly retained by the tribes when they agreed to settle on reservations with definite boundaries (373 U.S. at 595–601).

Other key developments in the Law of the River include a 1944 treaty with Mexico (Treaty for the Utilization of Waters of the Colorado, Tijuana and Rio Grande Rivers, Feb 13, 1944 (U.S.-Mex.), 59 Stat. 1219), under which the United States committed to deliver 1.5 maf of Colorado River water to Mexico at the border (about a bit less than 10% of the river's expected annual flow); a 1973

---

[1]283 U.S. 423 (1933); 292 U.S. 341 (1934); 298 U.S. 558 (1936); 373 U.S. 546 (1963); 376 U.S. 340 (1964); 383 U.S. 268 (1966); 439 U.S. 419 (1979); 460 U.S. 605 (1983); 466 U.S. 144 (1984); 531 U.S. 1 (2000).

decision by the International Boundary and Water Commission committing the U.S. to meet specified salinity targets for the waters reaching Mexico (International boundary and water commission, Permanent and Definitive solution to the International Problem of the Salinity of the Colorado River (Minute 242), done at Mexico City, D.F., Mexico, August 30, 1973); and a 1948 compact among the Upper Basin states apportioning 51.75% of the Upper Basin's 7.5 maf to Colorado; 23% to Utah; 14% to Wyoming; and 11.25% to New Mexico, after Arizona gets 50,000 acre-feet annually for its small share of the Upper Basin (Upper Colorado River Basin Compact of 1948, 63 Stat. 31 (1948), codified in state law, e.g., Colo. Rev. Stat. 37-602-101 to -106).

Numerous other agreements, statutes, court rulings, agency rules and guidelines, administrative orders by the Secretary of the Interior, and miscellaneous instruments further fill out the Law of the River; the details are beyond the scope of this Chapter.

It is notable that the Law of the River is primarily concerned with blunting and managing competition for scarce water resources among the seven Colorado River Basin states (Getches 1985). Relatively little of it is aimed at improving water quality or ecosystem health, or even at preventing conditions from further deteriorating (Adler 2008). Indeed, environmental considerations came as an afterthought, after severe environmental degradation had already occurred, though some environmental protection measures are now in place, and some argue these, too, should now properly be considered part of the Law of the River (Getches 1997). Federally operated desalination facilities just north of the border with Mexico aim to reduce salinity to levels specified by treaty, in order to ensure that the limited quantities of water reaching Mexico are usable for irrigated agriculture. Farther upstream, federally funded cost-sharing programs work with farmers to limit the salinity of irrigation return flows. A largely voluntary and collaborative endangered fish species recovery program works to improve fish habitat, maintain minimum in-stream flows, combat invasive non-native species, and restock native species in the Upper Basin, pursuant to the federal *Endangered Species Act* (ESA), and a parallel effort is underway in the Lower Basin, although these efforts are narrowly tailored to meet specific requirements of the ESA and must operate within the tight constraints of the Law of the River (Adler 2008). More recently, under a new five-year agreement between the United States and Mexico (known as Minute 319) and with active collaboration and support from non-governmental organisations, an experimental release of stored water from dams on the U.S. side of the border in 2014 restored spring pulse flow to the Colorado River Delta in Mexico, historically a critical habitat area for migratory birds along the Pacific Flyway (International Boundary and Water Commission, Minute 319 Colorado River Limitrophe and Delta Environmental Flows Monitoring Interim Report, May 19, 2016). The eight-week pulse was widely hailed as an ecological success, stimulating plant growth and attracting birds and other wildlife at levels not seen in decades. The agreement also provides for increased annual base flows to the Delta using water rights purchased from willing sellers upstream—a rare use of water markets to achieve environmental objectives. Minute 319 expired in 2017, but late in 2017 the

U.S. and Mexico agreed to expand environmental flows to the Delta and extend them to 2026, made possible largely through new investments in water conservation in Mexico (International Boundary and Water Commission, Minute 323 Extension of Cooperative Measures and Adoption of a Binational Water Scarcity Contingency Plan in the Colorado River Basin, September 21,2017).

The description thus far presented should suggest the degree of rigidity that pervades the governance arrangements for the Colorado River. It reflects a series of ad hoc, piecemeal decisions made by powerful decision-makers in response to *seriatim* pleas by self-interested parties—principally, the U.S. states—apparently with little thought given on any side to how the pieces fit into a coherent whole (Getches 1997). The result is a mind-numbingly complex set of legal requirements that serve no one's interests particularly well, bought at the price of enormous lobbying and litigation costs, coupled with much uncertainty until the latest definitive legislative, judicial, or administrative judgment comes down.

At its core, the Law of the River is a system of rules that more or less presumes that all the water in the Colorado River will be 'used up', i.e., put to consumptive uses elsewhere rather than kept in the river for ecological, aesthetic, recreational, or other in-stream uses (Glennon and Culp 2002). Perhaps this should not be surprising, given that it reflects the core ethos of the prior appropriation doctrine that has long dominated water law in the American West (Reisner 1986). The central question for the Colorado River then becomes, what is each state's legal entitlement to how much of the total flow of water, and secondarily, how do we put in place and manage the infrastructure to effectuate those allocations?

The Law of the River is extraordinarily rule-based, rights-based, and legalistic. It does not create regionally-scaled governance institutions with ongoing authority to make decisions and actively manage the Colorado, nor does it attempt to promote horizontal cooperation among the Basin states or to coordinate and harmonize their policies. Instead it puts in place hard and inflexible legal rules that attempt to shield each state's allocative share against potential predation by its sister states (Getches 1997). To that extent, it is also a largely non-participatory system of governance, insofar as neither the states nor residents of the Basin have a voice or a vote in most ongoing decisions of regional significance, except as parties to each successive round of litigation and as lobbyists trying to influence decisions made by Congress and the Secretary of the Interior. A partial exception is the various compacts, which are both federal legislation (they must be approved by Congress) and legally binding interstate contractual undertakings, akin to treaties in the international arena. But the various Colorado River compacts essentially establish legal rules and boundary conditions on state behavior, rather than creating regional forums for ongoing joint decision-making; and they are negotiated only rarely and sporadically, often under the expressed or implied threat of federal intervention if the states cannot agree.

Finally, the Law of the River is largely federal law, and to that extent top-down in orientation (Getches 1997). The Basin states are neither its authors nor its owners, again with the partial exception of the compacts. Management of the Colorado River depends heavily on the constitutional supremacy of federal law, the implied threat of federal lawmaking and enforcement, and the practical control the Federal Government exercises over the spigot by virtue of federal ownership and management of key pieces of infrastructure—dams, reservoirs, and diversion works—necessary to deliver actual 'wet water', as it is known in the American West (as opposed to property rights or contractual rights to water, which do not always result in 'wet water' actually flowing).

In sum, then, there is a system of governance over water resources in the Colorado River Basin. It is an extremely complex and thorny system known as the Law of the River, consisting primarily of legally binding rules, mostly federal, aimed principally at blunting and managing conflict and competition among the seven Basin states. Not much attention is paid to promoting cooperation and policy coordination among the Basin states; indeed, collaboration, cooperation, and policy coordination are concepts largely foreign to the management of the Colorado River's water resources (Getches 1997).

3 The Great Lakes: The Limitations of International Law in Transboundary Natural Resource Management

Straddling the international boundary between the United States and Canada, the Laurentian Great Lakes comprise the world's largest freshwater system with nearly 20% of the planet's fresh surface water. Covering an area of some 245,000 km^2 and with more than 16,000 km of coastline, the Great Lakes are the beating heart of North America's heavily populated industrial interior, an abundant source of water for the region's forty million people and a wide variety of commercial and industrial uses, a major shipping route for both intra-continental and ocean-going carriers, and a priceless aesthetic, recreational, and ecological resource (U.S. EPA 1995). The Great Lakes are also one of the world's most important shared transboundary freshwater systems, raising complex and difficult issues of transboundary governance.

From the outset, international governance of the Great Lakes has been amicable —though in fairness, amicable relations are probably more easily achieved amidst water abundance than under conditions of scarcity. In 1909, the United States and Canada adopted a bilateral Boundary Waters Treaty committing the parties to observe freedom of navigation and commerce in the Great Lakes and other boundary waters; to regulate obstructions, diversions, and artificial elevations or diminutions of natural lake levels; and to resolve management questions and disputes amicably (Treaty Between the United States and Great Britain Relating to Boundary Waters, and Questions Arising Between the United States and Canada, Jan 11, 1909, U.S.-Gt. Britain, 36 Stat. 2448, T.S. No. 548 (1909)). Toward that end, the Treaty established the International Joint Commission (IJC), a binational body whose members are appointed by the respective governments but by tradition act independently of the political and policy preferences of the governments in power. The IJC is empowered to regulate dams, diversions, and obstructions, to investigate and make recommendations to the governments on questions they refer to it ('references'), and to mediate disputes between the parties.

A major impetus for negotiation of the Boundary Waters Treaty was the completion in 1900 of an audacious engineering feat, reversing the course of the Chicago River to carry sewage effluent and industrial pollutants away from Lake Michigan, the source of Chicago's public water supply. Although Lake Michigan lies entirely within the United States, it is hydrologically at one with Lake Huron which straddles the international boundary, the two lakes separated only by a narrow straight. The new Chicago diversion reduced the levels of the two lakes by six inches, and it was feared that additional large-scale diversions could adversely affect navigation, hydropower production, and other uses of the Great Lakes (Walker 2014). Apart from establishing mechanisms to regulate such large-scale diversions, however, the Boundary Waters treaty made no provision for allocation of water supplies in the Great Lakes, no doubt because the ample volumes of water available obviated the need for such measures.

The 1909 Boundary Waters Treaty also included a crucial, and at the time novel, anti-pollution provision. Under Article IV of the Treaty, the parties contract to ensure that 'boundary waters and waters flowing across the boundary shall not be polluted on either side to the injury of health or property on the other.' This commitment was the earliest statement in treaty law of what would later become the central organising principle of customary international environmental law: the obligation to prevent serious transboundary harm by pollution (Knox 2008). As early as 1912, after an alarming rise in typhoid mortality in the Great Lakes region, the parties referred the question of transboundary bacterial pollution in the Great Lakes to the IJC (Read 1999). The reference led to the largest bacteriological investigation of its kind anywhere in the world up until that date, culminating in an IJC recommendation in 1918 to install sewage treatment works, especially in communities around the Detroit and Niagara Rivers where bacterial concentrations were highest (Final Report of the International Joint Commission on the Pollution of Boundary Waters Reference 1918). The IJC further urged the establishment of a single binational authority to set effluent standards, and volunteered itself for that task. In response, the governments asked the IJC to draft a treaty to implement these recommendations, but the draft treaty was never adopted (Read 1999). The city of Detroit did complete its first wastewater treatment plant by 1925, but only to a standard of 'primary treatment,' with the consequence that Detroit continued to be the principal source of pollution to Lake Erie for the next several decades (Reitze 1968). The city of Buffalo, the principal source of bacterial pollution to the Niagara River, did not complete a sewage treatment facility until 1938 (Rossi 1995).

After that, the anti-pollution provision of the Boundary Waters Treaty was largely ignored until the 1960s, when pollution in the Great Lakes became so severe that it compelled further action (Botts and Muldoon 2005). Acting on recommendations from the IJC under a reference on eutrophication in Lake Erie, the United States and Canada undertook the first Great Lakes Water Quality Agreement (GLWQA) in 1972, pledging to 'restore and maintain the chemical, physical, and biological integrity' of the Great Lakes by 'reduc[ing] to the maximum extent practicable the discharge of pollutants into the Great Lakes system' (Great Lakes

Water Quality Agreement of 1972 (U.S.-Can.), 23 U.S.T 301, T.I.A.S. No. 7312 (GLWQA 1972 or 1972 Agreement)).

For both the United States and Canada, the GLWQA represented the first major initiative in modern international environmental law. There had been bilateral, regional, and multilateral agreements on wildlife and other natural resources prior to 1972 (see, e.g., Convention Between the United States and Great Britain (for Canada) for the Protection of Migratory Birds, U.S.-Gr. Brit., Aug. 16, 1916, 39 Stat. 1702), and in limited ways pollution control had begun to creep into international law well before the 1970s—for example, through the *Trail Smelter* arbitration (Trail Smelter Arbitration (U.S. v. Can.), 3 R.I.A.A. 1905 (1938); further proceedings 3 R.I.A.A. 1938 (1941)) and the pollution control provision of the Boundary Waters Treaty itself. But the GLWQA represented something new—a binational agreement dedicated exclusively to water pollution control in a transboundary setting.

The 1972 Agreement was primarily a pollution control agreement, albeit an unusually broad one that pledged the parties to ensure adequate control of all sources of pollutants. Its stated goal was to restore and enhance water quality in the Great Lakes by establishing water quality standards for nutrients, toxic substances, materials that produce colors, odors, or other nuisance-like effects, as well as other pollutants. Notwithstanding these broad commitments, however, the principal and most urgent focus was a narrower one: controlling phosphorus pollution, which had been identified as the main culprit in Lake Erie eutrophication (Botts and Muldoon 2005).

The phosphorus problem was rapidly mitigated through regulatory bans on phosphate detergents and the construction of modern sewage treatment plants in the Lake Erie basin, coupled with strict regulatory controls on other major point sources of pollution, implemented on the U.S. side through the *Clean Water Act* which had been enacted roughly contemporaneously with the first GLWQA, and in Canada by regulatory requirements promulgated by the province of Ontario (Botts and Muldoon 2005). The parties soon recognised, however, that environmental problems in the Great Lakes ecosystem went beyond excessive phosphorus. Research pointed to the persistence of concentrated 'hotspots' of toxic contaminants in sediments on the lake beds, especially in ports and near industrial outfalls; the buildup of bioaccumulative toxic contaminants in the flesh of fish and other aquatic life; airborne deposition of pollutants not controlled by the new water pollution control laws; and the contributions of non-point pollution sources along the shores of the Great Lakes and in their tributaries (Botts and Muldoon 2005).

The 1972 Agreement also set in motion processes that would lead to further legal and institutional changes. Under the 1972 Agreement, the IJC was required to issue periodic reports on progress toward meeting the Agreement's water quality objectives and to make recommendations to the governments, building on the IJC's traditional role as impartial adviser to the governments by making it an independent 'watchdog' in the pollution control arena. The original Agreement also committed the governments to undertake a five-year review of the Agreement's effectiveness with an eye toward making such revisions as would be necessary—thus launching a dynamic, iterative review and revision process, not only of the Agreement's overall goals and objectives but also of the management approaches and institutional arrangements that might be necessary to achieve environmental improvements. Finally, among the institutional arrangements that emerged out of the 1972 Agreement were several measures that dramatically opened the process to citizen participation, creating additional pressure on the governments to address these broader problems (Botts and Muldoon 2005).

In 1978, the parties signed a revised Great Lakes Water Quality Agreement, committing to an 'ecosystem approach' to integrated management of the entire suite of environmental stressors and natural resources that made up the Great Lakes Basin Ecosystem, defined in the 1978 Agreement to include 'the interacting components of air, land, water and living organisms, including humans, within the drainage basin of [the international portion of] the St. Lawrence River' (Great Lakes Water Quality Agreement of 1978 (U.S.-Can.), Ottawa, 22 Nov., 1978. 30 U. S.T 1383, T.I.A.S. No. 9257 (GLWQA 1978)). The 1978 Agreement was the first anywhere to embrace the ecosystem approach—an approach that subsequently has been widely emulated elsewhere (Martin 1999).

The broad ecosystem restoration goals enunciated in the revised 1978 Agreement were largely kept intact when a new 1987 Protocol was negotiated (Protocol Between the United States of America and Canada Amending the Agreement of November 22, 1978, U.S.-Can., Nov. 18, 1987, T.I.A.S. No. 11551), and these goals remain foundational to the Great Lakes management regime today. But the 1987 Protocol added several important wrinkles. Recognising the important role of airborne deposition of pollutants, air pollution control was explicitly added to the Agreement's list of objectives. While maintaining overall basin-wide ecosystem restoration goals, the parties also committed to develop Remedial Action Plans for identified 'areas of concern' (toxic hotspots) throughout the Great Lakes Basin, and launched a process to develop lake-specific management plans for each of the Great Lakes. The 1987 Protocol thus incorporated a 'nested' management scheme at multiple interconnected scales, yet another important innovation and one more significant evolutionary step in the dynamic unfolding of Great Lakes governance (Regier 1999).

A subsequent 2012 Protocol makes additional commitments, recognising aquatic invasive species, discharges from ships, climate change, and habitat and species loss as priority concerns (Protocol Amending the Agreement Between Canada and the United States of America on Great Lakes Water Quality, 1978, as Amended on October 16, 1983, and on November 18, 1987, done at Washington, Sept. 7, 2012). The 2012 Protocol places special emphasis on restoration and maintenance of nearshore areas, where stressors tend to be greatest. The governments pledge to involve key subnational actors including states, provinces, municipalities, Tribal Governments, First Nations, watershed management agencies, and the public in Great Lakes management and restoration. The parties also pledge to use an adaptive management approach. The Protocol contemplates an enhanced role for the IJC and its subsidiary bodies, including the Great Lakes Water Quality Board, Science Advisory Board, and Great Lakes Regional Office, though the IJC's role remains largely one of information-gathering and advising. Indeed, the Protocol underscores that operational responsibility for implementation of all these commitments remains the sole responsibility of the national governments, relegating both the IJC and subnational actors to advisory and consultative roles.

The Boundary Waters Treaty and the Great Lakes Water Quality Agreement, through its multiple iterations and amendments have been path-breaking agreements, establishing a model of successful transboundary cooperation in the management of a critically important shared watercourse and to that extent, a worthy benchmark for the rest of the world (Robinson 2002). They have also been progenitors of much of modern international environmental law—the first to articulate the principle against transboundary harm by pollution, the first comprehensive free-standing transboundary pollution control agreement, and the first to adopt an ecosystem approach.

Yet for all that, the Great Lakes remain a deeply troubled system, hanging in a delicate balance between collapse and recovery (Bails 2005). To be sure, there has been substantial progress on some fronts. Overall Great Lakes water quality has improved since the 1960s when Lake Erie was proclaimed 'dead' and some of the other Lakes were thought to be not far behind (Dempsey 2005). We no longer dump untreated sewage into the Lakes or their tributaries (Andreen 2004)—at least, not usually. Water pollution from industrial point sources has been brought substantially under control, and the populations of some species have stabilised, and appear to be making a comeback in some parts of the Great Lakes system where they had been all but locally extirpated (IJC 2013).

Scientific understanding of the ecology, hydrology, and biogeochemistry of the Great Lakes system is also better than it ever has been, thanks to the cumulative efforts of independent scientists, academic institutions, nongovernmental organizations, and government agencies at both the federal and state/provincial levels (Botts and Muldoon 2005).

Despite such progress, however, the binational project of Great Lakes restoration sometimes appears to be on a treadmill. Among the major problems that remain:

- The most severely contaminated toxic 'hotspots' have been identified and designated as 'Areas of Concern' meriting priority remediation, but only a handful of clean-ups have been completed and progress has stalled on many (IJC 2013).
- Non-point source water pollution continues largely unabated and in some areas appears to be growing worse (Kerr et el. 2016).
- Notwithstanding the adoption of a Binational Toxics Strategy, the governments have made little headway against airborne deposition of both toxic and conventional pollutants (Andreen 2004).
- There is no real strategy for managing land use within the basin so as to protect the Great Lakes and their tributaries (IJC 2003).
- Invasive species carried by ships' ballast or infiltrating through rivers and canals continue to wreak ecological havoc, displacing native species and disrupting the food webs and ecological relationships that define aquatic life in the Great Lakes (National Research Council 2008).

Not only are these problems real, severe, and persistent, but the legal and institutional mechanisms capable of addressing them are not yet in place.

So we have an apparent paradox. On the one hand, Great Lakes governance arrangements might fairly be described as one of the most successful and durable models of binational cooperation in transboundary natural resource management the world has ever seen. Those legal and institutional arrangements are bolstered by a genuine political will on both sides of the border (at least within the Great Lakes basin) to commit real resources toward the project of protecting and restoring the Great Lakes ecosystem. Juxtaposed against that, however, we see widespread failure at the level of substantive policy.

At its core, the problem may be that the institutions are mismatched to the nature and scale of the problems to be addressed in the Great Lakes basin (Karkkainen 2006). The United States and Canada were quick to recognise that neither could manage the Great Lakes alone, so that some level of international cooperation and coordination was essential. That led to the Boundary Waters Treaty and a series of bilateral Great Lakes Water Quality Agreements spelling out an innovative vision of integrated ecosystem management and an ambitious set of environmental objectives—leaving implementation of those objectives to the good offices of the respective national governments, each on its own side of the international boundary. Thus, the regime for management of the Great Lakes essentially reverts at the implementation phase to the traditional assumptions of Westphalian public international law: the GLWQA is an agreement between national sovereigns, the federal governments of the United States and Canada, which remain the only real players. Their obligations run to each other, and each is exclusively responsible for implementing the Agreement within its own territory. Within this familiar institutional framework, policy failure is seen as a failure of implementation at the national level, borne of a failure of each party to hold its own feet to the fire.

An alternative view would hold that the failure runs deeper than a failure of implementation. An institutional arrangement in which the only two relevant players are the federal governments of the United States and Canada might be regarded as a flawed institutional design—however consonant that approach might be with the standard assumptions of public international law. This view suggests that until the transboundary governance institutions are realigned and, where necessary, redesigned into a new institutional architecture better fitted to the scope and nature of the task at hand, pouring more money through the same old institutional funnels may not get us much closer to providing effective solutions. Perhaps it is time to shift our focus away from thinking of management of the Great Lakes as an international problem requiring an international law solution, a binding contractual agreement between sovereign nation states. Instead, we might think of it as a transboundary problem, requiring a new form of effective transboundary governance, scaled to the resource we seek to manage and protect.

The recently adopted Great Lakes-St. Lawrence River Basin Water Resource Compact, (122 Stat. 3729 (2008)), and the Great Lakes-St. Lawrence River Sustainable Water Resources Agreement (2005) on water allocation suggest alternative institutional possibilities. These instruments are aimed at the rather modest goal of limiting out-of-basin diversions of water. More specifically, the legally binding Compact among the eight Great Lakes Basin states, and its mirror-image companion document, the good-faith Agreement between the same eight states and two Canadian provinces, provide for:

- A ban on new out-of-basin diversions, subject to narrowly limited exceptions for 'straddling' communities that are partly within the basin and partly outside it and for certain intra-basin transfers (e.g., a diversion from the watershed of one Great Lake to the watershed of another Great Lake).
- Establishment of uniform regional standards for evaluating and permitting proposed water withdrawals, including requirements that return flows must be to the source watershed; individual or cumulative adverse impacts on water quality or quantity must be prevented or curtailed; all withdrawals and consumptive uses must be implemented so as to incorporate environmentally sound and economically feasible water conservation measures; and each permitted withdrawal or consumptive use must be 'reasonable' as determined by reference to a multi-factor balancing test.
- Requirements that each state and province develop a comprehensive water resources inventory and contribute to a common database on water resources and withdrawals; adopt a state or provincial water management conservation and efficiency plan and submit it for regional review; establish a program to regulate water withdrawals and diversions in accordance with basin-wide standards set forth in the Compact and Agreement; and report at five year intervals on how the Compact and Agreement are being implemented in each respective jurisdiction.
- Establishment of a regional governing body called the Great Lakes Water Resources Council, consisting of the governors of each of the states (or their representatives), and a parallel body called the Regional Body consisting of the governors and the premiers of the two provinces. The Council and Regional Body meet concurrently and are jointly empowered to promulgate and enforce basin-wide regulations; to develop and implement region-wide water management conservation and efficiency plans; to review the water management plans and implementation reports of the basin states and provinces; to make recommendations to the states and provinces regarding implementation of the Compact and Agreement; and to exercise 'regional review' permitting authority over proposed withdrawals or diversions deemed to be of region-wide significance or of a precedent-setting character.

The Compact and Agreement apply not only to water within the Great Lakes and St. Lawrence River proper, but to all surface and groundwater within the Basin. In a controversial compromise, the Compact and Agreement classify shipments of water out of the basin in containers smaller than 5.7 gallons as not constituting 'diversions'. Also exempted is the longstanding diversion at Chicago, long governed by the United States Supreme Court's decree in *Wisconsin v. Illinois* (281 U.S. 696 (1930)).

Some critics within the Great Lakes Basin question whether these instruments will be effective in achieving their stated goal (Squillace 2006). Others question the goal itself, arguing that locking up twenty percent of the world's fresh surface water at a time of growing water shortages and an uncertain water future in an era of global climate change is a dubious undertaking (Squillace 2006). Both critiques raise important questions about the Compact and Agreement that are beyond the scope of this chapter. Still others have suggested that the Compact and Agreement were put forth as a solution to a remote and speculative, or even non-existent, problem (Tarlock 2008).

The focus here is not on the effectiveness of the Compact and Agreement themselves, nor on the wisdom of what these instruments are trying to achieve, but rather on what the Compact and Agreement represent as a novel kind of transboundary governance mechanism in federal systems. They provide a model in which the states and provinces did not wait for the national governments to act. Nor did the states and provinces assume that because questions of Great Lakes water allocation had a transboundary dimension, decisions about their management properly fell within the exclusive foreign affairs powers of their respective national governments, to be treated as questions of international law, the exclusive domain of national sovereigns and, according to classical Westphalian theory, no place for subnational actors. Instead, the states and provinces seized the initiative and crafted their own solution—a Compact among the eight states that became legally binding by virtue of Congressional approval, and a legally non-binding but morally compelling parallel good-faith Agreement between the eight U.S. states and two Canadian provinces, committing the provinces to the exact same provisions to which the U.S. states are legally bound by the Compact, and giving the provinces an equal seat at the table alongside the states in the regional governing body created by the instruments. The Compact and Agreement are then given further legal and practical effect by legislative ratification in each state and province, coupled with implementing legislation in each state and province to put the procedural and substantive commitments called for in the Compact and Agreement into effect. Through this ingenious device, the effect of the Compact and Agreement is to create an actual transboundary governance regime, complete with real transboundary decision-making institutions and backed by the force of law in each of the states and provinces with a stake in the resource, each harmonising its domestic laws with the common transboundary regulatory scheme.

That all this could take place without a sovereign-to-sovereign international treaty specifically authorising it might seem remarkable. And so it is, but it gives us a sense of the possibilities. These transboundary governance arrangements do not fit the familiar contours of international law and international law-making. Yet neither are they unlawful. Indeed, on the U.S. side at least, they come now with the formal blessing of the Federal Government, in the form of Congressional ratification of the Compact and acquiescence by silence with respect to the Agreement. It suggests there is space for more of this sort of thing, even in the Great Lakes Basin where similar institutional arrangements addressing fully integrated management of the shared water resources is a tantalising conceptual possibility, albeit not on anyone's policy agenda at the present time.

Still, there is a major disconnect at the heart of Great Lakes governance. Through the Compact and Agreement, the states and provinces have taken charge of managing water allocations, setting uniform Basin-wide standards for diversions and withdrawals, harmonising policies and procedures across jurisdictions to implement shared Basin-wide commitments, and establishing and empowering regional institutions to oversee water allocations and to make permitting decisions of regional significance. Yet an entirely separate governance regime exists outside and parallel to the water allocation system, consisting chiefly of binational commitments by the federal governments of the U.S. and Canada through a series of Great Lakes Water Quality Agreements.

4 How Does the Murray-Darling Compare?

We are now in a position to compare the governance regime in Australia's Murray-Darling system to those of its North American counterparts.

Integration of Environmental Policies into Water Resources Management
Tensions over the proper balance between environmental protection and uses of water as an input in human economic enterprises—principally irrigated agriculture—are rife in the Murray-Darling Basin. Although the adoption of the Basin Plan assuaged much palpable outrage, both environmentalists and irrigators frequently complain that their side is getting short shrift, and this has led to policy reversals, seeking to rebalance the equation in favour of one set of interests or the other. Such tensions are perhaps inevitable under the conditions of water scarcity that pervade in Australia's interior. To its credit, however, the Murray-Darling governance regime at least recognises the need to balance these competing interests, and to do so within a unified governing framework that seeks to accommodate both (Docker and Robinson 2014).

Like the Murray-Darling, the Colorado River is a major freshwater resource in an otherwise arid region, but environmental protection and human economic uses of water are not nearly as well integrated on the Colorado. As noted above, environmental protection came as an afterthought on the Colorado. Most of the governance regime is devoted to defining and protecting state-by-state water allocations, and environmental protection measures stand somewhere outside the main features of the institutional architecture, attempting to mitigate problems (such as excess salinity, species endangerment, and lack of flow to the Delta) after they have been created and institutionalised.

In contrast to the other two systems, the Great Lakes are characterised by water abundance rather than scarcity, and conflicts and trade-offs between economic uses of water and environmental protection are not prominent. Indeed, water is so superabundant in the Great Lakes system that almost any presently imaginable level of in-basin consumptive use seems easily accommodated without significantly compromising environmental quality or ecological integrity, and some of the most important economic uses of water—to support navigation and commerce as well as water-based recreation and tourism—are fully compatible with environmental goals to keep water abundant and in place. This is not to say that the Great Lakes are lacking in environmental problems, however. Climate change, invasive species, persistent toxic hotspots left from past industrial activity, and eutrophication and toxic algal blooms caused by land-based non-point source runoff are major threats, and a great deal of time, effort, and money has gone into diagnosing and attempting to remediate these problems. Although not entirely successful, environmental protection efforts in the Great Lakes Basin are impressive in their conceptual sophistication, calling for an integrated ecosystem approach long before that concept gained widespread adoption. Notably, however, environmental protection efforts are not well integrated with water allocation mechanisms, perhaps because such integration is not perceived as essential.

Some critics of environmental protection efforts in the Murray-Darling have argued that a singular focus on maintaining minimum in-stream flows has had a straightjacketing effect, elevating water quantity concerns over broader considerations of environmental quality and ecological integrity. Mimicking the historic seasonal pulses to which native flora and fauna are adapted may be more important than maintaining a specified level of base flow, and a broader ecological view would also address issues like invasive species, water quality, and the restoration of both in-stream and riparian habitats (Byron 2011). If these criticisms have validity, it suggests the Murray-Darling could learn much from environmental protection and ecological restoration efforts in the Great Lakes system and even on the Colorado, where recent efforts to return seasonal pulse to the Colorado Delta may contain valuable lessons. At the same time, however, both the Great Lakes and the Colorado have much to learn from the Murray-Darling about integrating environmental protection and water allocation within a unified governance regime.

Multi-level Governance All three governance regimes arise in federal systems in which both sovereign national governments and quasi-sovereign states and provinces play significant roles in both environmental protection and water allocation. What is notable in all three systems, however, is the degree to which current governance arrangements are top-down, with the federal governments of the United States and Canada and the Commonwealth government of Australia taking on a dominant role, largely relegating state and provincial governments to secondary status. In the Murray-Darling system, however, the role of a Murray-Darling Ministerial Council consisting of representatives of each of the Basin states plus the Commonwealth goes beyond consultation and advising; the Ministerial Council has some limited policy and decision-making authority over certain decisions affecting the interests of the Basin states, although its role is ultimately subsidiary to that of the MDBA, an agency of the Commonwealth government. In the Great Lakes, there are formal mechanisms for states/provinces to play an advisory and consultative role to the Binational Executive Committee that oversees implementation of the GLWQA. No formal mechanisms exist for states to play even an advisory and consultative role in federal decisions affecting the Colorado River.

An important exception to this pattern is the Canadian side of the Great Lakes system. Because Canada's federalist system grants the provinces primary responsibility over natural resources and overlapping federal-provincial jurisdiction over key aspects of environmental protection, the Canadian national government has found it necessary to negotiate a series of Canada-Ontario Agreements on Great Lakes Water Quality and Ecosystem Health in order to align Ontario's environmental protection policies and priorities with the national government's commitments under a series of Canada-U.S. Great Lakes Water Quality Agreements (Inscho and Durfee 1995). Arguably this has resulted in a deeper and more seamless level of national-subnational policy coordination on Canada's part than is seen elsewhere. Fortuitously, Ontario is the only Canadian province fronting directly onto the Great Lakes, bordering on four of the five lakes (excepting only Lake Michigan which lies wholly within the U.S.). This vastly simplifies matters, with Canada needing to reach agreement with only one subnational government, specifically one with a direct and immediate stake in maintaining environmental quality and restoring ecosystem health throughout the Basin.

States and provinces play a leading role in water allocation matters in the Great Lakes. As previously noted, however, the Great Lakes water allocation system is not well integrated with environmental protection efforts which are largely led by the federal governments, creating parallel governance regimes, one for water allocation at the state and provincial level, and one for environmental protection at the level of the national governments.

Horizontal Policy Coordination Horizontal coordination among states (and provinces in the Great Lakes Basin) is not a strong suit of any of the three management regimes, perhaps due in part to the heavily top-down nature of the governance arrangements currently in place. An important exception is the impressive level of horizontal coordination and policy harmonisation among the eight Great Lakes states and two Canadian provinces in implementing the Water Resources Compact and Agreement. However, a similar level of horizontal coordination and policy harmonisation does not extend to the wholly separate regime for environmental protection in the Great Lakes Basin, where the national governments of the United States and Canada are dominant. Horizontal coordination in the Colorado River system is minimal at best. Nor does horizontal coordination appear to play much of a role in the Murray-Darling Basin, except insofar as all the subnational governments are committed to carrying out common Basin-wide goals, objectives, and programs defined largely by Commonwealth legislation and the Murray-Darling Basin Plan promulgated by the regional Murray-Darling Basin Authority.

Regional Institutions Scaled to the Resource Perhaps the most distinctive feature of the Murray-Darling governance regime is the robust central role played by the Murray-Darling Basin Authority (MDBA), a powerful regional body created by Commonwealth legislation and empowered with extensive regulatory authority over water use throughout the Basin. No comparable regional institution exists on the Colorado, where key decisions are made variously by Congress, the U.S. Supreme Court, the federal Secretary of the Interior, and individual states. The

Great Lakes are a mixed bag in this regard. New regional institutions created under the Water Resource Compact and Agreement do have significant powers both to oversee state and provincial implementation and to make binding decisions on water allocation matters of basin-wide significance. On the environmental protection side, however, regional institutions are weaker. A Great Lakes Executive Committee chaired by the heads of the U.S. EPA and Environment Canada is empowered to review the progress of the parties in meeting commitments under the GLWQA, but its role is largely confined to advising the respective governments on additional steps that might be taken. Thus, despite its exalted name, the Great Lakes Executive Committee lacks the kind of decisive operational authority exhibited by the MDBA.

In-basin Versus Out-of-basin Uses One of the starkest contrasts among the three systems examined here is the degree to which water is used in-basin versus out-of-basin. Nearly half of the Colorado River Basin's water is used out-of-basin, most notably in California—the largest user—where only a small and thinly populated sliver of the state lies within the Basin. This is consistent with the broader water use ethos of the American Southwest, where fresh water is scarce and prior appropriation doctrine places no restrictions on out-of-basin diversions. This is in contrast to the riparian law doctrine that prevailed in the more water-abundant Eastern portions of the U.S., where historically only uses on riparian lands were lawful; indeed, prior appropriation doctrine developed out of practical necessity principally to facilitate out-of-basin diversions to benefit mining operations during the California Gold Rush, in derogation of the riparian law doctrine then prevailing in California. At the other extreme from the Colorado lie the Great Lakes, where almost no water is used out-of-basin and the chief goal of the Water Resources Compact and Agreement is to keep it that way by strictly limiting out-of-basin diversions. The Murray-Darling system lies somewhere in between. Most water is used in-basin, the principal exception being the City of Adelaide which lies just outside the Basin near the mouth of the Murray but draws most of its public water supplies from the nearby river (Richter 2013). Thus, apparently there are no restrictions in principle on out-of-basin diversions from the Murray-Darling system, but through a fortunate accident of geography, southeastern Australia's other major population centers—Sydney, Melbourne, and Brisbane—lie outside the Basin and are able to provide public water supplies from more local sources.

Use of Water Markets as a Policy Tool The Murray-Darling system is a recognised global leader in the use of water markets as a policy tool. In principle, water markets enhance economic efficiency by allowing water to flow to its 'highest and best use', and in addition allow for greater planning flexibility in a system with high variability in annual flow. Both permanent entitlements and annual allocations are actively traded. Market mechanisms are also used to buy back water entitlements in order to guarantee minimum environmental flows (Horne 2013). In contrast, water markets have been slow to catch on in North America. Although long advocated by economists and economics-minded reformers, water markets have faced stiff resistance in the Colorado River Basin (Getches 1997) mainly out of fear by agricultural users that urban and industrial interests in the Southwest's large and rapidly growing population centers will always be able to outbid the agricultural sector for water supplies, with the potential to squeeze out irrigated agriculture and agriculture-dependent communities entirely (Howe 2000). Similar pressures are not evident in the Murray-Darling system, where the vast majority of water use is for in-basin irrigated agriculture, so that most trading takes place between irrigators (Wheeler 2014). Notwithstanding the resistance to water markets in the American Southwest, some important experiments have taken place. In the early 2000s a series of high-profile and highly controversial trades permanently shifted large volumes of water from willing sellers in California's Imperial Valley, a highly productive agricultural region, to urban users in San Diego (Richter 2013). More recently, market acquisitions of water rights have facilitated partial restoration of base flows to the Colorado River Delta for environmental purposes. Water markets are somewhat more common in other parts of California, especially for short-term transfers, though they still represent a small fraction of water use (Park 2017). Water markets have never been a factor in the Great Lakes Basin where, due to abundant natural precipitation, irrigation is not widespread and, where it is used, almost always relies on groundwater pumping rather than surface diversions. The ample water resources of the Great Lakes themselves easily accommodate in-basin urban and industrial demands. Indeed, the very concept of water markets is anathema in the Great Lakes Basin where markets are perceived as a tool to move water out of the Basin for the benefit of out-of-basin users.

5 Conclusion

There is arguably no single, universally applicable model for management of freshwater systems. Context matters, and there is such wide variation in hydrological, climatic, biotic, legal, institutional, social, economic, and other critical elements of context that it would be a fool's errand to seek a 'one size fits all' solution. Consider, for example, the very different challenges that arise under the water-abundant conditions of the Great Lakes, as opposed to water-starved regions like the Murray-Darling Basin and the American Southwest.

In addition, efforts to manage freshwater systems are rarely, if ever, written on a blank slate. Path dependency plays a large role in shaping the expectations of interested parties and the range of available options for institutional design. For example, governance arrangements for the Colorado River are predicated upon a foundation of state-by-state quantitative entitlements to the use of water. If starting from scratch, we might view this approach as less than ideal, and under conditions of increasing water scarcity occasioned by population growth, drought, and climate change, it may become necessary to modify or curtail some of those entitlements. Yet it is almost unthinkable that the Basin states or their residents would ever agree to give up their state-by-state entitlements entirely. Expectations, and economies and societies built upon those expectations, are too deeply entrenched for that.

With these caveats, however, it is fair to hold up the Murray-Darling as one model of reasonably successful multi-jurisdictional governance of a critically important freshwater system—one from which valuable lessons may be drawn in such areas as integrated management of the environment and water use, horizontal and vertical policy coordination, the empowerment of resource-specific regionally-scaled institutions, and the use of markets to allocate water efficiently. But that is not to say that we should view the Murray-Darling governance regime uncritically. It has its warts, not least unresolved tensions between irrigators and environmentalists, and ecological conditions that many would regard as far from optimal. Thus rather than holding up the Murray-Darling as an approximation of a Platonic ideal, we should perhaps do better to examine it critically and empirically as an ongoing experiment from which both positive and negative lessons may be learned. Some of those lessons may find valuable application elsewhere ... or not. But such critical evaluation and lesson-drawing is not a one-way street, as there is also much to be learned from other efforts at multi-jurisdictional governance of freshwater systems.

In short, governance arrangements in the Murray-Darling Basin are a model, and an important one, for the management of freshwater systems in other complex multi-jurisdictional settings. But they are not *the* model. Local conditions vary, and some elements of the Murray-Darling approach may not translate well to some other settings. At the same time, many Australians would argue that the Murray-Darling approach has many imperfections even on its home ground—muddles within the model, so to speak.

References

Adler, R. W. (2008). The Colorado River compact: Time for a change? *Journal of Land, Resource and Environmental Law, 28,* 19–47.
Andreen, W. L. (2004). Water quality today–has the clean water act been a success? *Alabama Law Review, 55,* 537–593.
Bails, J. (2005). Prescription for Great Lakes ecosystem protection and restoration: Avoiding the tipping point of irreversible changes. Great Lakes Restoration.
Botts, L., & Muldoon, P. (2005). *Evolution of the Great Lakes water quality agreement.* Michigan: Michigan State University Press.
Byron, N. (2011). What can the Murray-Darling basin plan achieve? Will it be enough?. In D. Connell, & R. Q. Grafton (Eds.), *Basin futures: Water reform in the Murray-Darling basin* (pp. 367–384). Canberra: ANU Press.
Dempsey, D. (2005). *On the brink: The Great Lakes in the 21st century.* Michigan: Michigan State University Press.
Docker, B., & Robinson, I. (2014). Environmental water management in Australia: Experience from the Murray-Darling Basin. *International Journal of Water Resources Development, 30,* 164–177.
Getches, D. H. (1997). Colorado river governance: Sharing federal authority as an incentive to create a new institution. *University of Colorado Law Review, 68,* 573–658.
Getches, D. H. (1985). Competing demands for the Colorado river. *University of Colorado Law Review, 56,* 413–479.

Glennon, R. J., & Culp, P. W. (2002). The last green lagoon: How and why the Bush administration should save the Colorado River Delta. *Ecology Law Quarterly, 28,* 903–992.

Horne, J. (2013). Economic approaches to water management in Australia. *International Journal of Water Resources Development, 29,* 526–543.

Howe, C. W. (2000). Public values in a water market setting: Improving water markets to increase economic efficiency and equity. *Denver Water Law Review, 3,* 357–372.

Hundley, N. (2009). *Water and the west: The Colorado River Compact and the politics of water in the American West.* Berkeley: University of California Press.

Inscho, F. R., & Durfee, M. H. (1995). The troubled renewal of the Canada-Ontario agreement respecting Great Lakes water quality. *The Journal of Federalism, 25,* 51–69.

Karkkainen, B. C. (2006). Managing transboundary aquatic ecosystems: Lessons from the Great Lakes. *Pacific McGeorge Global Business & Development Law Journal, 19,* 209–240.

Knox, J. H. (2008). The boundary waters treaty: Ahead of its time, and ours. *Wayne Law Review, 54,* 1591–1607.

Kerr, J. M., DePinto, J. V., McGrath, D., Sowa, S. P., & Swinton, S. M. (2016). Sustainable management of Great Lakes watersheds dominated by agricultural land use. *Journal of Great Lakes Research, 42,* 1252–1259.

Lochhead, J. S. (2001). An Upper Basin perspective on California's claims to water on the Colorado River part 1: The law of the river. *University of Denver Water Law Review, 4,* 290–330.

Martin, T. (1999). Great Lakes water quality initiative. *Natural Resources & Environment, 14,* 15–19.

Park, D. (2017). California water reallocation: Where'd you get that? *Natural Resources Journal, 57,* 183–218.

Read, J. (1999). 'A sort of destiny': The multi-jurisdictional response to sewage pollution in the Great Lakes, 1900–1930. *Scientia Canadensis, 22,* 103–129.

Regier, H. A., Jones, M. L., Addis, J., & Donahue, M. (1999). Great-Lakes St. Lawrence River Basin assessments. In K. N. Johnson, F. J. Swanson, M. Herring, & S. Greene (Eds.), *Bioregional Assessments: Science at the Crossroads of Management and Policy* (pp. 133–165). Washington: Island Press.

Reisner, M. (1986). *Cadillac desert: The American West and its disappearing water.* New York: Viking.

Reitze, A. W., Jr. (1968). Wastes, water and wishful thinking: The battle of Lake Erie. *Case Western Law Review, 20,* 5–86.

Richter, B. D., Abell, D., Bacha, E., Brauman, K., Calos, S., Cohn, A., et al. (2013). Tapped out: How can cities secure their water future? *Water Policy, 15,* 335–363.

Robinson, N. A. (2002). Befogged vision: International environmental governance a decade after Rio. *William & Mary Environmental Law & Policy Review, 27,* 299–364.

Rossi, M. C. (1995). The history of sewage treatment in the City of Buffalo, New York. *Middle States Geographer, 28,* 9–19.

Squillace, M. (2006). Rethinking the great lakes compact. *Michigan State Law Review,* 1347–1374.

Tarlock, A. D. (2008). The international joint commission and great lakes diversions: Indirectly extending the reach of the boundary waters treaty. *Wayne Law Review, 54,* 1671–1694.

Walker, G. (2014). The Boundary Waters Treaty of 1909–A peace treaty? *Canada-United States Law Journal, 39,* 170–186.

Wheeler, S., Loch, A., Zhuo, A., & Bjornlund, H. (2014). Reviewing the adoption and impacts of water markets in the Murray-Darling Basin, Australia. *Journal of Hydrology, 518,* 28–41.

Environmental Water Transactions and Innovation in Australia

Katherine Owens

Abstract A process of water distribution is underway in the water-stressed and over-allocated Murray-Darling Basin. It remains to be seen, however, whether the *Murray-Darling Basin Plan 2012* (Cth) can gain enough momentum to restore the Basin's health. Large-scale government buybacks have also proven to be unpopular politically, and will play a more limited role in water recovery strategies going forward. In this context, models of impact investment provide a potentially effective option for improving system resiliency for the benefit of a range of uses. Through an examination of the Murray-Darling Basin Balanced Water Fund, the chapter will analyse the benefits and risks of this approach in the legal context of the MDB, as well as the legal supports and safeguards required to support these initiatives in meeting their environmental objectives. The analysis will draw on theories of innovation, governance perspectives on social learning and the concept of smart regulation in order to evaluate the potential of the Fund within a market-based water allocation framework. It will argue that, when accompanied by appropriate legislative safeguards, private initiatives of the kind represented by the Fund have the potential to function as an effective complementary measure, and improve system resiliency for the benefit of a range of uses, including the environment.

This chapter is based in part on work appearing previously in Owens (2016) Reimagining water buybacks in Australia: Non-governmental organisations, complementary initiatives and private capital. *Environmental and Planning Law Journal* 33: 342. Since this chapter was written in May 2017, the Australian media has reported on a number of serious allegations of water theft and government mismanagement of monitoring and compliance in the Barwon-Darling river system in the Murray-Darling Basin. Those reports have resulted in a number of inquiries, including by the New South Wales Ombudsman and an independent inquiry led by Ken Matthews AO, as well as the prosecution of a number of large irrigators by the New South Wales Government in the New South Wales Land and Environment Court. These developments do not alter the writer's conclusions as to the potential of the Murray-Darling Basin Balanced Water Fund as a complementary measure in the Basin.

K. Owens (✉)
The University of Sydney Law School, University of Sydney, Sydney, Australia
e-mail: kate.owens@sydney.edu.au

© Springer Nature Singapore Pte Ltd. 2018
C. Holley and D. Sinclair (eds.), *Reforming Water Law and Governance*,
https://doi.org/10.1007/978-981-10-8977-0_4

1 Introduction

In the context of considerable uncertainty and variability in water supply, water market frameworks have been implemented in Australia and other parts of the world. The tools and institutions that comprise these frameworks can provide an important means of reducing risk and building resilience into water management frameworks, allowing individual users and communities to adjust to changing conditions each year. When combined with environmental caps and the ability to hold water entitlements for environmental purposes, the use of water markets to recover water for the environment can also provide an important means of protecting ecosystem values. Despite these potential benefits, market-based models for environmental water recovery are also vulnerable in practice. The frameworks are implemented in interconnected hydrological systems, which serve multiple stakeholders for multiple purposes. Transactions to recover water for the environment can have detrimental impacts on rural communities in particular contexts, and general resistance from irrigation communities to water buybacks has resulted in government policy shifts in Australia from water 'buybacks' to other more expensive, and potentially less certain, recovery measures such as infrastructure upgrades and other 'environmental works and measures' (Adamson and Loch 2014; Pittock et al. 2013, p. 114).[1]

The political retreat from buybacks in Australia indicates a need to reimagine the legal institutions, actors and combinations of instruments required to mobilise environmental water transactions, in order to ensure that those transactions are sufficiently responsive to the economic, social and cultural values at play. Of particular relevance is the Murray-Darling Basin Balanced Water Fund ('the Fund'), launched in 2015, which is supported by private capital and will test a legal and institutional concept developed by The Nature Conservancy. The concept is designed to address systemic risks and foster a notion of shared, or 'multi-purpose' water within the Murray-Darling Basin (the 'MDB' or 'the Basin') (Gilmore 2016). This chapter will analyse the benefits and risks of this approach in the legal context of the MDB, as well as the legal supports and safeguards required to support the Fund in meeting its environmental objectives. It will argue that, when accompanied by appropriate legislative safeguards, private initiatives of the kind represented by the Fund have the potential to function as an effective complementary measure, and improve system resiliency for the benefit of a range of uses, including the environment. The analysis will draw on theories of innovation, governance perspectives on social learning and the concept of smart regulation in order to evaluate the potential of the Fund within a market-based water allocation framework. Part 4.2 will examine the potential role of private finance and impact investment initiatives in the context of environmental water transactions. Part 4.3 will outline the legal

---

[1]See also the Productivity Commission, which considers water buybacks to be the most cost-effective and precise mechanism for water recovery in the Murray-Darling Basin (Australian Government 2010b).

context in which environmental water transactions have emerged in the Murray-Darling Basin, before conceptualising the new developments represented by the Fund from both an innovation and governance perspective. Part 4.4 will evaluate the corporate governance arrangements that have been established for the Fund, in order to contribute to an understanding of the legal supports and safeguards that are likely to be required in this context. Part 4.5 will then focus on those aspects of the Fund and its legislative context that are likely to be instructive for international jurisdictions in devising water recovery strategies, and for non-government organisations looking to contribute to or broaden the scope of environmental watering activities.

2 Private Finance, Impact Investment and Environmental Water Transactions

Water markets can improve the resilience of water governance systems and the capacity of individual users to manage scarcity. Markets encourage the reallocation of water resources between a variety of consumptive uses, and can also create opportunities to rebalance water use and return water to the environment through acquiring and applying water rights to a range of environmental watering needs. In the MDB, these needs are expansive and include water for wetland habitat, floodplain vegetation and threatened native fish and waterbird populations (Australian Government 2016b). The framework established under the *Water Act 2007* (Cth) provides scope for participation in recovery measures by a wide range of public and private actors, functioning in a variety of institutional settings. The bulk of environmental water recovery has occurred, however, through the Australian Government's large-scale program of water buybacks, under which the government has committed AU$3.2 billion to purchase water entitlements in order to serve a variety of environmental watering needs (Australian Government 2017a). At present, only isolated examples exist at a global level of private initiatives and private funding of environmental water transactions (Culp et al. 2015, p. 6). In the United States, for example, philanthropic foundations and non-government organisations ('NGOs') have provided operational and funding support for environmental water transactions to supplement a range of government programs for instream flows (Culp et al. 2015, p. 112; King 2004, pp. 515–520). It is considered that water trusts and NGOs engaging in water recovery activities are more prevalent in places like the United States on account of the fact that Australian federal and state government activity has 'crowded out' private participation (O'Keefe and Dollery 2011, p. 128).

Public and philanthropic sources may not, however, be sufficient to fund and maintain recovery at the scale required to restore water systems in the long term (Patterson et al. 2016, p. 13; Yonavjak 2014; Martin and Williams 2016, p. 623). In the context of global conservation efforts, analysts have begun to refer to a 'conservation financing gap', with a 2014 report by Credit Suisse, WWF and McKinsey estimating annual conservation funding needs in the order of $300–400 billion annually, but annual funding levels comprising approximately $50 billion of government, multilateral agency and philanthropic funding (Credit Suisse et al. 2014, pp. 11–12). The analysts consider, however, that '[t]here would be sufficient financial capital available to meet conservation investment needs if the main investor segments [g]lobally allocated 1% of their new and reinvested capital to conservation'. In this context, 'impact investment' structures and other forms of social entrepreneurship could play a key role in facilitating private investment in environmental water recovery (Credit Suisse et al., p. 6; Patterson et al. 2016, p. 29). In the context of conservation, impact investment can be defined as:

> [I]nvestments intended to return principal or generate profit while also driving a positive impact on natural resources and ecosystems—specifically, decreased pressure on a critical ecological resource and/or the preservation or enhancement of critical habitat. In addition, conservation impact must be an important motivation for making the investment. Conservation impact cannot be simply a byproduct of an investment made solely for financial return (The Nature Conservancy and EKO 2014, p. 3).

The Nature Conservancy, through its investment unit NatureVest, for example, seeks to 'create and transact investable deals that deliver conservation results and financial returns for investors', and is working to create a portfolio of investment products as proof of concept (The Nature Conservancy, 'NatureVest' 2017; The Nature Conservancy 2016, p. 48). This potential for 'win-win' outcomes is an attractive proposition for both policy-makers and commentators, and is seen to respond to the need for new forms of innovation that can increase the resilience of water governance. These new initiatives can also carry certain risks, however, given the competing imperatives inherent in these structures, and in particular the need to reconcile the generation of financial returns with potentially competing environmental objectives (Culp et al. 2015, p. 24). Part 3 will examine the legal supports that have been used to balance these considerations in the context of the MDB and the Fund.

3 Legal Supports for the Emergence of Private Environmental Water Initiatives in the Murray-Darling Basin

Water management systems in the MDB have been developed to address, among other matters, the historical over-allocation of water entitlements to consumptive uses, degradation of water resources in the region and to return water to the environment without compromising the productivity of irrigated agriculture (The Nature Conservancy 2016, pp. 10, 74–77). The legal framework governing the Basin sets specific and enforceable parameters for environmental watering requirements through water planning measures, and has established 'the environment' as a market player that relies on financial support by government (Owens 2017, p. 160;

Water Act 2007 (Cth), Pt 2). The groundwork for these reforms was laid in the 1994 Council of Australian Governments Agreement and the 2004 National Water Initiative (NWI), and the resulting governance framework provides a number of legal, institutional and policy supports for environmental water transactions within the Basin (Owens 2017, pp. 83–115). Water rights have been unbundled, and are monitored and enforced. Both permanent water entitlements and seasonal allocations can be traded among a range of users across connected valleys and state borders within the Murray-Darling Basin, on a large and sophisticated trading market (Australian Government 2017c). The *Water Act 2007* (Cth) created the MDB Authority (MDBA), and a process for the development and implementation of a legally binding plan (*Water Act 2007* (Cth), Pt 2 and 9.). The *Murray-Darling Basin Plan 2012* (Cth) (MDB Plan) has set legally enforceable limits on surface and groundwater use, known as sustainable diversion limits (SDLs), which must represent a 'sustainable level of take' (*Water Act 2007* (Cth), ss 4 and 23). Those recovery targets have been set at 2,750 GL/yr below the baseline diversion limit in order to achieve a measure of water recovery for the environment, and will come into force in 2019 through the aggregation of state-level water resource plans (*Murray-Darling Basin Plan (Cth)*, cl 6.04.). At the end of 2016, more than 2,028 GL, or about 75% of this amount, had been recovered through a mix of government purchases of water licences ('water buybacks') and taxpayer-funded infrastructure upgrades (Australian Government 2017b). The water buybacks have been conducted by the Australian Government under its Restoring the Balance program and transferred to the statutory Commonwealth Environmental Water Holder (CEWH), who must manage the Commonwealth's environmental water holdings in order to satisfy the watering requirements of various environmental assets (*Water Act 2007* (Cth) Pts 2 and 6). The scheme relies on voluntary irrigator participation, or willing sellers, and the government's environmental watering efforts under the program have focused on the largest, publicly-owned floodplain wetlands (Australian Government 2014).

The Australian reforms can therefore be seen as an attempt to recast MDB water resources as *shared* resources, and achieve a high degree of certainty in relation to the delivery of environmental water requirements (Owens 2017, pp. 248–249). However, the MDB Plan was undermined by politics during its development process, which compromised the recovery targets that have been set through SDLs. The MDBA's own modelling had set the minimum level of recovery at 3,200 (not 2,750) GL/yr, and even this figure did not incorporate available modelling in relation to projected climate change impacts (Pittock et al. 2015; Australian Government 2010a; Wentworth Group of Concerned Scientists 2012; Australian Government 2012). The *Water Act* and MDB Plan also allow the MDBA to propose adjustments to the SDLs within certain limits, provided those measures come into operation by mid-2024 (*Water Act 2007* (Cth), ss 23A and 23B and *Basin Plan 2012*, Chap. 7). This adjustment mechanism may permit as much as 650 GL of water recovery to be obtained as virtual water through 'environmental works and measures', which are measures that are deemed by the MDBA and the Minister to be equivalent to water returned to the river (Commonwealth of Australia 2017,

p. 773; *Murray-Darling Basin Plan 2012* (Cth), Chap. 7). Observers note that the consideration of these works and measures has created opportunities for 'blurry metrics' and that the equivalency of these projects to a volume of water is difficult to measure in practice (Crase 2016; Carmody 2016). Irrigation communities have also resisted the Australian Government's water buyback program, criticising the government's approach on the basis that it has adopted a 'just add water' approach, which took advantage of farmers already stressed from the ongoing Millennium Drought and undermined rural communities (see, for e.g., Murray Irrigation (undated); Schulte 2016; Commonwealth of Australia 2011). This resistance led the Abbott government in 2015 to cap buybacks at 1,500 GL/year and to abolish the independent National Water Commission (*Water Amendment Act 2015* (Cth); *National Water Commission (Abolition) Act 2015* (Cth). The current policy shift towards water recovery through irrigation upgrades and environmental works and measures is considered to be maladaptive, and likely to lead to 'lock-ins' that are inappropriate in the context of increasing climate variability (Adamson and Loch 2014; Pittock et al. 2013, p. 114). In this context, further consideration should be given to the potential for private recovery initiatives, which are not subject to the same political pressures and which have the potential to complement the processes established under the Basin Plan.

From an innovation perspective, private environmental watering initiatives can be considered in terms of both the processes and outcomes of innovation (Bitzer and Hamann 2015, p. 9). The innovation *process* is one of learning and knowledge creation, which involves the collective generation of ideas, problem definition, and the generation of new knowledge by participants who participate collectively to meet recovery challenges (Bitzer and Hamann 2015, p. 9). This perspective also has much in common with theories of governance that emphasise the potential of particular initiatives to foster processes of social learning, when non-state and private corporate actors participate in the formulation and implementation of institutional innovations (Pahl-Wostl et al. 2008; Pahl-Wostl 2007, pp. 56–58; Davidson and de Loe 2016). These processes of social learning are likely to be particularly useful in the context of complex social-ecological systems, where there is a need to build adaptive cross-sectoral capacities, as well as flexible governance systems (Cole 2015; Pahl-Wostl et al. 2008, p. 24; Pahl-Wostl 2007, pp. 57–58; Bitzer and Hamann, p. 6). Concrete examples of private environmental watering initiatives are now being developed by The Nature Conservancy, through its investment unit NatureVest, in recognition of the potential of these processes to build capacity for ongoing cooperation through establishing relationships among investors, non-government organisations, government officials and other stakeholders. Impact investment structures also benefit from the functional advantages of NGOs in this context, which include the provision of environmental expertise and analysis to identify large-scale conservation opportunities, and the certification of conservation investments by using 'pragmatic measurement systems' (Credit Suisse et al. 2014, p. 27).

These processes of innovation and social learning are, in turn, more likely to be effective when supported by an appropriate mix of regulatory instruments, or 'smarter' regulation. The concept of 'smart regulation' describes a form of regulatory pluralism that supports 'flexible, imaginative and innovative forms of social control' under which, in the majority of circumstances, 'the use of multiple rather than single policy instruments, and a broader range of regulatory actors, will produce better regulation' (Gunningham and Sinclair 2017, p. 133; Gunningham and Grabosky 1998). In the context of environmental watering initiatives, smart regulation would support the implementation of private initiatives as 'complementary' measures, provided those measures are targeted towards addressing the environmental issues at hand (Gunningham and Sinclair 2017, p. 139). This more pluralistic approach to regulation also applies to regulatory participants, and seeks to empower non-government actors with the requisite skills and expertise (Gunningham and Sinclair (undated), p. 3), therefore 'freeing up' scarce regulatory resources, with a 'greater range of actors, including commercial third parties, such as banks, insurers, consumers, suppliers and environmental consultants, and non-commercial third parties... taking the weight off government intervention' (Gunningham and Sinclair (undated), p. 3). It is hoped that the more flexible, and less interventionist approach, envisaged by these theorists, will provide additional efficiency and flexibility to participants in developing and implementing responses, and the ability to provide a context-specific response (Gunningham and Sinclair undated).

There is, accordingly, strong theoretical support for the effectiveness of collaborations in water governance, and for the private provision of complementary measures in order to improve the resilience of existing water governance systems. In the context of water recovery, however, the success of these impact investment structures will depend on how effectively the initiatives are able to balance the competing values and logics of the stakeholders involved in the collaboration. The measures are based on market mechanisms, which allocate water to the 'highest and best' use and can therefore have social implications in the absence of appropriate legal safeguards (Martin 2016, p. 392; Owens 2017, p. 8). Investors will ultimately be profit seeking, which can raise tensions and conflict with environmental objectives unless appropriate checks and balances are in place (Bitzer and Hamman 2015, p. 7). Impact investment initiatives have also generated criticism on the basis that reports such as those issued by Credit Suisse et al. (2014) create undue pressure on NGOs, such as conservation organisations, to make their programs and projects 'investable', and for the various actors involved in these projects to be 'entrepreneurial subjects' (Dempsey and Suarez 2016, pp. 662, 666). This is regarded as problematic, as it requires or 'conditions' these actors to structure their affairs in accordance with the logic of the market and draws attention away from 'broader structural forces as the locus of problem solving, resulting in broad-scale yielding of consent to rules of the game once challenged and powers once resisted' (Dempsey and Suarez 2016, p. 666). In the context of ensuring water sustainability within the MDB, these initiatives may be considered to draw attention away from the inadequacy of the recovery targets that have been set under the MDB Plan, the potential for those targets to be incrementally undermined by the adjustment mechanism and the broader choices that need to be made within the Basin as to the irrigation activities that are likely to be appropriate in an increasingly variable climate.

At the same time, government spending on natural resource management is unlikely to significantly increase in the near future. Although operational funding for key public actors within the MDB, such as the MDBA and Commonwealth Environmental Water Office (which supports the CEWH), has now been confirmed on a continuing basis (Australian Government 2016a), the earmarked funding is in the order of tens of millions of dollars, and is not likely to be increased significantly in the short term (Martin et al. 2017; Martin and Williams 2016). In addition, market-based frameworks in the MDB have already replaced a bureaucratic process for water allocation, and private initiatives of this nature can therefore be seen as a logical extension of the reform processes that have been underway since the mid 1990s. While the competing institutional logics of those participating in these investment structures will be challenging, with appropriate regulatory settings this tension may provide an opportunity for innovation, which emerges out of a participatory process between private corporations and other non-government stakeholders, and which reduces the gap between what we currently spend and ecosystem needs (Bitzer and Hamann 2015, p. 5; Credit Suisse et al. 2014).

4 The Murray-Darling Basin Balanced Water Fund

Since 2015, impact investment concepts have emerged that promote a notion of 'shared water', and which are designed to address systemic risks to both consumptive users and the environment (Culp et al. 2015, pp. 26, 44, 297, 305). Peter Culp et al. advocate a form of 'next generation water trust', which is able to hold a portion of water that can be 'flexibly turned 'on' or 'off' and dedicated to a range of economic, social and environmental uses, and which would use a combination of philanthropic and private funding sources to invest in water resources 'that would be repaid through revenue streams generated via the strategic deployment of trust assets' (2015, pp. 44–45, 307). According to Culp et al., these new initiatives:

> ...will need to accomplish more than a simple reallocation of water resources from low-to-high value uses and the creation of reasonable investment returns; they will need to contribute to the management of growing systemic risk across sectors in the Basin, and they will also need to reflect a different kind of thinking about the management of water as a finite resource... (2015, p. 24).

The Nature Conservancy ('TNC') and partners have also developed an investable structure, known as a 'Water Sharing Investment Partnership' ('WSIP'). The concept relies on a market-based water allocation framework and, like the 'next generation water trust', creates revenue streams via the strategic deployment of water assets while supporting various social and environmental outcomes (Global Water Forum 2016):

Water is already changing hands in water markets, whether from one agricultural user to another or from agriculture to urban use. Nature and underserved communities are completely left out of the equation. The goal of a WSIP is to create a conscientious intermediary to help facilitate these water trades and to ensure a more balanced outcome by reserving a portion of the water for nature and underserved communities.

It is also intended that the WSIP will achieve these outcomes without an associated loss of agricultural productivity (The Nature Conservancy 2016, p. 48). Lauren Ferstandig, a Director of NatureVest, notes the added complexity in these structures, in that:

Investors expect a financial return whereas philanthropic donors do not… We can't just take all of the water and apply it to targeted conservation outcomes. We also have to put some of it to work, so to speak, leasing it the agricultural community or to cities to generate a financial return for the fund. In some cases, we sell water rights at the end of a defined period in order to pay back investors their initial capital (Global Water Forum 2016).

The concept developed by the TNC provides a high degree of flexibility in terms of funding sources, acquisition methods and the means by which revenue is to be generated. Funding sources can include impact investment, philanthropy and government support (The Nature Conservancy 2016, p. 50), and water may be acquired via either the voluntary acquisition of water rights or through 'water-saving measures' implemented by irrigators. The distribution strategy for each WSIP will vary depending on the particular needs of a water basin, and irrigators and other consumptive users will benefit from the operations of the WSIP in being able to lease or buy water rights held by the WSIP (The Nature Conservancy 2016, p. 50).

In order to test this concept TNC has, in partnership with the Murray-Darling Wetlands Working Group and the Kilter Group, recently launched the MDB Balanced Water Fund in the MDB (Kilter Rural 2015). The Fund has targeted the southern MDB market because:

[the] market retains transparent, robust legal and regulatory frameworks and is sufficiently liquid for the ongoing deployment and divestment of capital at a scale required to accommodate the Fund. These features combine to support what is arguably the most sophisticated water market in the world (Kilter Rural 2015, p. 12).

The Fund was launched in October 2015, with a $27 million investment that included $5 million of seed investment from TNC Australia, and targets wholesale investors to achieve the multiple objectives of securing water for agriculture, realising a financial return and restoring threatened wetlands.[2] With capital of $25 million, equity returns over a 10-year tenure are expected to be in the order of between 5 and 8% per annum (before tax) and, once the fund has raised $100 million of capital, that target return increases to between 7 and 9% per annum (before tax) (Kilter Rural 2015, p. 6). Investor returns are generated through the

---

[2]The Fund is open to 'wholesale clients' within the meaning of s 761G of the *Corporations Act 2001* (Cth), and investors to whom a product disclosure statement is not required to be provided under Div 2, Pt 7.9 of the Corporations Act (Kilter Rural 2015, p. 3).

annual lease of permanent water entitlements, the trade of annual water allocations and the long-term capital appreciation of the Fund's portfolio of water entitlements (The Nature Conservancy 2016, p. 53). The Fund's social objectives are to sustain irrigator access to water through the lease of water entitlements and the sale of water allocations, and to work with Traditional Owners to identify and to implement targeted culturally significant watering activities (Kilter Rural 2015, pp. 8, 19). The Fund's environmental objectives are to facilitate the conduct of environmental watering in wetlands on private and public land, to restore threatened wetlands and support threatened freshwater-dependent species across the southern Murray-Darling Basin, and to achieve biodiversity improvements in the Basin.[3]

Investment will be managed by Kilter Rural (the Portfolio Manager), which will acquire a portfolio of southern Murray-Darling Basin water entitlements across New South Wales, Victoria and South Australia, targeting 'high security' or 'high reliability' water entitlements (Kilter Rural 2015, pp. 19, 36). The Board of the Trustee (Kilter Investments) is responsible for the overall corporate governance of the Fund in accordance with the Fund's Constitution, including oversight and monitoring of the Fund's investment and environmental watering objectives (Kilter Rural 2015, pp. 36, 40). Those watering objectives will be implemented by the Environmental Water Trust ('EWT'), and the Trustee and TNC Australia have agreed that the Fund will make donations of water assets to the EWT over a ten-year term. Those water assets will be managed by the Murray-Darling Wetlands Working Group (WWG) (Kilter Rural 2015, p. 32; The Nature Conservancy 2016, p. 53),[4] a conservation organisation with more than 20 years of experience delivering environmental watering programs in the MDB, which will undertake environmental watering activities in accordance with a 'plan and strategy' approved by TNC Australia (Kilter Rural 2015, p. 32). The CEWH has focused primarily on restoring 'iconic' wetland areas, and it is intended that the Fund will complement government watering activities by focusing on the wetland systems located on private lands that are of high ecological significance and Aboriginal cultural significance (The Nature Conservancy 2016, p. 52). This includes:

> Aboriginal land or soon-to-be Aboriginal-owned [land]. We target those because scientifically it just makes sense because there's something like 30,000 wetlands in the Basin and probably 90 percent of them are on private property. The government is doing a really great job of managing those river icon sites; those really big-picture places, the state parks and things like that, and they're shifting massive volumes of water—we will never have that kind of massive volume of water.

---

[3]Targeted objectives of the Fund include: improved waterbird breeding and movement; improved health of native plants including several rarer species; maintenance of refuges for fish and other species; and increased wetland plant growth. In addition to environmental outcomes, the restoration of environmental flows will help to conserve sites of important Aboriginal cultural and spiritual value (Kilter Rural 2015, p. 32).

[4]The Nature Conservation Water Fund (NCWF), the trustee for the EWT, is now jointly owned by TNC Australia and WWG.

So we tend to focus more on smaller areas, more targeted, but which involve a lot of community engagement and that's very intensive work that the government often doesn't have the resources to do.

So that's kind of our niche, it's where we sit—we're able to bring together massive numbers of partners and stakeholders in a process that hopefully keeps them engaged and involved and collaborative and hopefully keeps everybody on side, and in the meantime we get a lot of outcomes from that. Each one is a completely different event but it can involve 10, 15, 20 different stakeholders and that's what we're good at (Environmental Stakeholder 2016).

An important element of the Fund's corporate governance framework, from an environmental perspective, is the role of TNC Australia which, as sponsor and cornerstone investor, led the development of the Fund and its objectives, and reserved certain matters in respect of which the Trustee requires TNC Australia's consent (Kilter Rural 2015, p. 40). TNC Australia also has certain rights to wind-up the Fund or replace the Trustee in limited circumstances (Kilter Rural 2015, pp. 40, 51, 52), and can require that the Trustee retire as trustee where there has been a 'persistent failure by the Trustee to make donations in accordance with Schedule 7 of the Constitution for a specified period and attempts to rectify the breach and achieve a resolution have failed' (Kilter Rural 2015, p. 51, 53, 56).[5] Accordingly, while the trustee has overall corporate responsibility for the Fund, the EWT and TNC will be responsible for environmental impact objectives under a distinctive hybrid approach:

[H]ardwired into the documentation is the necessity for the Fund to make water available to the environment based on the set criteria that can only be changed with a 75 percent majority vote of all the unit holders and a reserved right of The Nature Conservancy, and there are very few of those things where the TNC has a reserved right and they're pretty much all around ensuring the long-term conservation outcomes of the Fund because we want to make sure that those things are enshrined in the Fund, as opposed to being open to interpretation or change by the portfolio manager. ...So we've hardwired a lot of those rules into the Fund but given a lot of discretion to the portfolio manager about how they construct and manage the portfolio, and a lot of discretion to the Wetlands Working Group about how the environmental watering gets done practically on the ground, because TNC sees itself as being the organisation who is there to oversee the balanced outcomes of the Fund, not just the financial outcomes and environmental outcomes, and we always try to work with the best-in-class organisations who have the expertise that we don't have (Gilmore 2016).

These controls are reinforced by environmental monitoring and reporting obligations, under which WWG will provide annual reports regarding the conservation outcomes of its activities (Kilter Rural 2015, p. 39).[6] Performance metrics must also

---

[5]Schedule 7 (Donations for Environmental Watering) of the Constitution specifies the conditions of cash donations and water assets for environmental purposes. The Schedule may not be amended or replaced without TNC Australia's consent (Kilter Rural 2015, p. 46).
[6]MWWG's other responsibilities are expected to include: identification, scoping and prioritisation of environmental watering projects; securing appropriate permits and approvals for watering activities; coordination with government bodies, including the CEWH, to maximise environmental outcomes of watering activities; and liaising with key Basin stakeholders.

be set against each watering event (Kilter Rural 2015, p. 32), with examples of possible metrics provided in the Fund's Information Memorandum (Kilter Rural 2015, p. 33). The EWT must then report on performance against stated outcomes for each watering event undertaken (Kilter Rural 2015, pp. 32–33). The Trustee is also required to provide an environmental report on an annual basis that outlines the environmental outcomes achieved through water donations in terms of the EWT's objectives (Kilter Rural 2015, p. 33). All of these reporting measures will be made available to the investors in the Fund, and the public (Environmental Stakeholder 2016; Gilmore 2016).

Although the governance structure of the Fund contains a number of controls designed to safeguard the Fund's environmental objectives, a key question is how decision-making under the Fund will reconcile the need to generate financial returns with the potentially competing objective of producing environmental outcomes (Patterson et al. 2016, p. 31). The proponents of the Fund stress that the Fund has been structured on a 'counter-cyclical' basis, to the effect that a greater distribution of water assets will be made to environmental and cultural watering activities in periods of high rainfall, which is consistent with both natural patterns of wetland inundation and periods of low demand for irrigation uses (Kilter Rural 2015, p. 21; The Nature Conservancy 2016, p. 52). According to Richard Gilmore of TNC Australia:

> In Australia the variability of the river systems is often a threat, but in the case of the Fund creates an opportunity because those wetlands have evolved over millennia to variable flows and we can leverage the variability of those flows in order to make water available to agriculture when it needs it and to wetlands when they need it, and the happy coincidence for the Fund is those are largely different times. You don't have that seasonal variability in a lot of places—I mean, the Murray is something like ten times more variable than the Danube and the Darling is something like 50 times more variable than the Danube (Gilmore 2016).

The Fund implements its counter-cyclical approach through annual determinations by the Trustee (Kilter Investments) as to the relevant water classification for the upcoming irrigation year, which will range from Very Dry to Very Wet (Kilter Rural 2015, p. 31):

> Strategy 1—Wet or Very Wet years:
>
> During a Wet to Very Wet year scenario, the entire environmental watering allocation will be diverted onto floodplains or very large wetlands. These environmental water donations will be added to natural flows to augment and extend the inundation of floodplain wetlands.
>
> Strategy 2—Moderate years:
>
> During a Moderate year scenario, water will be diverted into prioritised wetlands and creeks where appropriate infrastructure exists or can be developed efficiently. This strategy will target specific, smaller individual wetlands and creeks rather than broad floodplain inundation.
>
> Strategy 3—Dry or Very Dry years and drought:
>
> During a Dry Year scenario, small diversions of environmental water may be made into wetlands that provide critically important bird or other animal habitats, sustain areas of

drought refuge for aquatic species, provide benefits of cultural significance, or improve water quality. In a Dry Year or Very Dry Year, environmental watering will be limited and may not occur.

Particular percentages of water have been earmarked to those classifications, ranging from a minimum of 10% (Very Dry year) to a maximum of 40% of water allocations (Wet/Very Wet year) received by the Fund (Kilter Rural 2015, p. 31). It is anticipated that, on average, approximately 20% of the Fund's water entitlements will be committed to environmental watering purposes in the southern Murray-Darling Basin wetlands over the life of the Fund (Kilter Rural 2015, p. 21). Accordingly, a minimum of 60% of the Fund's portfolio of water entitlements will be made available to irrigators in the southern Murray-Darling Basin, but in dry conditions that proportion is likely to rise to 90% of Fund water entitlements (Kilter Rural 2015, p. 21). In crisis or emergency situations, the Constitution also sets out the expectation of the parties that the Trustee may decide to *not* commit Fund water to environmental watering activities, in specifying that in a Dry Year or Very Dry Year, environmental watering 'will be limited and may not occur' (Kilter Rural 2015, p. 31). A prolonged drought would, therefore, stress test the controls established under the Fund's Constitution.

The Trustee's expectation is that environmental water donations to Basin wetlands within the first 10 years of the Fund's operation will be between 55 and 79 GL of water (Kilter Rural 2015, p. 32):

> [which] demonstrates how an impact investment solution can generate environmental and social benefits at a far greater rate than could be achieved under a traditional philanthropy-funded model. Current projections indicate that the WSIP will be able to restore thousands of hectares of wetlands over the coming decade to the great benefit of people, frogs, fish, turtles and waterfowl. Investment of AUD$100 million will enable the purchase of approximately 40 gigaliters (40 million cubic meters) of water entitlements. The WSIP will be entirely self-funding, including coverage of all ecological monitoring and other operational costs, while projecting a return of 7 to 9 percent to wholesale investors (The Nature Conservancy 2016, p. 53).

An environmental stakeholder considers that the rules are 'pragmatic', and make the scheme workable in the sense that those rules are sufficiently clear for investors, who need to be able to form 'clear expectations' of what the Fund would deliver (Environmental Stakeholder 2016). However, the Determination Framework also highlights the potential limitations of an investment structure in the context of emergency conditions, which are foreseeable in the inevitable event of a drought, and underlines the need for private environmental watering initiatives to build on or supplement relevant public programs that are designed to serve critical watering needs.

5 Broader Lessons from Australia's Experience in Relation to Private Initiatives, Impact Investment and Environmental Water Recovery

The Fund therefore includes a number of regulatory innovations that are likely to be enduring, or even transformative, and are instructive for other jurisdictions in devising water recovery strategies, and/or NGOs looking to participate in environmental watering activities. The relatively sophisticated water market in the MDB, as well as market and political forces, have supported the emergence of the Fund and dictated its structure and objectives. However, it is important to recognise that it is the legal framework under the *Water Act* and the MDB Plan that establishes the ground rules for environmental protection within the MDB, and therefore determines the environmental limits within which a private initiative like the Fund may operate. The Fund has been deliberately conceived as an 'additional' or 'complementary' measure; it is not envisaged, or indeed advocated, that the Fund or the market take the place of the Australian Government in determining appropriate environmental baselines. The Fund will instead seek to increase the pool of investment funds available and provide private citizens with an opportunity to manage environmental assets. According to an environmental stakeholder:

> We really support the work of CEWH and we think that our Fund is not a better way to do it —it's a different and complementary way to do it…If the investors felt their investments were just doing the work that the Commonwealth was required to do anyway, we would find it very difficult to raise the funds because we wouldn't be able to demonstrate the additionality (Gilmore 2016).

The Fund's water holdings will also be additional to those held by the Australian Government. The Information Memorandum advises that:

> Water Entitlements acquired and held by the Fund cannot at the present time be counted towards the Basin Plan's environmental water recovery targets. The Trustee is prepared to work collaboratively with relevant Commonwealth entities to consider accredited proposals for delivery of environmental water in the Murray-Darling Basin (Kilter Rural 2015, p. 32).

The roles of the various public and private actors in relation to environmental watering activities, therefore, remain well delineated, and the model accords with the broad vision of those theorists who advocate the implementation of 'smarter' and complementary combinations of instruments that are targeted to particular environmental issues.

The Fund's structure also turns on the ability of private actors to hold and trade environmental water rights in the MDB, and form collaborative partnerships. Providing private actors with this legal capacity is desirable from an environmental perspective, as private entities will not be constrained by the political considerations and constraints outlined in Part 2 of the Chapter in providing environmental water (Owens 2017, pp. 260–261). Impact investment initiatives are also a collaborative enterprise and, in the context of water recovery, it is envisaged that those initiatives will rely on collaborative partnerships between a range of NGOs, which will provide opportunities for 'social learning' and innovation, bring into play all relevant knowledge and resources and enable communications and interactions that are likely to build the capacities of its key participants (Cole 2015, p. 115). These opportunities are explicitly recognised by TNC, which has outlined the Fund's potential to develop mutually beneficial arrangements between investors and the NGO community (The Nature Conservancy 2016, p. 60):

> Many NGOs have experienced, knowledgeable water experts on staff that can provide the data, knowledge base and political insights that can enable investors to better understand opportunities and risks around water markets. NGOs can help investors understand how much water is needed to serve various objectives, and the legal and physical mechanisms through which water transactions can be achieved. In turn, given that many NGOs lack sophisticated understanding of investor perspectives and even the language of investment, collaboration should prove to be mutually beneficial. Investors can help attune NGOs to investor concerns or risks that need to be addressed in the design of a Water Sharing Investment Partnership, or other creative financial solution, and the likelihood of attracting sufficient investment to the project.

Through collaboration, the Fund's partners have already established a strong consensus on the goals and values of the Fund, and formalised their ongoing functional relationships. This represents a departure from previous collaborative efforts in the MDB, which have not involved the same degree of systemic and organisational integration between the various stakeholders in their common goals (Owens 2016, p. 355; Margerum and Robinson 2015, p. 54). The Fund's investment structure incorporates robust corporate governance measures, making it highly attractive to private investors, while also building capacity to meet a range of environmental watering objectives. Through TNC's transnational networks, and the reporting measures provided for under the Constitution, the Fund is also likely to provide opportunities for learning, and emulation, through the dissemination of environmental reports and monitoring results (Owens 2016, p. 355). In combining a range of complementary competencies, it is hoped that these 'unlikely allies' will achieve a range of economic, social and environmental goals, and generate environmental benefits on a far greater scale than would ordinarily be achieved on a non-profit basis (Owens 2016, p. 355).

Another key innovation likely to be instructive for international development is the Fund's triage system of allocation, which reconceptualises the Fund's water holdings as 'shared', or 'multi-purpose', water to be applied to for the purposes of tackling systemic risks within the MDB. According to Richard Gilmore of TNC Australia:

> if we were to go into the market and buy 10 gigalitres of water just for conservation purposes we would be reinforcing the perception in the agricultural community that there's a conflict between agriculture and nature, and we didn't want to perpetuate that assumed conflict. We think that conflict is not only unhelpful but it's unnecessary (Gilmore 2016).

Instead, the water is considered to be 'productive water', for both agriculture and the environment, which is 'productive in different ratios', and which 'flexes around what the conditions are on the ground' that change 'from season to season' (Gilmore 2016).[7] The Fund therefore provides a new framing for activities involving environmental water recovery, and will hopefully achieve a measure of environmental improvement without polarising consumptive and environmental uses. In other jurisdictions, similar opportunities may be constrained by existing water allocation frameworks (The Nature Conservancy 2016, p. 51). Prior appropriative water rights in the western United States, for example, have particular components, which must be adhered to in order for the water right to retain its validity, including a specified beneficial use (Owens 2017, pp. 94–95). A formal change process is required in order to alter the appropriation's purpose, and downstream water users who use the return flows of upstream users acquire a legal claim to those return flows, which reduces the flexibility to amend the location and purpose of water rights under those systems (Owens 2017, p. 95). Observers are therefore considering how best to facilitate short-term, temporary, trades in these jurisdictions, which would not require a fundamental reform of the water allocation system (see, for e.g., Culp et al. 2015).

6 Conclusion

The experience within the MDB has demonstrated the difficulty of optimising social, economic and environmental outcomes within a single water management system. A process of water redistribution is currently underway; however, it remains to be seen whether the MDB Plan can gain enough momentum to restore the health of water-stressed and over-allocated water basins. Large-scale government buybacks, although favoured by both economists and environmentalists, have proven to be unpopular politically, and will play a more limited role in water recovery strategies going forward. In this context, models of impact investment provide a potentially effective option for improving system resiliency for the benefit of a range of uses, including the environment. Impact investment under the Fund should not be considered a panacea for the water problems experienced in the MDB, or as a replacement for current government programs. However, shifts in government policy to restrict the activities of the CEWH necessitate a reconsideration of environmental water recovery initiatives within the MDB, from a largely government-led process to one that will involve a flexible mix of competition and

---

[7]The notion of water that is held under a water access right for a range of possible uses does not, however, appear to have been contemplated by the regime under the W*ater Act 2007* (Cth), which is based on water rights that are administered by state and territory governments. For example, the definition of 'held environmental water' under s 4 of the *Water Act 2007* (Cth) does not deal specifically with water held for multiple purposes. Water Access Licences issued under the *Water Management Act 2000* (NSW) are not limited to a particular use (s 56); however, a water use approval must be obtained, in order to use water for a particular purpose at a particular location (s 89).

collaboration between government and non-government actors (Esty and Geradin 2000, p. 255).[8] The success of the Fund will turn, of course, on the myriad of operational and implementation decisions that must now be made. It is anticipated, however, that the new partnerships created under the Fund should improve the resilience of water governance within the MDB. International jurisdictions seeking to encourage similar initiatives will need to ensure that a range of legal supports are in place, including the delimitation of the market through environmental caps, the ability of private actors to hold and trade water and the ability for that water to be transferred on a seasonal basis between a variety of uses.

References

Adamson, D., & Loch, A. (2014). Possible negative feedbacks from 'gold-plating' irrigation infrastructure. *Agricultural Water Management, 145,* 134–144.

Australian Government, Department of Agriculture and Water Resources. (2017a). *Commonwealth water purchasing in the Murray-Darling Basin.* Retrieved April 3, 2017, from http://www.agriculture.gov.au/water/markets/commonwealth-water-mdb.

Australian Government, Murray-Darling Basin Authority. (2017b). *Progress on water recovery.* Retrieved April 3, 2017, from http://www.mdba.gov.au/managing-water/environmental-water/progress-water-recovery.

Australian Government, Murray-Darling Basin Authority. (2017c). *Interstate water trade.* Retrieved April 3, 2017, from https://www.mdba.gov.au/managing-water/water-markets-trade/interstate-water-trade.

Australian Government. (2016a). *Budget 2016–17: Australia's Mid-Year Economic and Fiscal Outlook.* Retrieved April 3, 2017, from http://www.budget.gov.au/2016-17/content/myefo/html/.

Australian Government, Murray-Darling Basin Authority. (2016b). *Annual watering priorities 2016–17.* Retrieved April 3, 2017, from http://www.mdba.gov.au/publications/mdba-reports/annual-watering-priorities-2016-17.

Australian Government, Murray-Darling Basin Authority. (2014). *Basin-wide environmental watering strategy.* Retrieved April 3, 2017, from http://www.mdba.gov.au/sites/default/files/pubs/Final-BWS-Nov14_0816.pdf.

Australian Government, Murray-Darling Basin Authority. (2012). *Hydrologic modelling to inform the proposed basin plan: Methods and results.* Retrieved April 3, 2017, from https://www.mdba.gov.au/sites/default/files/.../proposed/Hydro_Modelling_Report.pdf.

Australian Government, Murray-Darling Basin Authority. (2010a). *Guide to the proposed Basin plan: Technical background.* Retrieved April 3, 2017, from https://www.mdba.gov.au/sites/default/.../guide.../Guide-to-proposed-BP-vol2-0-12.pdf.

Australian Government, Productivity Commission. (2010b). *Market mechanisms for recovering water in the Murray–Darling Basin: Final research report.* Retrieved April 3, 2017, from http://www.pc.gov.au/inquiries/completed/murray-darling-water-recovery.

---

[8]Trends towards alternatives to buybacks, such as infrastructure projects, are also becoming increasingly popular in the western United States. See for example, Oregon's Allocation of Conserved Water Program (see https://www.oregon.gov/owrd/pages/mgmt_conserved_water.aspx . Accessed 3 April 2017).

Bitzer, V., & Hamann, R. (2015). The business of social and environmental innovation. In V. Bitzer & R. Hamann, M. Hall, E. W. Griffin-EL (Eds.), *The business of social and environmental innovation* (pp. 3–24). Switzerland: Springer International.

Carmody, E. (2016). Changes in Murray-Darling Basin laws and policies between 2007 and 2016. *Bridging, 19*, 6–7.

Cole, D. (2015). Advantages of a polycentric approach to climate change policy. *Nature Climate Change, 5*, 114–118.

Commonwealth of Australia. (2017). *Parliamentary debates: House of Representatives: Official Hansard.* 13 February. Retrieved April 3, 2017, from http://www.aph.gov.au/Parliamentary_Business/Hansard?wc=13/02/2017.

Commonwealth of Australia. (2011). *House of representatives, standing committee on regional Australia. Of drought and flooding rains: Inquiry into the impact of the guide to the Murray-Darling Basin plan.* Retrieved April 3, 2017, from http://www.aph.gov.au/parliamentary_business/committees/house_of_representatives_committees?url=ra/murraydarling/report.htm.

Credit Suisse AG, World Wildlife Fund, Inc and McKinsey and Company. (2014). *Conservation finance: Moving beyond donor funding toward an investor-driven approach.* Retrieved April 3, 2017, from https://www.cbd.int/financial/privatesector/g-private-wwf.pdf.

Crase, L. (2016). Latest Murray-Darling squabble sheds light on the plan's flaws. *The Conservation.* Retrieved April 3, 2017, from https://theconversation.com/latest-murray-darling-squabble-sheds-light-on-the-plans-flaws-69484.

Culp, P., et al. (2015). *Liquid assets: Investing for impact in the Colorado River Basin.* Retrieved April 3, 2017, from http://www.ckblueshift.com/resources.

Davidson, S. L., & de Loe, R. C. (2016). The changing role of ENGOs in water governance: Institutional entrepreneurs? *Environmental Management, 57*(1), 62–78.

Dempsey, J., & Suarez, D. C. (2016). Arrested development? The promises and paradoxes of 'selling nature to save it'. *Annals of the American Association of Geographers, 106*(3), 653–671.

Esty, D. C., & Geradin, D. (2000). Regulatory co-opetition. *Journal of International Economic Law, 3*(2), 235–255.

Global Water Forum, Q&A. (2016). *Water markets and impact investment help address global water scarcity.* Retrieved April 3, 2017, from http://www.globalwaterforum.org/2016/09/19/qa-water-markets-and-impact-investment-help-address-global-water-scarcity/.

Gunningham, N., & Grabosky, P. (1998). *Smart regulation: Designing environmental policy.* Oxford: Oxford University Press.

Gunningham, N., & Sinclair, D. (2017). Smart Regulation. In P. Drahos (Ed.), *Regulatory theory: Foundations and applications* (pp. 133–148). Canberra: ANU Press.

Gunningham, N., & Sinclair, D. (Undated). *Designing smart regulation.* Retrieved April 3, 2017, from https://www.oecd.org/env/outreach/33947759.pdf.

Interview with Richard Gilmore, Country Director, The Nature Conservancy Australia, Telephone Interview, March 23 2016.

Interview with environmental stakeholder, Telephone Interview, March 23 2016.

Kilter Rural. (2015). *Information memorandum: The Murray-Darling Basin balanced water fund.* Retrieved April 3, 2017, from http://kilterrural.com/news-resources/murray-darling-basin-balanced-water-fund-information-memorandum.

King, M. A. (2004). Getting our feet wet: An introduction to water trusts. *Harvard Environmental Law Review, 28*, 495–534.

Margerum, R. D., & Robinson, C. J. (2015). Collaborative partnerships and challenges for sustainable water management. *Current Opinion in Environmental Sustainability, 12*, 53–58.

Martin, P. (2016). Creating the next generation of water governance. *Environmental and Planning Law Journal, 33*(4), 388–401.

Martin, P., & Williams, J. (2016). Next generation rural natural resource governance: A careful diagnosis. In V. Mauerhofer (Ed.), *Legal aspects of sustainable development: Horizontal and sectorial policy issues* (607–628). Berlin: Springer.

Martin, P., et al. (2017). The environment needs billions of dollars more: Here's how to raise the money. *The Conversation*. Retrieved April 3, 2017, from https://theconversation.com/the-environment-needs-billions-of-dollars-more-heres-how-to-raise-the-money-70401.

Murray Irrigation Ltd. (Undated). *Response to the guide to the proposed Murray-Darling Basin plan*. Retrieved April 3, 2017, from http://www.murrayirrigation.com.au/media/2503/Submission-to-the-MDBA-re-Guide-to-the-Proposed-Murray-Darling-Basin-Plan.pdf.

O'Keefe, S., & Dollery, B. (2011). Collaborating and coordinating disparate interests: Lessons from water trusts. In L. Crase & S. O'Keefe (Eds.) *Water policy, tourism and recreation* (pp. 120–132). RFF Press.

Owens, K. (2016). Reimagining water buybacks in Australia: Non-governmental organisations, complementary initiatives and private capital. *Environmental and Planning Law Journal, 33*, 342–355.

Owens, K. (2017). *Environmental water markets and regulation: A comparative legal approach*. London: Routledge: Earthscan Studies In Water Resource Management.

Pahl-Wostl, C. (2007). Transitions towards adaptive management of water facing climate and global change. *Water Resources Management, 21*(1), 49–62.

Pahl-Wostl, C., Mostert, E., & Tàbara, D. (2008). The growing importance of social learning in water resources management and sustainability science. *Ecology and Society, 13*(1), 24.

Patterson, L., et al. (2016). *Conservation finance and impact investing for U.S. Water: A report from the 2016 Aspen-Nicholas Water Forum*. Retrieved April 3, 2017, from https://www.aspeninstitute.org/publications/2016-aspen-nicholas-water-forum-report/.

Pittock, J., Max Finlayson, C., & Howitt, J. (2013). Beguiling and risky: 'environmental works and measures' for wetland conservation under a changing climate. *Hydrobiologia, 708*(1), 111–131.

Pittock, J., Williams, J., & Grafton, R. Q. (2015). The Murray-Darling Basin Plan fails adequately to deal with climate change. *Water, 42*(6), 28–34.

Schulte, S. (2016). *More than flow for the Murray-Darling Basin Plan. King's Water*. Retrieved April 3, 2017, from https://blogs.kcl.ac.uk/water/2016/11/17/more-than-flow-for-the-murray-darling-basin-plan/.

The Nature Conservancy and EKO. (2014). *Investing in conservation*. Retrieved April 3, 2017, from http://www.naturevesttnc.org/reports/.

The Nature Conservancy, NatureVest. (2017). *About us*. Retrieved April 3, 2017, from http://www.naturevesttnc.org/about-us/.

The Nature Conservancy. (2016). *Water share: Using water markets and impact investment to drive sustainability*. Retrieved April 3, 2017, from https://global.nature.org/content/water-share.

Wentworth Group of Concerned Scientists. (2012). *Does a 3,200Gl reduction in extractions combined with the relaxation of eight constraints give a healthy working Murray-Darling Basin River System?* Retrieved April 3, 2017, from http://wentworthgroup.org/wp-content/uploads/2013/10/Wentworth-Group-Evaluation-Modeling-with-Constraints.pdf.

Yonavjak, L. (2014). *Conservation finance is gearing up for wall street*. Retrieved April 3, 2017, from http://conservationfinance.org/news.php?id=232.

Part II
Water Markets—Property, Regulation and Implementation

Water Entitlements as Property: A Work in Progress or Watertight Now?

Janice Gray and Louise Lee

Abstract This chapter considers a seemingly simple question: are water entitlements property? Yet this question is deceptive in its simplicity. The chapter's focus is water entitlements and licences in the context of water trading in Australia, but it also references other international jurisdictions and suggests that the concerns raised in the Australian context may have relevance for other jurisdictions where water trading is either being relied on, or being considered as a governance tool. In exploring the property question, the chapter briefly outlines three different theories of property and highlights some existing tensions between different analyses of property in order to demonstrate that property is a complex and nuanced concept which may take different forms depending on which justificatory and analytical approaches are employed. The chapter argues that despite property's popularity in neo-liberal politics and its convenience as a regulatory tool, the impacts of propertisation should be considered very carefully before the proprietary route is embraced. Accordingly, the chapter also discusses statutory property generally and specifically the characterisation of water entitlements/licences under legislation, before considering whether water entitlements need to be characterised as property to support trading. It then explores the possible effects of keeping water entitlements/licences outside the proprietary frame, arguing that such an approach may open up opportunities for a wider range of governance tools.

J. Gray (✉)
Faculty of Law, University of New South Wales, Sydney, Australia
e-mail: j.gray@unsw.edu.au

L. Lee
Corrs Chambers Westgarth, Sydney, Australia
e-mail: louise.lee@corrs.com.au

1 Introduction

Towards the turn of last century, many Australian states and territories embarked on an extensive program of water reform. That program was reinforced and expanded when, in 2004, states and territories agreed upon the National Water Initiative (NWI) (Council of Australian Governments 2004) a national blueprint for water reform. Recognition of the need for reform had been gathering momentum throughout the last decade of the twentieth century, but the seriousness and urgency of Australia's water problems only became obvious to many when the ill-effects of water over-allocation were highlighted by the Millennium Drought (late 1996–mid 2000) (ACIL Tasman and Freehills 2004). That drought ravaged land and waterscapes. It saw the introduction of water-use restrictions, water-use targets, increased dam building plans, the construction of desalination plants, and the growth of sewer mining and recycled water operations.

Water reform policy in this era was driven by a range of issues, two of which were: (a) a desire to address widespread natural resource degradation resulting from both natural causes and heavy water use; and (b) the availability of sufficient water in places where it was needed. The strategic frameworks which were developed as a response to these concerns were shaped largely along efficiency and competition lines. The frameworks covered issues such as 'water pricing, the appraisal of investment in rural water schemes, the specification of, and trading in, water entitlements, [adaptive] resource management (including recognising the environment as a user of water via formal allocations), institutional reform and improved public consultation' (National Competition Council 2017).

More broadly, the reforms introduced across the various Australian jurisdictions represented (and continue to represent) a new, hybridised form of governance which relies on the inter-dependency of markets, regulation, and integrated and collaborative water planning. However, little legal attention has been given to the question of the *nature* of the newly styled water entitlements in that new governance model. Are they property and indeed, do they need to be property in order for them to form the subject of trade under the new governance model?

The answer to the question of whether water entitlements constitute a form of private property is important. It may have implications for water law and governance (particularly water trading) but it may also be relevant to other legal sub-disciplines which interact with water law including succession, compulsory acquisition, taxation, trusts and restitution, for example. The implications of entitlements' classification are, therefore, potentially extensive and may impact on the effectiveness, efficiency, fairness and resilience of not only water governance itself, but also social and legal relationships more widely.

Nevertheless, in many Australian jurisdictions, the legal nature of water entitlements remains unsettled. This has not, however, stopped certain stakeholders, groups and even some scholars taking a 'business as usual' approach. Some have proceeded on the basis that the answer does not really matter. It is just a legal nicety. Others have proceeded as though the issue of entitlements classification has been resolved already. Indeed, some of the literature, particularly that of economists and government agencies, refers to 'water property rights' without equivocation (see Hughes 2015; Grafton et al. 2010; National Water Commission 2014). Such a position is possibly based on the view that if something is tradeable, it must be property—a view strongly embraced by scholars in the Law and Economics sub-discipline. However, to many others, including scholars from disciplines such as sociology, cultural studies, environmental science, philosophy and even law, this approach is limited and problematic (on the pyschology of property (see Babie et al. (2017)). It forecloses the diverse justifications for property and arguably takes a reductionist view of the conditions which must be met for a 'thing' to be propertised.

Whilst it is beyond the scope of this chapter to explore in detail the diverse justifications for property, the chapter does outline three key theoretical positions and flags the potential significance of the range of property justifications. It also discusses statutory property and its applicability in the water sector, focussing in particular on the law of New South Wales, Australia.

The chapter then questions the trend to propertise, observing that propertisation has both positive and negative impacts. Propertisation may (but will not necessarily) improve water management. Indeed, propertisation may cause as many problems as it solves (see Freyfogle 2007; Singer 2000a).

With this in mind, consideration is given to some of the practical and perhaps unforeseen consequences of water entitlements' propertisation which may emerge if governments seek to reset their policy and legislative directions. Such changes in direction may arise if, for example, governments conclude that trading and markets do not sufficiently help steer water management in a sustainable direction nor yield desired outcomes in terms of equity, fairness and efficiency. Hence, we consider whether a proprietary classification may lock future generations into path dependency. Once something is propertised it is very difficult to 'unpropertise' it, as the cost of doing so is often prohibitive. Accordingly, we argue that high levels of prudence should be exercised before the propertisation course is embarked upon.

In probing these and other issues we seek to tease out some broad lessons for policy-makers, lawyers and theorists both in Australia and beyond.

2 Understanding Property

In order to better understand whether water entitlements are, or should be, property, it is helpful to explore the concept of property. There is no single, unitary definition of property. As Rose (1994) and others observe, property remains a contested term (Waldron 1985, 1998). The diversity of the work of Locke, Blackstone, Bentham, Voltaire, Singer, Smith, Rose, Posner, Heller, Alexander, Penner, Cooke and Kevin Gray, for example, speaks to the range of different conceptual and theoretical positions available.

Despite such differences it is, however, generally agreed that property is a social institution with a legal dimension. It is the institution responsible for regulating access to material resources or things (Singer 2000b; Edgeworth et al. 2012). It is, therefore, a concept which is commonly employed in domains such as politics, economics, geography, environmental science and philosophy, as well as law.

To lawyers, property consists of the legal and technical rules for delineating the rights and duties of owners, as well as the formal mechanisms which effectuate fragmentation and transfer. It also consists of the values underpinning those rights, duties and the choice of mechanisms. In turn, those values form the basis of the various justifications for property.

Indeed, judicial and legislative elaboration of property's doctrinal rules commonly interacts and engages with (a) conceptual questions as to the nature of property and (b) evaluative assessments as to the appropriate limits or boundaries of property rights (Gray et al. 2017). Cases such as *ICM Agriculture Pty Ltd v Commonwealth* demonstrate this.

2.1 Different Approaches to Property

The following briefly outlines three different approaches to property, in order to highlight that property is a complex concept and that calls (particularly uninformed calls) for the propertisation of various things, including water entitlements, should be approached with caution.

Economic Approach to Property One prominent understanding of property which has been re-invigorated in the neo-liberal era is an economic understanding. Penner observes that '[t]he economic view of property is very simple. Property comprises any valuable resource in respect of which an individual has exclusive entitlement' (Penner 1997). Seen this way, property has virtually endless possible permutations. It may be tangible or intangible. It may cover a range of 'things' or just a few, but its advocates are confident of one thing–that (private) property and the trading of it has a flow-on beneficial impact on society.

From an economic perspective, the justification for the creation and recognition of property rights stems from the ability to develop incentives for the efficient use of resources (Posner 1992), to prevent over-exploitation of resources (Lueck 2003), and to internalise externalities (Demsetz 1967). Additionally and importantly, private property (with as little State-based regulation as possible) is thought to promote opportunities for wealth maximisation. From this perspective, the more private property there is, the better. Hence 'things' (such as water access entitlements) not previously brought within the private property paradigm should be re-envisioned as property so as to improve society more generally.

Labour and Property This theory asserts that individuals have a property right in their own person and by extension are entitled to own what they produce by employing their own efforts and what they have laboured on (Gray et al. 2017). In other words, it is through labour (rather than through trade, for example) that one creates and acquires property. Locke was the first philosopher to arrive at such a view; a view which is grounded in individualism (Laslett 1964).

Hence, under Lockean theory, if an individual caught a wild animal, then the labour involved in that act resulted in the wild animal belonging to (or becoming the property of) the individual. Aside from the gendered nature of Locke's theory (men's labour has traditionally been more easily recognisable than women's), it does not seem to be a particularly good fit for water access entitlements. A right of access to water, as opposed to a right of use or a right to install works, may, for example, involve little or no actual labour. Further, Locke's restriction on the acquisition of property imposed by his requirement that individuals cannot diminish the opportunities of others to acquire property, would seem to generate further issues for the applicability of the theory in relation to water entitlements. The more shares there are in the pool of available water (entitlements are calculated on shares or percentages), the less water each shareholder is able to hold.

Property and Social Vision Singer is the key proponent of this theory, but he is far from alone in arguing that property is linked to democracy and social vision more broadly. Singer conceives of property not just as a mechanism for coordination; but as a 'quasi-constitutional framework for social life' (Singer 2014). In other words, property is a tool which simultaneously reflects and promotes (even defends) underlying social values. In Singer's view, each system of property is built upon the context and values that stimulate its creation in the first instance. This is evident in the emergence of new forms of property and ownership concepts such as moral rights for artists, intellectual property including rights pertaining to cyber-space games (Walpole and Gray 2010), the potential ownership of one's genetic information (Martin and Verbeck 2002; Bennett Moses 2008), and even the ownership of sewage (Gray and Gardner, 2008) and unbundled elements of fugacious resources such as water (Gray 2011).

To Singer, cultural, moral, religious and legal traditions help inform what is meant by property (Singer 2000a, b). On this reasoning, property is not a concept that is best approached in positivist terms (see Hart 1958 on positivism). Nor is it a concept that should be interpreted along Posner-type lines which emphasise ownership and permit people to do with their land (or other assets) what they will as long as their conduct does not hurt others (Posner 1992). By contrast, Singer sees society as having the ability to construct property in such a way that helps shape the type of community in which people would like to live, and he sees dependence on the ownership model alone as getting in the way of that potential, a point which he ably demonstrates through his discussion of the U.S. case *Friendswood Development Co v. Smith-Southwest Industries Inc.*

There are, of course, many additional theories of property, but space prevents further elaboration. Those theories include:

- utilitarian theories (involving judgements of 'goodness and badness', see Alexander and Penalver 2012);
- feminist theories (Rose 1992; Dickenson 1997; see also Wilson 2008; and Carr and Wong in Bright and Blandy (eds) 2016);
- Marxist theories (involving state-owned property);
- Hegelian theories (emphasising rights-based personhood);

- Kantian theories (involving individual freedom as an end in itself); and
- Indigenous theories or understandings (see Marshall 2017 referring to Arnold).

Tensions Between Conflicting Analyses of Property Tensions exist between conflicting views and analyses of property, potentially impacting on decisions about what should be propertised and how property should be allocated or shared between members of society. Those 'what' and 'how' questions are relevant to the water sector.

One of the key modern tensions is the tension between the 'bundle of sticks' approach to property and more essentialist approaches. Hohfeld (1913), who did not actually use the term 'bundle of sticks' in his writing, is often credited with being the founder of this approach. Honore (1999) later elaborated, spelling out examples of the different types of sticks or rights which may comprise the bundle that is property. They include: the right to possess; the right to use; the right to manage; the right to income; the right to capital, the right to security; and the power of transmissibility (Honore 1999; see Gray et al. 2017). Under the 'bundle of sticks' approach, no one right is individually necessary to establish property, but a variable collection of rights may be sufficient to do so. Hence, lawyers, scholars and theorists often speak of the 'indicia of property' to suggest that the existence of some of the elements to which Honore and others refer will point in the direction of a property characterisation.

The 'bundle of sticks' approach has, however, been criticised for being too malleable and for too readily permitting the admission of 'things' into the pantheon of property. For example, the approach does not stipulate how many rights (or sticks) should be in the bundle. Nor is it clear whether each of the rights in the bundle should be equally weighted, and which rights, if any, are essential to establishing the existence of property. It is around these (and other) questions that several theoretical tensions have emerged. Penner (1997), Merrill (1998) and Smith (2012), for example, posit that the right of exclusion is paramount. To these scholars, without the right to exclude, there is no property. Merrill (1998) goes so far as to state that:

> [t]he right to exclude others is more than just 'one of the most essential' constituents of property—it is the *sine qua non*. Give someone the right to exclude others from a valued resource, i.e. a resource that is scarce relative to the human demand for it, and you give them property. Deny someone the exclusion right and they do not have property.

Merrill's approach is sometimes described as a 'means' based approach because it focuses on the qualities or means that property uses (Baron 2014). Such means emphasise 'modularities' and 'boundaries' and include exclusion and *in rem* rights (see Merrill 1998; Smith 2012). Meanwhile 'ends' theorists focus on the ends property is designed to serve; ends such as democracy (Singer 2014) and human flourishing (Alexander 2009).

Merrill's formulation is clearly at odds with several other understandings of property, including that of Kevin Gray and in particular, Gray's notion of 'regulatory property', which Gray describes as an expanded right of access, arising as a corollary of the curtailment, by way of statute, of the conventional property right of exclusion (Gray 2010). Regulatory property as conceived by Gray is a democratic concept, but it simultaneously attacks the heart of what Penner, Smith, Merrill and others consider essential for the establishment of property because it cuts back the right of exclusion in order to open up the right of access. According to Gray, property is 'a continuum along which varying kinds of proprietary status may fade finely into one another' (Gray 1991). This open and fluid conceptualisation again departs from the more essentialist approach of Penner (1997), Smith (2012) and Merrill (1998).

Presumably to 'means' theorists such as Merrill (1998) and Smith (2012), Gray's approach would manifest other problems, too. It may, for example, be said to lose sight of the *in rem* dimension of property. Arguably, loss of that dimension is very significant in the natural resources domain. Natural resources, including water, necessarily involve a strong degree of materiality. Put another way, 'thingness' matters.

Conceiving of property as a flexible institution invites all manner of questions including: how far can the right of exclusion be curtailed before it undermines the very existence of property? Clearly, the right of exclusion has already been diminished in numerous ways. For example, each time a plane flies over one's land an infringement of the *cujus est solum* doctrine occurs. This and other examples suggest that some level of curtailment will be tolerated, even by essentialists, when it comes to the right of exclusion. The issue raises the question: at what point would property in water infrastructure be lost by way of a diminution of the right to exclude in favour of expansion of a right to access (particularly for competitors seeking to use that same water infrastructure)?

Despite property's many benefits, including its wide sphere of enforceability, the powerful remedies available for breach of a property right and property's capacity as a tool of social organisation, propertisation may also cause problems. It may, for example, deepen disputes. Opposing parties may equally rely on their different sets of property rights to assert that they should succeed. At other times, parties may rely on the same rights in the property bundle but, viewed from different perspectives, each party may see those same rights as serving different purposes. Propertisation may, therefore, exacerbate, rather than solve, disputes. For example, one party may argue that she should succeed because she is exercising her property rights in a lawful manner by abstracting groundwater, while a second party, her neighbour, may argue that his right to use and enjoy the adjoining land has been interfered with by the exercise of the first party's water abstraction rights. Hence, merely enveloping things and relationships within the proprietary frame will not necessarily provide a solution. In some cases it may entrench positions and exacerbate disagreements (Freyfogle 2007).

Another property tension which may exist is between positivists and other theorists including, but not limited to, social realists and natural law theorists. To many positivists, who separate moral rights from legal rights and who rely on logic and deduction as the cornerstone of legal reasoning, the theoretical positions of scholars Singer, Kevin Gray and Natural Law theorist, Fuller (Fuller 1958), for example, reveal weaknesses. Singer's approach to property, which is concerned with democracy building and social vision realisation is, for example, criticised as being too flexible. It allows property systems and their component rights and entitlements to be left open to interpretation because social values are subjective. The approach also embraces the view that social values are fluid. They change over time and in response to external pressures and circumstances. Hence in periods of drought, there may, for example, be more value and emphasis placed on the propertisation of water resources as compared with periods of plenty.

Yet, while positivists reject the role of subjectivity and morality in legal decision-making, judicial decisions historically demonstrate that the context in which the existence of a property right is mediated, and the values and perspectives each party brings to the table, *are* important to the final determination (see, for example, *Mabo v Queensland*, a native title case dealing with rights in, and relationships to both water and land).

What emerges from the above discussion is that schisms in property literature exist, reflecting different schools of thought on the analysis, philosophy and doctrine of property. At times, the different positions are in competition with each other but at other times, they coalesce over blurred boundaries. When there are calls to propertise new things such as water access entitlements, policy makers, judges and legislators in particular, need to respond, bearing in mind the complexities of property discussed above. Not to be aware of how property may be used as a tool of social organisation and/or democracy (Singer 2000a, b); how it may help some people flourish but not others (Alexander 2009); how it may be a reward for labour (Locke in Laslett 1964) or as an example of theft (Proudhon 1840), or how it may be a gendered concept (Rose 1992), for example, would be to fail to appreciate its impact on society and how it may privilege certain members of society over others. Property is a nuanced institution which should be handled with a respectful awareness of the implications for which a property characterisation may be responsible. Property is not a static concept, locked in time without a relationship to the wider world.

A failure to appreciate the analytical, philosophical and doctrinal aspects of property may lead to reification of the concept. Property may end up being treated simply as a tradeable object. Such an outcome would be to diminish and misunderstand property's potential and may ultimately obscure the impacts of propertising resources such as water entitlements.

2.2 Two Key Questions

What Should Be Propertised? The fundamental question of which things should be propertised is relevant to all societies. The way different societies respond to that question reflects their attitudes to both different theories of property and the reconciliation of tension between conflicting approaches to, and analyses of, property. However, most societies do not seek to propertise resources if they are in abundance (Hume 1739; Dickenson 2007; Gray and Gardner 2008; Cooke 2012). While resources are plentiful, common practice is to treat them as common property or a common pool resource (see Plato, who argued in *Republic* that collective/common ownership was necessary to serve the common good by promoting the common interest (Plato (trans) 1993). Purported abundance (along with its fugacious nature) has meant that water in flow, for example, has not been propertised. Additionally, in some jurisdictions, policy reasons may have kept a thing outside the proprietary paradigm (see Kant 1964). Hence until recently, sewage (despoiled water) arguably fell into this category (Gray 2011) along with human bodies (Fagot-Largeault 1998), body parts and corpses (Dickenson 2007; *Moore v. Regents of University of California*; Boyle 2002). Conversely, what has traditionally brought resources into the proprietary frame is desire or acquisitiveness. When there is demand for a limited resource, such as water, propertisation is one method of regulating that resource.

How Should Property Be Allocated? Once something is propertised, the next and interlinked question which commonly arises is, how should that property be shared or distributed? The answer to that question is connected to the justifications for property on which the relevant society relies. The property theories outlined above reveal several different justifications at play. Different theories will privilege different interests such as public or private, male or female, individual or common, democratic or autocratic, socio-economic, racial and/or religious interests, for example.

Justifications for property based on concerns about where the boundaries between public and private domains should lie raise particular, associated questions such as: should an individual be able to do whatever he/she likes with his/her property simply because it is his/hers, or does a property holder have a responsibility to the greater heritage of humankind, particularly the *environmental* heritage of humankind? In this regard, Chief Justice Taney of the US Supreme Court in *Charles River Bridge v. Warren Bridge* stated that '[w]hile the rights of private property are sacredly guarded, we must not forget that the community also have [sic] rights, and the happiness and well-being of every citizen depends on their faithful preservation'.

Joseph Sax captured the dilemma ably when he considered whether one has the right to play darts with a Rembrandt or whether one is under a duty to preserve the Rembrandt for the greater good (Sax 1999). In the water context, we might similarly ask whether a private individual or company should have the right to seriously or unsustainably deplete water resources from a watercourse or whether that individual/company has an obligation to preserve water for the benefit of the wider community. Should individual or communal rights be privileged? (Hannam 2017).

3 Implications of Propertisation

Statutory Property Several modern forms of property are examples of statutory property—that is, they have been created under, and have an existence that is entirely dependent upon, statute. Copyright provides one such example. Water access entitlements potentially provide another.

Academic literature and judicial decisions have considered the question of whether statutory rights are property both from a general perspective (Hepburn 2010; Story 2006) and with reference to specific (water) statutory schemes (McKenzie 2009). It is clear from analysis of this work that statutory rights are capable of constituting property. Whether a statutory right is proprietary in nature will ultimately depend on close examination of the provisions in the particular statute in question. The High Court in *Wik Peoples v State of Queensland* saw this as the key difference between statutory property and common law property (see also *Telstra Corporation Limited v The Commonwealth of Australia*; *Davies v Littlejohn*).

Accordingly, statutory property does not necessarily equate with common law property, although in certain circumstances it may be a close approximation. This depends on the nature of the rights conferred under the relevant statute and the context and purpose for which the issue is being examined. For example, where a diminution or extinguishment of particular property rights amounts to 'an acquisition of property' by the Commonwealth, s 51(xxxi) of the Commonwealth Constitution requires that statutory property, as well as common law property, be the subject of compensation on just terms (*Attorney General NT v Chaffey*). Thus the High Court has held previously that s 51(xxxi) of the Constitution applies to copyright (*Australian Tape Manufacturers Association Ltd v Commonwealth of Australia*), which, as observed above, is an entirely statutory creation (see *Pacific Film Laboratories v Commissioner of Taxation*).

The English decision of *Armstrong DLW GMBH v Winnington Networks Ltd* acknowledged that it may be difficult to draw a clear distinction between common law and statutory property because the relevant property terms used in statutory definitions are themselves derived from common law concepts.

In *Winnington*, the nature and character of carbon emission allowances was considered by the High Court of England and Wales; the nature being determinative of the remedies available to the appellants. At issue was whether either a claim in conversion or a proprietary restitutionary claim could be brought in relation to the fraudulent act of scammers, who had logged onto the online accounts of holders of carbon emission allowances and then on-sold those allowances to third party purchasers. Ultimately, the Court held that the allowances were a form of intangible property over which there could be no claim in conversion, but for which a proprietary restitutionary claim was available.

It is acknowledged that the mere fact that a statutory right may be diminished or extinguished by subsequent enactment does not mean that it is not a property right (*Attorney General v Chaffey*; *Commonwealth v WMC Resources Ltd*). After all, common law forms of property may also be diminished or extinguished by enactment but still preserve their proprietary status, as the intrusion of planning law's statutory restrictions on property interests such as fee simple estates reveals. However, rights and interests created by statute may possess certain additional characteristics which make them more difficult, or even impossible to be recognised as property.

One such characteristic is an inherent susceptibility to administrative variation and/or revocation. Although the kind of vulnerability to variation which will have this effect is a matter of degree (*ICM Agriculture v Commonwealth of Australia*), it is clear that courts have historically looked for the existence of some degree of permanence or stability before characterising a right as proprietary. In *National Provincial Bank Ltd v Ainsworth*, for example, Lord Wilberforce observed:

> Before a right can be admitted into the category of property, or of a right affecting property, it must be definable, identifiable by third parties, capable in its nature of assumption by third parties, and have some degree of permanence or stability.

Two things are clear from Lord Wilberforce's words: (1) that such a description does not emerge from a vacuum. It is shaped by attitudes towards the range of understandings of, and justifications for, property; and (2) that inherent variability of the kind noted by Lord Wilberforce has precluded several types of statutory rights from being protected by s 51(xxxi) of the Constitution (*Attorney General v Chaffey*; *Commonwealth v WMC Resources Ltd*). Hence, in *Minister for Primary Industry and Energy v Davey*, amendment to a management plan which lessened the value of the appellant fishermen's fishing units and disqualified the fishermen from operating did not constitute an acquisition of property by the Crown justifying compensation under s 51(xxxi). In that case, Burchett J found that the fishing units established under the *Fisheries Act 1952* (Cth) were property to the extent that 'the units may be transferred, leased and otherwise dealt with as articles of commerce' but that 'nevertheless, they confer only a defeasible interest subject to valid amendments to the [plan of management] under which they are issued. The making of such amendments is not a dealing with property; it is the exercise of powers inherent at the time of its creation and integral to the property itself'.

In *ICM Agriculture Pty Ltd v Commonwealth*, the High Court considered whether the conversion of a bore licence held under the former *Water Act 1912* (NSW), into an aquifer access licence under the *Water Management Act 2000* (NSW) (WMA), constituted an acquisition of property for the purpose of s 51(xxxi). Although the WMA is a State Act, the claim was, for various reasons, able to be brought under the Commonwealth Constitution (see Gray 2012).

When the WMA was passed it provided for the development of water sharing plans covering the relevant groundwater systems. Replacement of the plaintiffs' bore licences with aquifer access licences under the WMA resulted in a reduction in the water available to the plaintiffs for consumptive use.

Ultimately, French CJ, Gummow and Crennan JJ found that there had been no 'acquisition'. Accordingly, it was unnecessary to decide the subsequent issue of whether the bore licences constituted property. However, those same judges noted

Mason J's view as expressed in *R v Toohey; Ex parte Meneling Station Pty Ltd* that a licensing system subject to Ministerial control or forfeiture may be insufficiently permanent or stable to be treated as proprietary in nature. Meanwhile other members of the High Court in *ICM,* Hayne, Kiefel and Bell JJ 'readily accepted that the bore licences that were cancelled were a species of property' because the entitlements attaching to the licences could be traded or used as security. However, their Honours also noted that, because the rights given by the licences were statutory rights, they were 'inherently susceptible to change or termination' and were 'inherently fragile'. This analysis suggests that statutory property may be something less permanent and less stable than common law property.

Of course, these views were provided by way of *obiter dicta* only, and even if the High Court had decided the issue of legal classification, the utility of the decision would have been limited as *ICM* deals with bore licences, which were features of the pre-NWI system in NSW.

Another case of some relevance, but which also concerns pre-NWI styled water access licences, is that of *Australian Rice Holdings Pty Ltd v Commissioner of State Revenue*. In that case, the Victorian Supreme Court considered whether the rights conferred by water licences under the *Water Act 1969* (Vic) amounted to property for the purposes of the full payment of *ad valorem* stamp duty on assignment. Harper J held that the water licences did confer proprietary rights, on the basis that the licences displayed the key indicia of property: they had value, could be renewed and transferred, were well defined and had a degree of stability.

More recently and in relation to NWI styled water entitlements, the NSW Supreme Court in the case of *Martin v Martin* considered whether a water access licence (WAL) issued under the WMA was property in the context of succession. At issue was whether the bequest of rural land, under a will, to the plaintiff (the deceased's son) included a WAL held in the name of the deceased, or whether the WAL was part of the residual estate bequeathed to the defendant (the deceased's widow). Following consideration of the statutory characteristics of WALs, White J held that "[a]n access licence is personal property separate from the ownership or rights to occupy land which might have the benefit of the licence" and "personal property capable of being transferred independently of land without the need for ministerial consent".

The judgments considered above reveal that the answer to the question of whether water entitlements and WALs are property, absent statutory clarification, may ultimately depend on the context and purpose for which the question is being asked. The answer may also be influenced by the justifications for, and understandings of, property that the court favours. Hence, a court may have less trouble concluding that there are property rights in water entitlements in a succession (as in *Martin v Martin*) or taxation (as in *Australian Rice Holdings*) context than in relation to S 51(xxxi) of the Constitution (such as in *ICM*). The factual circumstances, pleadings in the case and the potential for broad precedent to be set, all play a role in the judicial outcome. Arguably, philosophical and justificatory proclivities concerning property also influence judicial decision-making and provide further explanation for the reluctance of courts to address squarely the property issue in relation to cases concerning the Constitutional guarantee for compensation, on just terms, for loss of property. Such a proposition brings us back to our earlier discussion on how different understandings of, and justifications for property may influence the role property plays in society, including in the water sector.

A clear body of jurisprudence has yet to emerge in the water sector, making it difficult to predict whether a court would consider a water entitlement to be property in any given context if statute did not deem it to be so. Further, even where legislation does deem water entitlements to be property, how that proprietary characterisation may impact on society deserves attention.

Property and Water Entitlements The NWI, which is, of course, a non-legislative instrument, requires states to separate 'water property rights' from land title and provide 'clear specification of entitlements in terms of ownership, volume and reliability'. Yet it was not so prescriptive as to require that reforming legislation specify the exact nature of the water rights or entitlements to be created, although Fisher (2004) points out, in the development of NWI styled water rights there were references to 'trade in water', 'water entitlements', 'property rights' and 'full property rights'. Accordingly, the implementation of the 'property objectives' of the NWI has varied across Australian jurisdictions, with some jurisdictions such as Western Australia still not having unbundled their water rights.

Most states' water legislation has not gone so far as to declare water entitlements, nor the WAL in which the entitlements are commonly packaged, to be personal property. The exceptions are South Australia and Tasmania. In both cases those states have expressly declared that a water licence is the personal property of the licensee (*Natural Resources Management Act* 2004 (SA) s 146(8) and *Water Management Act* 1999 (Tas) s 60)). Section 146(8) of the *Natural Resource Management Act 2004* (SA) for example, provides:

> A water licence is personal property and may pass to another in accordance with the provisions of this Act or, subject to this Act, in accordance with any other law for the passing of property.

A similar provision has been included in s 59B of the *Water Act 2003* (UK), which provides that an abstraction licence 'shall be regarded as property forming part of the deceased's personal estate, whether or not it would be so regarded apart from this subsection, and shall accordingly vest in his personal representatives'. The UK provisions are limited to the context of succession and in that sense, are to be contrasted with the Chilean Constitution which explicitly guarantees water rights as private property irrespective of context.

While it does not necessarily follow that the mere assertion of a property right will cause a right to be proprietary (*Wik v State of Queensland*), application of the ordinary rules of statutory interpretation allows us to infer that, in making an express declaration of personal property, the South Australian and Tasmanian parliaments intended to create a form of statutory personal property.

In jurisdictions other than South Australia and Tasmania, the question of whether water entitlements are property is less clear. As we have observed above, in interpreting the WMA, the NSW Supreme Court has concluded that water access licences are property in the context of succession, while the High Court of Australia has arguably left the door ajar for an alternative finding, in the context of compensation (for loss of property). Examination of the language of the statutory provisions and the substance of the rights conferred under relevant regulatory provisions casts some light on the nature of water entitlements.

The WMA, for example, provides that a WAL may be subject to a registered security interest (WMA s 71D(1)) and that the Minister may record in the NSW Water Register a caveat on the title of a WAL (WMA s71E). The holder of the security interest may sell the burdened WAL pursuant to an exercise of the power of sale (WMA s 71X). The holder of a WAL may transfer the water entitlements conferred by the WAL to another person for a specified term, resembling a leasehold arrangement and known as a term transfer (WMA s 71 N). These features demonstrate that WALs are treated as property in many ways. Do these features, considered cumulatively, amount to a property characterisation? s 71D(1) of the WMA contemplates that a WAL held by multiple persons may be held as a tenancy in common, which itself reveals an assumption made by parliament that WALs are proprietary, because traditionally at least, it has not been possible to hold a non-proprietary interest as a tenancy in common. Similarly s 101 of the *Water Act 2000* (Qld) provides that all licences owned by multiple persons are owned as tenants in common. Meanwhile, under the *Water Act 1989* (Vic), water shares may be held by any legal person or persons as joint tenants or tenants in common.

On the other hand, as discussed above, courts have historically looked for the existence of some degree of permanence or stability before characterising a right as proprietary. NWI styled water entitlements are inherently variable and susceptible to administrative variation. WALs in NSW, for example, may be varied (WMA s 66 (3)), subjected to conditions (WMA s67), suspended (WMA s 78(1)) or cancelled (WMA s 78(1)) by the Minister administering the WMA. Further, because NWI styled water entitlements are dependent upon water quality and availability, which fluctuate both temporally and spatially, they are intrinsically variable. Entitlements are also vulnerable to sudden and unexpected changes as a result of suspension, cancellation or amendment of water plans, brought about by climatic and hydrological factors which are unpredictable and unrelated to the behaviour of the entitlement-holder.

Plans are, in turn, able to be amended or repealed at any time (WMA ss 42 and 45) and may be suspended if the Minister considers that there is a severe water shortage in relation to any water management area or water source (WMA s 49A). Specifically, water allocations (which reflect the actual volume of available water in an entitlement) are subject to change each year depending on the volume of water the Minister makes available for sharing via a determination and as set out in a water sharing plan.

Given their inherent variability, is it possible that WALs (and/or the entitlements they house) could be considered to lack the degree of permanence or stability required to afford them proprietary status? White J in *Martin v Martin* did not think so, but perhaps another court in a superior jurisdiction might.

Attention has been given to these considerations in other jurisdictions, such as the United States of America. With respect to the water rights regime of the Colorado River, Hennessy (2004) concludes that 'private parties hold property rights, albeit imperfect ones, in water' because they are 'use' rights, subject to forfeiture if the use is not 'reasonable or beneficial'. Further, although they include exclusion rights, transfer rights are limited because strict transfer rules have been developed to address the interdependency of water rights. Such rules act as a serious restraint on alienability yet interestingly, freedom of alienability has long been considered 'a key and enduring principle' of property law (Gray et al 2017).

In summary and turning back to Australia, while the statutory provenance of water entitlements will not of itself preclude proprietary status, their susceptibility to administrative variation suggests that in many contexts water entitlements may not meet the threshold for a proprietary characterisation (see pro and contra views in McKenzie 2009). Further, if Parliament intended WALs to be property, why did the NSW legislature, for example, not simply deem them to be so, as in South Australia and Tasmania? It is notable that, despite calls for 'property right arrangements formally to be put in place' (Commonwealth of Australia 1995) and 'full property rights which will form a sound basis for inter-jurisdictional trade where applicable' (Agriculture and Resource Management Council of Australia and New Zealand 1995) to be implemented, the NSW legislature assiduously avoided using the word 'property' in the WMA when referring to WALs and entitlements (Gray 2006). In light of this, can it necessarily be assumed that WALs were intended by the legislature to be property?

On the other hand, fitting a trend line to the present, it may be logical to expect (although it may not necessarily be advantageous in terms of outcomes) that NWI styled water entitlements will be considered property, given that a key aspect of the reforms was to make water entitlements more robust (McKenzie 2009). This perhaps accords with the broader view of property rights in water taken by Clarke and Malcolm (2017) with respect to the UK regime. Those authors argue that property rights in water, far from being confined to absolute, exclusionary, private property rights, exemplify the complexity of modern 'real-life' property rights systems, and that 'established and normative accounts of the role of property in water are based on an untenably narrow conception of property centred on private ownership which fails to take into account the spectrum of private, communal, public and state property interests' (Clarke and Malcolm 2017, 122). However, in practice, it *is* narrow, private property rights that many advocates of market-based water regimes tend to support because such rights are seen as readily transferable, secure and certain, and attract extensive remedies if breached.

Further, if the theory of property on which property characterisations is based is an economic one, then the very tradeability of water entitlements will position them in the proprietary category. However, if another theory of property which emphasises an alternative conceptualisation of property is relied upon, courts may guide the law in a different direction.

Do Water Entitlements or Licences Need to Be Property to Support Water Trading? We now turn to the question of whether water licences or underlying entitlements need to be, or should be, property to support water trading. We also explore what it would mean if water entitlements were not property. These questions are important because the NWI styled Australian water governance model is dependent on water trading and markets—both commonly thought to rely on property. Lee and Jouravlev (1998), for example, argue that 'well-defined property rights' are 'the essential foundation of any market allocation mechanism' and are a required condition for trade. If property rights are not well defined, they argue, the consequent uncertainty will reduce the expected value of the rights and therefore, the incentive to engage in trading. In support of this view, those authors suggest that the existence of secure property rights in Chile had a discernible correlation with the overall growth in the value of Chile's agricultural production since the 1980s. However, as has been hinted at by Bogojevic (in a different environmental trading context, that of emissions trading) the role of property is potentially complex (Bogojevic 2013). It needs careful attention. Lange and Shepheard provide some of this attention particularly when they consider the nexus between property and stewardship (Lange and Shepheard 2014).

If, however, entitlements could be traded without their propertisation being either a pre-condition for, or the result of that trade, further alternative governance options may open up.

One key problem with the propertisation of entitlements to support water trading is that once a thing is propertised, it is often very difficult to reverse that propertisation. Property is a valuable resource in the hands of its holder. Hence if property is lost, holders will (understandably) seek compensation. One impact of government's vulnerability to the payment of compensation may be that governments will be deterred from introducing alternative water governance models which are not based on trading. They may feel wedded to trading because it is too expensive to deviate from it.

In other words, a government may be locked into the trading and propertisation path (and be unable to dismantle the supporting framework) even if that government came to the view that the trading and propertisation path was not delivering desired outcomes. Hence, if a government wished to change its legislative and policy course and convert the private property rights of entitlement holders into common property or state-owned property rights, for example, it may be restricted in its capacity to do so. The propertisation of water entitlements may, therefore, lead to future path dependency.

In the United Kingdom (where water entitlements have not been wholly unbundled from land but abstraction licences are nonetheless valuable tradable administrative rights with a property flavour), the impacts of potential path dependencies have been lessened by amendments to the licensing scheme introduced under the *Water Act 2003*. These amendments have curtailed the rights to compensation triggered by limitations on abstractions through licence variations or revocations under section 61 of the *Water Resources Act 1991*, by introducing a requirement for new licences to be time-limited, and by removing any right to compensation where the regulator varies or revokes a non-time-limited licence where water abstraction is likely to cause serious damage to the environment (*Water Act 2003*, s 27). Of course in Australia, no amount of legislative intervention at the state level can overcome the potential for compensation to be payable as a result of an acquisition of property by the Commonwealth.

Rejecting the Proprietary Characterisation The above discussion begs the question—what if propertisation is not actually a pre-condition for, nor the product of, water trading? What if water entitlements could be de-coupled from property and regulatory mechanisms were responsible for transferring *sui generis* non-proprietary entitlements between parties? Then the result would be that water could still be moved between locations and holders, but the subject of the underlying trade would not be property.

If that were the case, then the question of compensation for loss of property would not arise. A non-proprietary characterisation of WALs may, therefore, liberate water governance from its present path. Fear of compensation would no longer be a disincentive for change. Choices other than water trading and markets may re-appear on legislative and policy agendas.

Whilst calls to dismantle the water trading model in Australia have, thus far, been few (and largely ignored), there have been vocal calls by eminent scholars to re-nationalise another resources sector; electricity generation and distribution (John Quiggin Opinion and Consulting 2014). Such calls may flag a different policy direction where alternative governance tools assume importance. It is too soon to tell, but similar calls (accompanied by political support) have also been evident in the UK in relation to the re-nationalisation of the railways (see Bring Back British Rail 2017; Elgot (2016).

If trade can be decoupled from property, and the objective of trade can be met without needing to rely on a proprietary characterisation of water entitlements, perhaps the NWI did not need to call for 'property rights' in water in the first place. Indeed, markets appear to be operating and water entitlements are increasingly being traded in the absence of clear definitions of property rights in many jurisdictions. Over the 2013/2014 financial year, rights in 2,421 gigalitres were traded across Australia, representing an increase of 80 percent on the 2012/2013 financial year (Morey et al. 2015). New derivative products have also emerged to form the basis of trade. Along with permanent entitlements and temporary allocations, water options are now being traded.

Yet if the unbundling of water entitlements from land and the introduction of a comprehensive regulatory framework based on *sui-generis* non-proprietary rights are themselves sufficient to achieve the NWI's primary aims, the importance of the property debate would fall away. The benefit would be additional flexibility for governments, who need not feel constrained by the compensatory aspect of propertisation, nor need to address the implications of different analyses of, and justifications for, property. A retreat from the trading model may even be possible.

4 Conclusion

Conscious that many of the calls for water propertisation were made in an era of desperation (drought and water mismanagement) and perhaps without great awareness of the implications of such a characterisation, this chapter has sought to unpack some of the different theories and analyses of property in order to examine whether property is necessarily an effective, efficient and fair tool of governance leading to resilience in the water sphere.

It argued that there are many different ways of understanding and conceptualising property beyond an economic approach to property. Indeed other theoretical positions may be more sympathetic to the idea that regulatory tools are able to transfer *non-proprietary* interests in the form of water entitlements and licences. On this reasoning, property and trade would be de-coupled. If this proves possible, additional governance options may open up because governments would no longer need to fear that retreating from a governance model based on propertised tradeable entities would be prohibitively expensive in terms of compensation.

If, however, governments, legislatures and courts either have embarked on, or are considering opting for the propertisation path, they should be very mindful of the implications of such a decision. Property's wide sphere of enforceability, the robust nature of the remedies available in cases where a property right is breached, and the capacity of property to help shape societies are all attractive features but property also has its limitations which should not be overlooked. Propertisation may cause as many disputes as it solves. It may also encourage path dependency, particularly in the water trading context and it may shape societies in unintended or negative, as well as positive ways, depending on how property is justified and conceived.

To propertise or not to propertise?—that is the question which governments, legislators and courts need to address in the water context, both in Australia and internationally. The answer will have implications not simply for the water entitlement holders themselves, but for the wider community, because the effects of holding or not holding property rights are far-reaching. In the interests of not only enhanced water management but of society more generally, it is time for clear and well considered responses to be developed, which reflect engagement with the theoretical underpinnings of the property concept.

References

ACIL Tasman and Freehills. (2004). *An effective system of defining water property titles*. Report to the Australian Government Department of Agriculture, Fisheries and Forestry and Land & Water Australia.

Agriculture and Resource Management Council of Australia and New Zealand. (1995). Water allocations and entitlements: A national framework for the implementation of property rights in water. Taskforce on COAG Water Reform, Commonwealth of Australia, Canberra.

Alexander, G. S. (2009). The social-obligation norm in American property law. *Cornell Law Review, 94*(4), 745–819.

Alexander, G., & Penalver, E. (2012). *An Introduction to Property Theory*. Cambridge, UK: Cambridge University Press.

Australian Government National Water Commission. (2014). *10 years of water wins: Australia's National water initiative*. Commonwealth of Australia.

Babie P., Burdon, P., & Da Rimini, F. (2017). The psychology of property. Paper presented at the 14th Australasian Property Law Teachers Conference, Curtin University, Perth, 26–28 September 2017.

Baron, J. (2014). Rescuing the bundle of rights metaphor in property. *University of Cincinatti Law Review, 82*(1), 57–102.

Bennett Moses, L. (2008). The applicability of property law in new contexts: From cells to cyberspace. *Sydney Law Review, 30*(4), 639–662.

Bogojevic, S. (2013). *Emissions trading schemes: Markets, states and law*. London: Hart Publishing.

Boyle, J. (2002). Fencing off ideas: Enclosure and the disappearance of the public domain. *Dædalus: Journal of the American Academy of Arts and Sciences, 131*(2), 639–662.

Bring Back British Rail. (2017). Bring Back British Rail website., Retrieved December 10, 2016, from http://www.bringbackbritishrail.org.

Carr, H., & Wong, S. (2016). Feminist approaches to property law research. In S. Bright, & S. Blandy (Eds.), *Researching Property Law* (pp. 164–179). London: Palgrave Macmillan.

Clarke, A., & Malcolm, R. (2017). The role of property in water regulation: Locating communal and regulatory property rights on the property rights spectrum. In C. Godt (Eds.), *Regulatory Property Rights: the transforming notion of property in transnational business regulation* (pp. 121–140), Leiden, Boston: Brill/Nijhoff.

Commonwealth of Australia. (1995). *Second Report of the Working Group on Water Resource Policy to the Council of Australian Governments, Canberra*.

Cooke, E. (2012). *Land law* (2nd ed.). Oxford: Clarendon Press.

Council of Australian Governments, Intergovernmental Agreement on a National Water Initiative Between the Commonwealth of Australia and the Governments of New South Wales, Victoria, Queensland, South Australia, Australian Capital Territory and Northern Territory. (2004). Retrieved January 24, 2017, from http://www.agriculture.gov.au/SiteCollectionDocuments/water/Intergovernmental-Agreement-on-a-national-water-initiative.pdf.

Demsetz, H. (1967). Toward a theory of property rights. *American Economic Review, 57*(2), 347–359.

Dickenson, D. (2007). *Property in the body*. Cambridge: Cambridge University Press.

Dickenson, D. (1997). *Property, women and politics*. Cambridge: Polity Press.

Edgeworth, B., Rossiter, C., Stone, M., & Connor, P. (2012). *Sackville and Neave, Australian property law* (9th ed.). Sydney: Lexis Nexis.

Elgot, J. (2016). Corbyn to launch transport campaign with rail pledges. Available via The Guardian. Retrieved March 1, 2017, from https://www.theguardian.com/politics/2016/aug/16/corbyn-to-launch-transport-campaign-with-rail-pledges.

Fagot-Largeault, A. (1998). Ownership of the human body: Judicial and legislative responses in France. In H. Ten Have, J. Welie, & S. Spicker (Eds.), *Ownership of the human body: Philosophical considerations on the use of the human body and its parts in healthcare* (pp. 115–140). Dordrecht: Kluwer Academic Publishers.

Fisher, D. E. (2004). Rights of property in water: Confusion or clarity. *Environmental and Planning Law Journal, 21*(3), 200–226.

Freyfogle, E. T. (2007). Private property: Correcting the half truths. *Planning and Environmental Law, 59*(10), 3–11.

Fuller, L. (1958). Positivism and fidelity to law: A reply to professor Hart. *Harvard Law Review, 71*(4), 630–672.

Grafton, Q., Squires, D., Fox, K. J. (2000). Private property and economic efficiency: a study of a common pool resource. *Journal of Economics and Law, 43*(2), 679–714, cited in Gray, J., &

Lee, L. (2016). National Water Initiative styled water entitlements as property: Legal and practical perspectives. *Environmental Planning and Law Journal, 33*(4), 284–300.

Gray, J. (2006). Legal approaches to the ownership, management and regulation of water rights from riparian rights to commodification. *Tranforming Cultures eJournal, 1*(2), 64–96.

Gray, J. (2008). Watered down: Legal constructs, tradable entitlements and the regulation of water. In G. Develeena, H. Goodall, & D. Stephanie (Eds.), *Water, sovereignty and borders in Asia and Oceania* (pp. 147–168). London: Routledge.

Gray, J., & Gardner, A. (2008). Legal access to sewage and the re-invention of wastewater. *AJNRL&P, 12*(2), 115–159.

Gray, J. (2011). Mine or ours? Sewage, recycled water and property. In K. Bosselman & V. Tava (Eds.), *Water rights and sustainability* (pp. 154–172). Auckland: New Zealand Centre for Environmental Law.

Gray, J., Foster, N., Dorsett, S., & Roberts, H. (2017). *Property law in New South Wales* (4th ed.). Sydney: Lexis Nexis.

Gray, J. (2012). The legal framework for water trading in the Murray Darling Basin: An overwhelming success? *Environmental, Planning and Law Journal, 29*(4), 328–348.

Gray, K. (1991). Property in thin air. *Cambridge Law Journal, 50*(2), 252–307.

Gray, K. (2010). Regulatory property and the jurisprudence of quasi-public trust. *Sydney Law Review, 32*(2), 237–267.

Hannam, P. (2017). High and dry: Adani seeks additional surface water to feed giant coal mine, SMH. Retrieved 6 April, 2017, from http://www.smh.com.au/environment/high-and-dry-adani-seeks-additional-surface-water-to-feed-giant-coal-mine-20170405-gve42a.html.

Hart, H. L. A. (1958). Positivism and the separation of law and morals. *Harvard Law Review, 71* (4), 593–629.

Hennessy, M. (2004). Colorado river water rights: Property rights in transition. *University of Chicago Law Review, 71,* 1661.

Hepburn, S. (2010). Statutory verification of water rights: The 'insuperable' difficulties of propertising water entitlements. *Australian Property Law Journal, 19*(1), 1–22.

Hohfeld, W. N. (1913). Some fundamental legal conceptions as applied in judicial reasoning. *Yale Law Journal, 23*(16), 28–59.

Honore, A. M. (1999). Ownership. In L. J. Coleman (Ed.), *Readings in the philosophy of law* (pp. 557–597). London: Routledge.

Hughes, N. (2015). Capacity sharing and the future of water property rights. Paper presented at ABARES, Crawford School, ANU seminar, Sydney 17 September 2015.

Hume, D. (1739). *A treatise on human nature.* In L. A. Selby-Bigge, & P. H. Nidditch (Eds.). Oxford: Clarendon Press.

John Quiggin Opinion and Consulting. (2014). *Electricity privatisation in Australia: A record of failure. Report commissioned by the Victorian Branch of the Electrical Trades Union.*

Kant, I. (1964). Lectures on Ethics, cited in Cohen, G. A. (1995). *Self-ownership, freedom and equality.* Cambridge: Cambridge University Press.

Lange, B., & Shepheard, M. (2014). Changing conceptions of the right to water: An eco-socio-legal perspective. *Journal of Environmental Law., 26*(2), 215.

Laslett, P. (Ed.). (1964). *Locke: Two treatises of government.* Cambridge: Cambridge University Press.

Lee, T., & Jouravlev, A. (1998). Price, property and markets in water allocation. Report prepared for the United Nations Economic Commission for Latin America and the Carribbean. Santiago, Chile.

Lueck, D. (2003). First possession as the basis of property. In T. L. Anderson & F. S. McChesney (Eds.), *Property rights: Cooperation, conflict and law* (pp. 200–226). Princeton: Princeton University Press.

Martin, P., & Verbeck, M. (2002). Property rights and property responsibilities. In *Property rights and responsibilities: Current Australian thinking.* Land and Water Australia.

Marshall, V. (2017). *Overturning aqua nullius: Securing Aboriginal water rights.* Canberra: Aboriginal Studies Press.

McKenzie, M. (2009). Water rights in NSW: Properly property? *Sydney Law Review, 31*(3), 443–463.
Merrill, T. W. (1998). Property and the right to exclude. *Nebraska Law Review, 77*(4), 730–755.
Morey, K., Grinlinton, M., & Hughes, N. (2015). *Australian water market report 2013–14*. Australian Bureau of Agricultural and Resource Economics and Sciences, prepared for the Department of Agriculture and Water Resources, Canberra.
National Competition Council, National Competition Policy. Retrieved January 27, 2017, from http://ncp.ncc.gov.au/pages/water.
Penner, J. E. (1997). *The idea of property in law*. Oxford: Clarendon Press.
Plato, C. 370 BC. *Republic*. (1993). Trans. R. Waterfield. Oxford: Oxford University Press.
Posner, R. A. (1992). *Economic analysis of law* (4th ed.). Boston: Little Brown and Co.
Poudhon, P.-J. *What is Property? Or, an inquiry into the Principle of Right and of Government*. (1840). Trans B. R. Tucker. Humboldt Publishing, c 1890.
Rose, C. M. (1992). Women and property: Gaining and losing ground. *Virginia Law Review, 78*(2), 421–459.
Rose, C. M. (1994). *Property and persuasion: Essays on the history, theory, and rhetoric of ownership*. Westview Press.
Sax, J. (1999). *Playing darts with a Rembrandt: Public and private rights in cultural treasures*. Ann Arbor: University of Michigan Press.
Singer, J. W. (2000a). *Entitlements: The paradoxes of property*. Yale: Yale University Press.
Singer, J. W. (2000b). *The edges of the field: Lessons on the obligations of ownership*. Massachusetts: Beacon Press.
Singer, J. W. (2014). Property as the law of democracy. *Duke Law Journal, 63*(6), 1287–1335.
Smith, H. E. (2012). Property as the law of things. *Harvard Law Review, 125*(7), 1691–1703.
Storey, M. (2006). Not of this Earth: The extraterrestrial nature of statutory property in the 21st century. *Australian Resources and Energy Law Journal, 25*(1), 51–64.
Waldron, J. (1998). *The right to private property*. Oxford: Clarendon Press.
Waldron, J. (1985). What is private property? *Oxford Journal of Legal Studies, 5*(3), 313–349.
Walpole, M., & Gray, J. (2010). Taxing virtually everything: cyberspace profits, property law and taxation liability. *Australian Taxation Review, 39*(1), 39–60.
Wilson, D. (2008). *Women, marriage and property in wealthy landed families in Ireland*. Manchester: Manchester University Press.

Case Citations

Armstrong v Winnington Networks Ltd (2013) Ch 156.
Attorney General (NT) v Chaffey (2007) 231 CLR 651.
Australia Rice Holdings Pty Ltd v Commissioner of State Revenue (2001) 48 ATR 498.
Australian Tape Manufacturers Association Ltd v Commonwealth of Australia (1993) 176 CLR 480.
Commonwealth v WMC Resources Ltd (1998) 194 CLR 1.
Charles River Bridge v. Warren Bridge 36 U.S. 420 (1837).
Davies v Littlejohn (1923) 34 CLR 174.
Friendswood Development Co v. Smith- Southwest Industries Inc 576 OS/ W/ 2d 21. 22 (Tex 1978).
ICM Agriculture Pty Ltd v Commonwealth of Australia (2009) 240 CLR 140.
Mabo v Queensland (No 2) 175 CLR 1.
Martin v Martin (2010) NSWSC 700.
Milirrpum v Nabalco (1971) 17 FLR 141.
Minister for Primary Industry and Energy v Davey (1993) 47 FCR 15.
Moore v Regents of University of California, 51 Cal, 3rd 120, 793, P. 2d, 271 Cal. Rptr. 146 (1990).

National Provincial Bank Ltd v Ainsworth (1965) 1 AC 1175.
Pacific Film Laboratories v Commissioner of Taxation (1970) 121 CLR 154.
R v Toohey; Ex parte Meneling Station Pty Ltd (1982) 158 CLR 327.
Telstra Corporation Limited v The Commonwealth of Australia (2008) 234 CLR 210.
Victoria Park Racing and Recreation Grounds Co Ltd v Taylor (1937) 58 CLR 479.
Wik Peoples v State of Queensland (1996) 187 CLR 1.
Zapletal v Wright (1957) Tas SR 211.

Regulatory and Economic Instruments: A Useful Partnership to Achieve Collective Objectives?

Adam Loch, C Dionisio Perez-Blanco, Dolores Rey, Erin O'Donnell and David Adamson

Abstract In this chapter we examine how water governance and demand management arrangements can be linked to economic instruments, such as water markets, to address the broad range of water reallocation problems that exist in many global contexts. The utilization of economic instruments is context-specific throughout the world and can take many forms. This chapter therefore lists the pros and cons of some more common instruments. While successfully combining regulatory and economic instruments is far from straightforward, policy-makers can learn from growing evidence of successful partnerships between these two approaches. It may be costly both in terms of political support and transaction investments to strip away existing arrangements in favour of more flexible and better-suited institutions to manage scarce water resources. However, it would be expected that ignoring the problems, and hoping they will resolve themselves, would be more harmful to private and public welfare outcomes in the long run.

A. Loch (✉) · D. Adamson
Centre for Global Food and Resources, University of Adelaide,
Adelaide, SA, Australia
e-mail: adam.loch@adelaide.edu.au

D. Adamson
e-mail: david.adamson@adelaide.edu.au

C. D. Perez-Blanco
Universidad de Salamanca, Salamanca, Spain
e-mail: dionisio.perez@cmcc.it

D. Rey
Cranfield Water Science Institute, Cranfield University, Cranfield, UK
e-mail: d.reyvicario@cranfield.ac.uk

E. O'Donnell
Centre for Resources, Energy and Environmental Law at Melbourne
Law School (CREEL), Melbourne, VIC, Australia
e-mail: erin.odonnell@unimelb.edu.au

1 Introduction

In the end, behaviour must be guided not by self-restraint alone, but also by a combination of regulation and market forces (Getches 2014).

Humans dominate changes to Earth's water resource systems, presenting challenges for scientists and public policy makers interested in the regulation and management of human-biophysical environmental interaction, and guiding that toward sustainable outcomes (Crutzen 2002). With regard to freshwater resource utilisation, evidence suggests that current and predicted growth rates are unsustainable. Potential conflicts surrounding water resources now constitute the single biggest global economic threat (World Economic Forum 2015). Many nations are also failing to meet sustainability criteria, where investments in human knowledge and institutional capital are insufficient to offset depletions in natural capital (Arrow et al. 2004). Moreover, the precautionary principle warns us that, even if that was the case, substitution between man-made and natural capital cannot be taken for granted.

Our ability to increase water through engineering projects (dams and storage) has become heavily constrained through a reduction in suitable sites coupled with the growing awareness of environmental and social costs associated with such construction. Current approaches to water reallocation therefore require radical change (Sivapalan et al. 2012), and now mainly revolve around demand-management institutions. In broad terms, three institutional approaches to demand-based reallocation of water resources are typically discussed in the literature: administrative, collective negotiation and economic instruments such as market-based transfers (Meinzen-Dick and Ringler 2008); where market-based measures constitute the most radical approach. This is because market-based solutions are complex and costly to implement, making them less suitable for developing nations—and even some developed-nation contexts. Some critics also contend that water is a basic human right, that should not be subject to market (re)allocation (Bakker 2007) or commodification. However, market-based transfer arrangements may naturally arise, and be sustained, where strong institutional arrangements are adopted in the pursuit of good water governance. Appropriate institutions thus include robust legal and planning structures, informal and formal markets, various government agencies, interpersonal networks and norms that guide water demand and use behaviour, and compliance/enforcement resources or expertise.

From a strengthened institutional base, vigorous public polices such as extraction and use regulations, taxes and the establishment of clear property rights can help with the economic pricing of natural and environmental resources to better approximate and include social costs; and improve matters along the sustainability dimension (Arrow et al. 2004). By way of example, the mix of top-down regulation, catchment water planning with stakeholder consultation, water levies and adoption of levies and market transfers in Australia suggest world-leading water resource governance and reallocation framework pillars (Godden and Foerster 2011). Yet the development of these pillars has been slow, taking considerable time and government investment to implement and grow (Holley and Sinclair 2016). Consequently, markets are far from a panacea for the reallocation of scarce water resources and progress toward sustainable water-use outcomes. Further, if market-based arrangements are specifically aimed at achieving the reallocation of water to meet environmental objectives, they will require on-going institutional capacity-building and adaptive governance arrangements for the best probability of success and sustainability (Garrick 2015).

In this chapter we therefore examine a range of economic instruments that link to water governance and demand management, including water markets, which may be of broader appeal and use in water reallocation contexts. Economic instruments have many types, just as water reallocation requirements occur in many different contexts. In our view, both good water regulation (governance) and the implementation of appropriate economic instruments must be achieved in the pursuit of effective water management arrangements.

2 Economic Instruments and Regulatory Processes

Given the focus on engineering solutions, early forms of water regulation typically created frameworks were based on required performance or technology standards. These 'command and control' regulations sought limited environmental pollution and water-resource degradation by imposing fines to enforce compliance (Gunningham 2009). By the 1980s, there was a distinct shift in focus to applying limits to resource extraction (Basin closures), and the use of economic incentives such as taxes, levies and market-based mechanisms in the belief that these approaches would enhance resource-use flexibility, lower costs, and drive continuous improvement (Stewart 1992). This historic construction of a dichotomy between 'regulation' (which was conflated with 'command and control' methods) and economic incentives continues to influence public policy development today (Driesen 2006; Gunningham 2009).

Although informal markets can emerge in some circumstances with limited government oversight, robust economic instruments that deliver efficient and effective management of natural resources critically depend on their regulatory environment. This is especially the case for fugitive resources (such as river flows) and public goods, where economic instruments rely on the capacity of the state to control access, or create and enforce property rights (Scott 2008). In these cases, economic instruments should be framed as a *type* of regulation, rather than an alternative to it (Driesen 1998). This approach enables regulatory reforms to be considered as part of the selection of the economic instrument, and supports policy-makers to avoid the potential pitfall of inadequate funding for creation and maintenance of the regulatory framework. In the particular case of markets, this approach also focuses attention on transaction costs. In Australia, for example, each individual decision by a rural water user to buy or sell (temporary) water allocations using a water exchange platform might be a relatively small investment of time and effort. However, this relatively inexpensive transfer is made possible by significant and ongoing government investment in defining and enforcing water rights, maintaining water registers, accounting for water use, and making this information publicly available (Fisher 2006; Holley and Sinclair 2016).

The OECD has framed water crises as 'primarily governance crises' (2015, pg. 2). Economic instruments are an essential component of the toolkit for responding to water crises, and enabling sustainable management of water in both urban and rural contexts, but their efficacy will depend on their regulatory and governance frameworks. This is especially true at the intersection between private and public benefits or the generation of positive welfare outcomes from natural resource uses.

3 Theoretical Foundations of Economic Instruments

Welfare economics lies at the heart of economic instruments. In 1890, Alfred Marshall successfully argued that economics is not the study of wealth alone, but that it must include a study of how individuals interact with society to gain material welfare. This identification of private and social aspects of the economy enabled Pigou (1920) to explore how negative production externalities (e.g. pollution) can reduce positive economic welfare. Pigou noted that an economy could be worse off if the private benefit gained from the consumption or production of goods was less than the costs borne by society from the development or their consumption of goods. To annul these negative externalities of production, Pigou suggested taxing the consumption of polluting goods as one reduction solution, and using the income generated as a redistribution mechanism to compensate negatively-impacted individuals.

Coase (1960) furthered the economic debate by arguing that taxes were inefficient for dealing with externalities. Coase suggested that public institutions often failed to define the correct level of tax (i.e. governance failure), and that their involvement introduced a level of transaction costs such as bureaucratic frictions that dealt inefficiently with the problem. As an alternative, Coase argued that by treating both the factors of production and the negative externalities generated during production as a property right, markets could be used to set a price for, and define the socially-acceptable quantity of, those negative externality rights. However, in time markets have proven their inability to deal with or incorporate negative externalities (i.e. market failure). As Bromley (1989) discusses, in reality the combination of market failure,[1] missing markets,[2] and governance failures such

---

[1]Market failure occurs when markets fail to allocate resources efficiently. In this case, no market may exist for the good.

[2]Missing markets deal with intertemporal externality problems that create information asymmetry for current generations, and exclude the concerns of future generations.

as the exclusion of policy, regulations and enforcement have prevented markets from combating negative externalities.

Thus, over time we have learned that markets do not organically spring into life, but rather they need strict regulatory boundaries and investments to be established and work effectively, efficiently and sustainably. These regulatory investments include *static transaction* costs to administer, monitor and enforce existing policies or arrangements, *institutional transition* costs to design, test and implement new arrangements to overcome barriers to dealing with externalities. Where those barriers prove too much for current institutions to address, we may also experience *institutional lock-in* transaction costs that reduce public/private flexibility to deal with externalities (Marshall 2013).

However, economic instruments can help reduce transaction costs by designing policy, establishing market-based or other incentives, and improving their efficiency. When incorporated into the policy design stage, economic instruments can be used to define: the number, nature and structure of property rights; how to allocate initial rights to alternative sectors in the economy; the rules and regulations required to develop incentives or working markets; and the benefit of alternative strategies for reallocating the share of rights based on changing social expectations. Further, economic instruments can draw from a diverse set of economic theory and concepts. For example, behavioural economics can be applied to explore how individuals respond to economic incentives designed to reallocate resources. Auction theory can be applied to help initially allocate resources and find prices. Policy analysis can assist with designing the rules and regulations of property rights and their trade, to prevent market failure and missing markets, while policy evaluation can also help identify where efficiencies in markets and regulations may need to be addressed. Finally, institutional economic theory can help design public frameworks to minimise transaction costs, while production economic concepts can illustrate how to maximise profits while minimising the production of externalities. Thus, we can clearly see that a model for dealing with externalities lies at the intersection between natural resource and environmental economics.

In summary, the development of regulatory policies aimed at dealing with negative resource-use externalities requires a clear understanding of the non-linearity impacts of individual behaviour-reactions in response to complex problems that are imbued with risk and uncertainty. Once policy objectives are set, economic instruments to change individual behaviour reactions work well, especially when the best possible science is used to parametrise economic models such that they align closely to those policy goals. Until these parameters are set, it is the role of the economist to highlight the benefits and costs of alternative economic instruments in achieving the best possible outcome for current and future generations. We therefore now turn to an examination of some of the more popular economic instruments below, with a discussion of some of their pros and cons.

3.1 Economic Instrument Types

There are a diverse range of economic instruments (Bernstein 1997), most of which are applicable to the (re)allocation of water resources. The major types are detailed in the following sections.

Water Levies The First Fundamental Theorem of Welfare Economics, which states that markets tend toward Pareto-optimality, is often regarded as an argument towards *laissez-faire* (re)allocation arrangements (Mendelsohn 2016). However, prerequisites for Pareto- and allocative-efficiency (i.e. no transaction costs, price-taking behaviour and local non-satisfaction of desires) rarely apply to the real world, and often result in market failures. Thus, Pareto-efficiency does not equate with individual desirability. It may be possible to mimic Pareto-efficient equilibria, shifting resources in a way that gains outweigh losses and, if necessary, compensating those that end up worse-off. Successive policy iterations may result in equilibria to achieve not only a Pareto- but also allocative-efficient status; meaning that the marginal benefit of every good is equal to its marginal cost. However, in the case of water, these policy iterations may include biased allocations of rights towards polluters or, as discussed, the exclusion of users (e.g. environmental flows) from the market (Hanemann 2006). In concert with regulation-policy, economic instruments can contribute to solving market failure and achieving allocative efficiency. Among these instruments, levies on water use, commonly (and wrongly) referred to as pricing, are possibly the most popular.

A water levy consists of a financial charge imposed on shareholders, in addition to or instead of the market price, to recover the costs of water supply (water works) and any negative externality associated with water use.[3] In the absence of water markets, a water levy should also convey the resource costs embedded in prices to prevent inefficient allotments and the exclusion of productive water uses. Water levies are preferably volumetric; although in places or sectors where metering is deficient, other arrangements may be used such as flat-rates based on the acreage or installed irrigation capacity. From a regulatory perspective, water levies depend on legal capacity to impose costs upon water users, which can be controversial where access to water is considered a human right. In addition, volumetric water levies depend on water metering and accounting frameworks, including the ability to access private property to read water meters. Finally, a regulatory framework must include provision for non-payment, and the ability to manage debts associated with water use in an equitable and effective manner.

Just like water extraction limits, levies can play an effective role in curbing water demand; particularly in agriculture, the largest water user worldwide.

---

[3]It is this point that can work against arguments for provision of water as a basic human right. The infrastructure involved to establish (and eventually replace) the means to cleanse, store and deliver water is significant—usually regardless of the volumes put through it. Thus, while water access is a right, it comes at a high economic cost which must be addressed in some meaningful way by policy-makers.

Evidence shows that the price elasticity of water is typically negative (Rogers et al. 2002), meaning that higher levies will lead to a reduction in water use. On the other hand, the price elasticity of water demand in agriculture is non-constant and can rapidly transition between elastic and inelastic states as water becomes scarce. In other words, trade-offs between water conservation and farm income can range from irrelevant to very high. Empirical studies show that in water-abundant basins sporadically exposed to drought events, higher levies can succeed in limiting water use with low to moderate farm losses, as farmers phase out irrigated crops on marginal land (Pérez-Blanco et al. 2016). However, further strengthening the irrigation constraint may threaten valuable crops and motivate inelastic water demand curves, leading to significant impacts on income. Eventually, increasing levies may force farmers to stop irrigating and suffer large (in the case of perennial crops) capital losses and farm exit (Wheeler et al. 2012). Information on the price elasticity of demand may also be of relevance in water policy planning (e.g. water works), where the revenue from water levies is earmarked (i.e. tariffs) and relevant to the budget. Understanding the price elasticity of water demand is thus of critical importance to anticipating any economic impacts of and behavioural-reaction responses to, water levies, the design of compensatory policies, and the planning of water management arrangements.

Notwithstanding their potential to address market failures and reallocate water and economic resources to enhance overall welfare, water levies in agriculture are consistently reported to be below desirable levels (e.g. Colby 1987). While realised water levies (partially) account for the costs of water works, the costs of negative externalities are rarely conveyed to the polluter. In addition, while we may seek to identify the true cost of water, the presence of subsidies or support payments to certain sectors of society may mean that we fail to optimally allocate or use water resources. Thus, the transition from cheap water to new arrangements adapted to the realities of many river basin-closure contexts remains largely incomplete.

Subsidies (Including Efficiency Incentives) Environmental subsidies are typically used to enhance the supply of positive externalities. The mechanics are simple: government subsidies can be used to reduce the costs of production, shift the supply curve downwards, and increase the amount of goods produced—potentially along with desirable externalities. Admittedly, water use is typically related to the generation of negative externalities that are better addressed via making polluters pay (Freeman 2003). But the effectiveness of water levies is often at odds with other relevant criteria, notably affordability, and this has led policy-makers to rely on multiple forms of subsidisation that can be applied to bypass unpopular levies.

Water-related subsidies can adopt different forms: some are explicit and implemented through price supports, subsidised loans and direct payments; others are implicit and involve reduced regulation and tax relief. As a result, the regulatory frameworks used to support subsidies are highly context-specific. However, subsidies do depend on being able to identify a desirable outcome, so they rely on a water planning process that can articulate a clear policy objective, and a framework to support the implementation of the particular subsidy.

Deficient environmental regulation (e.g. setting environmental flows) is a form of implicit subsidisation that loosens the supply constraint, reduces prices and may lead to water over-allocation. In extreme cases, loose regulation coupled with low withdrawal costs can lead to environmentally-degraded water bodies. Implicit subsidies can also be the result of low cost-recovery levels where, even if regulations are in place, inadequate signals such as low levies and/or sanctions translate into poor or misleading economic incentives. Implicit subsidies are widespread and may lead to outcomes characterised by poor environmental quality and water over-allocation. Explicit subsidies also abound. Subsidised loans are frequently used to drive the adoption of more efficient technology, which is the case of irrigation modernisation programmes in countries like Australia, Austria, Mexico, the Netherlands, Portugal, the United States or Spain. These 'green' investments are justified on the grounds that higher technical efficiency reduces the need for water inputs in agricultural activity. In practice, evidence shows that farmers adjust their cropping patterns to the new set of alternatives/constraints and may end up further depleting, rather than improving, the ecological state of water bodies (Gómez et al. 2014; Loch and Adamson 2015). Price supports that keep market outlays above competitive levels are also widespread in the agricultural sectors of many developed countries.[4] Be it through financial payments or price controls, price supports artificially inflate the amount of goods produced and inputs consumed, including water. Price supports have also been consistently criticised due to their negative impact on the trade balance of developing nations, and several developed nations have recently substituted them with direct payments decoupled from the amount of goods produced—the case of the EU. Decoupling subsidies from quantities produced to effect direct payments can help avoid the exceedance supply characteristic of price supports, and the negative environmental products that come with them (Galko and Jayet 2011).

It must be noted that neoclassical economics is notorious in its opposition toward environmental subsidies unless market failures arise; there may be better alternatives than subsidies to address these failure, such as levies. While affordability and redistributive concerns can motivate the adoption of subsidies, empirical evidence shows they often backfire and aggravate, rather than ameliorate, water governance challenges. This negative impact can be aggravated economy-wide, since subsidies reduce total factor productivity, increase interest rates and savings, and reduce resources available for environmental expenditure.

Compliance Incentives (Buybacks, Voluntary Agreements) Compliance incentives comprise a wider category of inducements that cannot be ascribed to the more common *taxa* of levies and subsidies. Compliance incentives rely on negotiated agreements to adopt mutually-settled practices that benefit private individuals, while also contributing to water policy objectives. Compliance incentives are strictly voluntary, and require the existence of welfare-enhancing opportunities.

---

[4]In contrast, governments in developing countries typically impose price controls to keep prices below competitive levels.

These opportunities can be reaped through pecuniary exchanges, also known as *Payment for Ecosystem Services* (PES), or without direct remuneration. While many examples of the former can be found in the literature, the latter are more difficult to find, especially in areas with fully-fledged water markets where competition for water leaves little room for unexploited welfare-enhancing opportunities.

The advantage of PES lies in its capacity to unblock policy or behavioural transition (Marshall 2013). PES are based on the *beneficiary-pays-principle*, meaning that polluters are compensated to reform their behaviour. In water resource management, this process hinges on a water accounting framework, where water use is metered and water rights are enforced, so that changes to water use in response to compliance incentives can be identified and accounted for. Examples of PES in water management can be found in the large water acquisition programs implemented in: Australia's Murray-Darling Basin (AU$2.2 billion for the period 2009–2012), south-eastern Spain (EUR829.9 million for the period 2007–2027) and the United States, notably California (US$547 million during 1987–2011, 55% of which occurred after 2003). Water acquisitions or *buybacks* are often coupled with costly flanking measures to mitigate micro- and macroeconomic impacts, often larger than the buyback programme itself. The foregone revenue inflows that could be attained through the use of a polluter-pays principle approach (i.e. levies) complete the cost of PES policies. There is debate in the literature whether non-pecuniary voluntary exchanges constitute an economic instrument or not. This misunderstanding stems from the definition of what constitutes a voluntary exchange in an economic context (Maziotis and Lago 2015). Some authors consider regulations, rewards or penalties to be valid motivations for a voluntary exchange and therefore include them in this category, yet this is not admissible from an economic standpoint where voluntary agreements result from non-coerced exchanges. More detailed accounts of non-pecuniary, voluntary agreements for water ecosystems restoration through public-private partnerships can be found in Gómez et al. (2014).

Marketable Permits Of the three demand-based water (re)allocation approaches defined at the beginning of this chapter, water markets are the most challenging to design, implement and sustain over time. Water trade can span informal (e.g. between neighbours) to formally assessed and recognised transfers that involve a variety of rules and processes designed to protect the interests of all parties (NWC 2014). Formal water trade thus entails a number of benefits, such as explicitly accounting for water use costs—including externality opportunity costs—across all users. Whilst countries such as Australia, the United States, Chile, Mexico, South Africa and China have made considerable progress toward water trade as a means to improve their use of water resources, the expanded application of formal water trade arrangements remains highly contentious. The complexity associated with trade in wider social-ecological systems means that transfers are not always able to comprehensively resolve all socioeconomic issues around water use (Meinzen-Dick 2007). The conditions for enabling effective water trade thus critically depend on local circumstances (Maestu and Gómez-Ramos 2013), and must generally be consistent with the principles of sound water management (Grafton et al. 2012).

Matthews (2004) raises ten questions relevant to any discussion about the establishment or reformation of a water rights system, which have significance for water managers interested in subsequent market adoption. These include: how any rights to water are currently specified, distributed and prioritised; whether existing rights are tradeable in nature, or if transformation would be required; how clear are current operational water use rules and can those rules assist/hinder transfers; how certain are we of our data on current water source, supply, usage and measurement; how might we enforce change or compensate losers in the modifications proposed, and who will achieve this; and are all aspects of the system (e.g. groundwater interaction, return flows, losses etc.) accounted for. Issues that might help to stimulate water markets are also discussed, such as adopting uniform rights across all uses (but with heterogeneous use-reliability or preferences), increased water pricing/charges, removing spatial limitations to use, and adopting a national registry system (Matthews 2004).

This highlights the need to embed water markets in water right reforms, as part of an integrated approach, if trade is the objective. If water markets are used to achieve these outcomes, then they will also require ongoing institutional capacity-building and adaptive governance arrangements for the best probability of sustainable implementation (Marino and Kemper 1998; Wheeler et al. 2017). Given this complexity, developing-nations may find less complicated and costly non-market reallocation approaches more suitable (Marston and Cai 2016). But in many developed-nations, growing water scarcity, greater environmental concern and limited supply-side options have driven an increased emphasis toward demand-side reallocation policies. Proponents of water markets argue they offer more efficient and effective approaches to reallocating water, preferably before it becomes scarce, and can also protect social and environmental values (e.g. Crase and O'Keefe 2009; Chong and Sunding 2006) when underpinned by appropriate regulatory and monitoring programs. Benefits such as efficient transfers of water to high-value uses (e.g. the environment in recent decades) and gradual changes in people's perceptions of water use values and sustainable behaviour can also be attributed to market-based transfer outcomes in some contexts (Table 1).

Administrative (re)allocation mechanisms, on the other hand, may lead to less-efficient allocations and weak incentives unless water is already scarce—although they are typically more equitable than trade mechanisms. Subsequent debate about water markets has thus focused on the privatisation of urban water supplies (Segerfeldt 2005; Goldman 2007), and consequently many researchers prefer that governments periodically reduce particular water uses so that it can be allocated to the environment and/or other users. Therefore, water markets exemplify the interdependence between policy, regulation, and economic instruments. It is unlikely that effective, efficient and sustainable markets will emerge in contexts where robust regulatory institutions—coupled with sound water governance principles—do not form a basis for structuring socio-ecological policy and objectives. However, the major advantage of properly implemented and maintained water

Table 1 Pros and cons of water trade to ensure efficient, equitable and sustainable water allocation

Economic efficiency	• Net irrigation revenue gains where trade can freely occur (Qureshi et al. 2007) • Lead to socially-optimal and efficient allocation (Möller-Gulland 2010) • Facilitation of water reallocation from low to high value uses (Maestu et al. 2010) • Efficient under low transaction costs (Pujol et al. 2006), but it can increase total use and thus decrease net welfare where negative externalities increase (Quiggin 1988) • Excessive regulatory control and subsides may result in inefficient markets (Möller-Gulland 2010)
Water use efficiency	• Most likely in the long term. Water efficiency measures stimulated by the market may make additional water available for the environment without reducing overall economic activity (Qureshi et al. 2007) • Should be secure to provide users incentives to invest in water conservation practices (Dosi and Easter 2000)
Equity	• Equitable if properly regulated and if the distribution of rights is fair (may work against equity in some cases) (Grafton et al. 2010) • In agriculture, water markets may lead to the concentration of water in more efficient and intensive farms (Pujol et al. 2006) • They can generate third-party effects or externalities if not properly regulated (Janmaat 2011) • They can generate third-party externalities, or unjustified profits on sellers, if not properly regulated (Janmaat 2011)

markets is that, where a resource is shown to be over-extracted/consumed and new information arises about the unsustainable nature of extractions, reductions in use can then be managed in a manner that does not necessarily negatively impact market confidence. Markets can similarly be used to affect behavioural change among users that can positively benefit both public and private objectives.

4 Taking Stock

It should now be clear that, in recent decades, governments have started to rethink the way water is managed and shared among competing users, changing from a supply management approach to a demand management perspective. In semi-arid climates where inter-annual water availability variation is extreme, large infrastructure may prove insufficient to mitigate the effects of water scarcity in an unstable and environmentally limiting context (Calatrava and Garrido 2005). Managing demand in a modern context involves implementing water conservation measures, providing economic incentives, reforming water charge schemes and re-defining/reallocating water rights. There is now also a growing consensus that greater reliance on economic principles in managing and allocating water is critical for more efficient and sustainable use.

As such, to address current and future water availability problems, there is a need for effective and flexible institutional arrangements and allocation mechanisms to mitigate and manage water scarcity (Grafton et al. 2010). Properly designed and regulated economic instruments can make a significant contribution to changing individual behaviour towards a more sustainable use of available water resources, water reallocation and the mitigation of water-related risks. However, if not carefully planned, negative externalities and unexpected impacts could arise from their implementation. In closed basins where all available water resources are already allocated, the only way to meet increasing demand is through the reallocation of available water resources among competing needs. Inter-sectoral reallocation is seen as one pillar of water demand management that can be achieved either through decentralised or other reallocation mechanisms. However, a current major problem is that water is often allocated based on institutions established when water was not considered to be a scarce resource (Frederick 2001), and this problem will be exacerbated in the future due to increasing water demand. Hence, developing robust water governance and reallocation frameworks takes considerable time and resources to design, implement and evaluate.

In many countries, increasing attention is now being paid to the need to improve water rights definition. As long as the resource is plentiful, there is little pressure to redefine or enforce water rights. When water becomes scarcer and competition increases, property rights can clarify expectations and reduce conflicts. Water rights are the heart of any allocation system, and essential for a successful reallocation (Meinzen-Dick and Bakker 2001). It then follows that 'the great virtue of creating property rights in water is that it can be bought and sold' (Getches 2004, pg. 12). Consistent with this view, market mechanisms to reallocate water resources are increasingly encouraged by many experts and organisations.

The European Commission, in its Blueprint to Safeguard Europe's Water Resources, considers water markets as a tool that could help improve water use efficiency and overcome water stress, if a cap for water use is first implemented (European Commission 2012). The Murray–Darling Basin in Australia is home to one of the world's most successful water markets, underpinned by well-defined and secured rights with set limits to extraction, as well as national regulatory frameworks for setting, monitoring and enforcing water market rules. As the market has matured, farmers have become more accepting of it over time as a risk-management strategy during times of scarcity. The market has also enabled regulators to assist rural communities and drive positive environmental outcomes (NWC 2014). Some deficiencies still exist in Australian water markets, such as unnecessary trade barriers, the need for improved water market and weather information, limited types of water trade products, inadequate understanding of return flow impacts, and possible future lock-in of some enterprises such as perennial production systems. However, many expected negative market outcomes such as reduced regional spending, higher unemployment, permanent water trades away from areas and underutilised water infrastructure have generally not emerged (Grafton et al. 2016).

Broad adoption of water markets in Australia has also not precluded adoption of other economic instruments (e.g. improved water pricing). The European Environment Agency (2012) considers that water levies and market-based instruments are essential for sustainable water management and efficient water allocation. The IPCC (2007) also believes water markets may play an important role in reducing water supply vulnerabilities. However, there are institutional, physical and social barriers (i.e. transaction costs) that impede the implementation of markets for water in that context. While markets and water prices have been used to manage demand, allocate water resources, and provide incentives to conserve and invest in new supplies and incentivise environmentally positive externalities (Frederick 2001), a great deal of disruption to existing institutions, infrastructure, political agreements and social-norms may be required to establish the fundamental conditions. This disruption can, in turn, motivate significant behavioural-resistance among stakeholder groups. In contrast, subsidies and compliance incentives have promoted behavioural change, aiming at a more sustainable use of water resources or to solve environmental issues. The main advantage of the latter is they are more popular than other instruments given their voluntary nature.

That said, water markets have been successfully created in different parts of the world—mainly in those areas with severe water scarcity problems. In most cases, the establishment of water markets has resulted in effective water conservation, rising awareness of the true value of the resource, and investment in water saving and water reuse technologies. Water infrastructure and allocation rules mitigate climatic cycles but do not completely eliminate supply risks. Spot water markets facilitate the efficient allocation of this resource and have some supply risk reduction properties, but do not provide efficient risk allocation mechanisms per se, which exploit differences in risk tolerance and exposure (Calatrava and Garrido 2006; Rey et al. 2016). This can be solved with secondary markets for water (i.e. option contracts (Rey et al. 2016)). Of course, markets also come with some disadvantages that should be taken into account. Economic criticisms of water markets may stem from arguments that transactions costs may be higher than those derived from other water allocation mechanisms (Pujol et al. 2006), although there is more recent evidence that transaction cost reductions can be empirically observed (Loch et al. 2018). Further, the adoption of markets may generate third-party effects or externalities, in some cases exceeding the social benefits derived from the exchange (Rosegrant and Binswanger 1994). Empirical evidence of these claims, though, is less prominent in the literature, and it appears unlikely that the adoption of water markets will limit regulatory or governance policy options in contexts with strong legislative, judicial and executive powers.

Finally, according to one of the basic principles of economics, anything scarce and in demand commands a price. However, setting a price that reflects the true value of water and incentivising individuals to use water efficiently, whilst simultaneously ensuring affordability, is not a simple task. In most cases, water levies are still low, not reflective of the full cost of the service and the negative externalities of water use, and they are sometimes not linked to actual extraction. While levies for municipal and industrial water use may increasingly reflect the full costs of service, agricultural water use remains heavily subsidised. Subsidies send inadequate price signals to water users, aggravating rather than solving the problem they were designed for. All water problems are policy problems. Getches (2014) expressed this sentiment clearly:

> Water issues are typically discussed as physical problems […]. But essentially all water problems have a policy nexus. It is rare that a water problem cannot be solved if public policy can be harnessed and directed effectively. […] Often these problems of competition can be privately resolved by payments from one party to another. But public policy must intervene if water is to serve more and varied interests, inasmuch as water is a public good. […] In the end, behaviour must be guided not by self-restraint alone, but also by a combination of regulation and market forces.

There is a lot we can learn from previous experiences of economic instrument implementation in different contexts. But also it is important to recognise that what may be working effectively in a particular setting may not be the most-appropriate option for another context with different water availability issues, regulation institutions and uses.

5 Conclusion

Standard policy approaches aimed at improving economic efficiency such as property rights, addressing externalities and achieving common water-use objectives are no guarantee of sustainability into the future. However, improvements in technological capital including our knowledge-base related to regulatory institutions and economic instruments may assist us to draw consumption and sustainability objectives closer together.

In this chapter we have sought to elaborate on the useful partnership between regulatory and economic instruments that can be utilised to achieve effective and efficient water governance arrangements. While this approach is far from straightforward—relying on a critical mass of institutional capacity and system understanding—policy-makers can take advantage of the growing evidence in many contexts of the successful marriage between these two approaches. It may be costly both in terms of political support and transaction investments to strip away existing arrangements in favour of more flexible and better-suited institutions to manage scarce water resources. But it would be expected that ignoring the problems and hoping they will resolve themselves would be more harmful to private and public welfare outcomes.

Acknowledgements Dr. Loch was funded via an Australian Research Council grant DE150100328, and the UNESCO grants program 2015–16. Dr. Adamson was funded via an Australian Research Council grant DE160100213. Dr. Perez-Blanco was funded via AXA Research Fund Post-Doctoral Fellowships Campaign 2015, and the European Institute of Technology via Climate-KIC Europe's AGRO ADAPT Project. Dr. Rey was funded via the Natural Environment Research Council (NERC) programme on Droughts and Water Scarcity, funded through the Historic Droughts (NE/L010070/1) and MaRIUS (NE/L010186/1) projects.

References

Arrow, K., Dasgupta, P., Goulder, L., Daily, G., Ehrlich, P., Heal, G., et al. (2004). Are we consuming too much? *The Journal of Economic Perspectives, 18*(3), 147–172.

Bakker, K. (2007). The "commons" versus the "commodity": Alter-globalization, anti-privatization and the human right to water in the lobal South. *Antipode, 39*(3), 430–455.

Bernstein, J. D. (1997). Economic instruments. In R. Helmer, I. J Hespanhol (Eds.), *Water pollution control—A guide to the use of water quality management principles*. Geneva: United Nations Environment Programme, The Water Supply & Sanitation Collaborative Council and the World Health Organization.

Bromley, D. W. (1989). *Economic interests and institutions: The conceptual foundations of public policy*. New York: Blackwell Publishing.

Calatrava, J., & Garrido, A. (2005). Modelling water markets under uncertain water supply. *European Review of Agricultural Economics, 32*(2), 119–142.

Calatrava, J., & Garrido, A. (2006). Difficulties in adopting formal water trading rules within users' associations. *Journal of Economic Issues, 40*(1), 27–44.

Chong, H., & Sunding, D. (2006). Water markets and trading. *Annual Review of Environment and Resources, 31,* 239–264.

Coase, R. H. (1960). The problem of social cost. *Journal of Law and Economics, 3,* 1–44.

Colby, B. (1987). Do water market prices appropriately measure water values? *Natural Resources Journal, 27,* 617–651.

Crase, L., & O'Keefe, S. (2009). The paradox of national water savings: a critique of 'Water for the Future'. *Agenda, 16*(1), 45–60.

Crutzen, P. J. (2002). Geology of mankind. *Nature, 415* (6867), 23–23.

Dosi, C., & Easter, K. W. (2000). *Water scarcity: Economic approaches to improving management*. Center for International Food and Agricultural Policy, University of Minnesota.

Driesen, D. M. (1998). Is emissions trading an economic incentive program? Replacing the command and control/economic incentive dichotomy. *Washington and Lee Law Review, 55,* 289–350.

Driesen, D. (2006). Economic instruments for sustainable development. In B. J. Richardson & S. Wood (Eds.), *Environmental law for sustainability* (pp. 277–308). Portland, Oregon: Hart Publishing.

European Commission. (2012). A blueprint to safeguard Europe's water resources. Brussels: COM (2012) 673 final.

European Environment Agency. (2012). Towards efficient use of water resources in Europe. Brussels: EEA Report, No 1/2012.

Fisher, D. E. (2006). Markets, water rights and sustainable development. *Environmental and Planning Law Journal, 23,* 100.

Frederick, K. (2001). water marketing: obstacles and opportunities. *Forum for Applied Research and Public Policy*, (Spring), 54–62.

Freeman, A. (2003). *The measurement of environmental and resource values: Theory and methods*. Resources for the Future: Washington D.C.

Galko, E., & Jayet, P.-A. (2011). Economic and environmental effects of decoupled agricultural support in the EU. *Agricultural Economics, 42*(5), 605–618.

Garrick, D. (2015). *Water allocation in rivers under pressure*. Cheltenham: Edward Elgar Publishing.

Getches, D. (2004). Water wrongs: Why can't we get it right the first time? *Environmental Law Reporter, 34* (1).

Getches, D. (2014). Water scarcity in the Americas: Common challenges—A Northern Perspective. In A. Garrido & M. Shechter (Eds.), *Water for the Americas, Challenges and Opportunities,* (pp. 15–39) London: Routledge.

Godden, L., & Foerster, A. (2011). Introduction: Institutional transitions and water law governance. *Journal of Water Law, 22*(2–3), 53–57.

Goldman, M. (2007). How "Water for All!" policy became hegemonic: The power of the World Bank and its transnational policy networks. *Geoforum, 38*(5), 786–800.

Gómez, C. M, & Pérez-Blanco, C. D. (2014). Simple myths and basic maths about greening irrigation. *Water Resources Management 28* (12), 4035–4044.

Gómez, C.M., Pérez-Blanco, C.D., & Batalla, R.J. (2014). Tradeoffs in river restoration: Flushing flows versus hydropower generation in the Lower Ebro River, Spain. *Journal of Hydrology, 518* (Part A), 130–139.

Grafton, R. Q., Landry, C., Libecap, G., McGlennon, S. & O'Brien, R. (2010). An integrated assessment of water markets: Australia, Chile, China, South Africa and the USA. In *NBER Working Paper Series*. Cambridge, MA: National Bureau of Economic Research working paper series, working paper 16203.

Grafton, R. Quentin, Gary Libecap, E., Edwards, R. O., & Candry, C. (2012). Comparative assessment of water markets: Insights from the Murray-Darling Basin of Australia and the Western USA. *Water Policy, 14*(2), 175–193.

Grafton, R. Quentin, Horne, J., & Wheeler, S. A. (2016). On the marketisation of water: Evidence from the Murray-Darling Basin, Australia. *Water Resources Management, 30*(3), 913–926.

Gunningham, N. (2009). Environmental law, regulation and governance: Shifting architectures. *Journal of Environmental Law, 21*(2), 179–212.

Hanemann, W. Michael. (2006). The economic conception of water. In P. Rogers & R. Llamas (Eds.), *Water crisis: Myth or reality?* (pp. 61–91). New York: Taylor & Francis.

Holley, C., & Sinclair, D. (2016). Governing water markets: Achievements, limitations and the need for regulatory reform. *Environmental & Planning Law Journal, 33*, 301–324.

IPCC. (2007). Climate Change 2007: Synthesis report. Summary for policymakers: An assessment of the intergovernmental panel on climate change. Intergovernmental panel on climate change.

Janmaat, J. (2011). Water markets, licenses, and conservation: Some implications. *Land economics, 87*(1), 145–160.

Loch, A., & Adamson, D. (2015). Drought and the rebound effect: a Murray-Darling Basin example. *Natural Hazards, 79*(3), 1429–1449.

Loch, A., Wheeler, S. A., Settre, C. (2018). Private transaction costs of water trade in the Murray–Darling Basin. *Ecological Economics, 146*, 560–573.

Maestu, J., & Gómez-Ramos, A. (2013). Conclusions and recommendations for implementing water trading. In J. Maestu (Ed.), *Water trading and global water scarcity: International experiences*. New York: RFF Press.

Maestu, J., Gómez, C. M., Garrido, A., & Llamas, M. R. (2010). Water uses in transition. *Water policy in Spain*, 39–48.

Marino, M., & Kemper, K. (1998). Institutional frameworks in successful water markets: Brazil, Spain, and Colorado, USA. In: *World Bank Technical Paper 427*. Washington: World Bank.

Marshall, G. (2013). Transaction costs, collective action and adaptation in managing complex social–ecological systems. *Ecological Economics, 88*(1), 185–194.

Marston, L., & Cai, X. (2016). *An overview of water reallocation and the barriers to its implementation*. Wiley Interdisciplinary Reviews: Water. https://doi.org/10.1002/wat2.1159.

Matthews, O. (2004). Fundamental questions about water rights and market reallocation. *Water Resources Research, 40*, W09S08.

Maziotis, A., & Lago, M. (2015). Other types of incentives in water policy: An introduction. In M. Lago, J. Mysiak, C. M. Gómez, G. Delacámara, & A. Maziotis (Eds.), *Use of economic instruments in water policy insights from international experience* (pp. 317–324). Switzerland: Springer.

Meinzen-Dick, R. (2007). Beyond panaceas in water institutions. *Proceedings of the National Academy of Sciences, 104*(39), 15200.

Meinzen-Dick, R., & Bakker, M. (2001). Water rights and multiple water uses—Framework and application to Kirindi Oya irrigation system Sri Lanka. *Irrigation and Drainage Systems, 15* (2), 129–148.

Meinzen-Dick, R., & Ringler, C. (2008). Water reallocation: drivers, challenges, threats, and solutions for the poor. *Journal of Human Development, 9*(1), 47–64.

Mendelsohn, R. (2016). Adaptation, climate change, agriculture, and water. *Choices, 31*(3).
Möller-Gulland, J. (2010). *The initiation of formal water markets—Global experiences applied to England*. Oxford: University of Oxford.
NWC. (2014). Australian water markets: Trends and drivers 2007–08 to 2012–13. Canberra, Australia: National Water Commission.
OECD. (2015). OECD principles on water governance. In *Directorate for Public Governance and Territorial Development*. Paris: OECD.
Pérez-Blanco, C. D., Standardi, G., Mysiak, J., Parrado, R., & Gutiérrez-Martín, C. (2016). Incremental water charging in agriculture. A case study of the Regione Emilia Romagna in Italy. *Environmental Modelling and Software, 78*, 202–215.
Pigou, A. (1920). *The economics of welfare*. London: Macmillan Press.
Pujol, J., Raggi, M., & Viaggi, D. (2006). The potential impact of markets for irrigation water in Italy and Spain: a comparison of two study areas. *Australian Journal of Agricultural and Resource Economics, 50*(3), 361–380.
Quiggin, J. (1988). Private and common property rights in the economics of the environment. *Journal of Economic Issues, 22*(4), 1071–1087.
Qureshi, M. Ejaz, Connor, J., Kirby, M., & Mainuddin, M. (2007). Economic assessment of acquiring water for environmental flows in the Murray Basin. *Australian Journal of Agricultural and Resource Economics, 51*(3), 283–303.
Rey, D., Garrido, A., & Calatrava, J. (2016). Comparison of different water supply risk management tools for irrigators: option contracts and insurance. *Environmental & Resource Economics, 65*(2), 415–439.
Rogers, P., Silva, R., & de Bhatia, R. (2002). Water is an economic good: How to use prices to promote equity, efficiency, and sustainability. *Water Policy, 4*, 1–17.
Rosegrant, M., & Binswanger, H. (1994). Markets in tradable water rights: Potential for efficiency gains in developing country water resource allocation. *World Development, 22*, 1613–1625.
Scott, A. (2008). *The evolution of resource property rights*. Oxford: Oxford University Press.
Segerfeldt, F. (2005). *Water for sale: How business and the market can resolve the world's water crisis*. Washington DC: Cato Institute.
Sivapalan, M., Savenije, H. H., & Blöschl, G. (2012). Socio-hydrology: A new science of people and water. *Hydrological Processes, 26*(8), 1270–1276.
Stewart, R. B. (1992). Models for environmental regulation: Central planning versus market-based approaches. *Boston College Environmental Affairs Law Review, 19*(3), 547–562.
Wheeler, S., Zuo, A., Bjornlund, H., & Lane-Miller, C. (2012). Selling the farm silver? Understanding water sales to the Australian government. *Environmental & Resource Economics, 52*(1), 133–154.
Wheeler, S., Loch, A., Crase, L., Young, M., Grafton, R. Q. (2017). Developing a water market readiness assessment framework. *Journal of Hydrology 552*, 807–820.
World Economic Forum. (2015). *Global risks 2015*. Geneva: World Economic Forum.

Water Markets and Regulation: Implementation, Successes and Limitations

Cameron Holley and Darren Sinclair

Abstract Although markets are widely promoted as an efficient tool for managing water, little critical attention has been directed to the legal and governance issues of water markets, including matters such as compliance and enforcement, water accounting, and the overall effectiveness of water trading. In response to these gaps, the chapter critically evaluates Australia's cap and trade instrument, drawing on a review of the literature and survey and interview data collected from government and non-government stakeholders. The findings reveal achievements, including flexible responses to past and future droughts; efficiencies that contribute to economic and environmental benefits; and increasing trade and market functionality.

We are grateful for the research assistance and editing of Amelia Brown and Trent Wilson. Parts of this chapter first appeared Cameron Holley and Darren Sinclair. 'Rethinking Australian water governance: successes, challenges and future directions' (2016) 33(4) *Environmental and Planning Law Journal* 275–283 and Cameron Holley and Darren Sinclair, 'Governing Water Markets: Achievements, Limitations and the need for regulatory reform' (2016) 33 (4) Environmental and Planning Law Journal 301–324. These articles were first published by Thomson Reuters in the Environmental and Planning Law Journal and should be cited as Cameron Holley and Darren Sinclair. 'Rethinking Australian water governance: successes, challenges and future directions' (2016) 33(4) EPLJ 275–283 and Cameron Holley and Darren Sinclair, 'Governing Water Markets: Achievements, Limitations and the need for regulatory reform' (2016) 33(4) EPLJ 301–324. For all subscription inquiries please phone, from Australia: 1300 304 195, from Overseas: +61 2 8587 7980 or online at legal.thomsonreuters.com.au/search. The official PDF version of this article can also be purchased separately from Thomson Reuters at http://sites.thomsonreuters.com.au/journals/subscribe-or-purchase.

C. Holley (✉)
UNSW Law School, Sydney, Australia
e-mail: c.holley@unsw.edu.au

C. Holley · D. Sinclair
Connected Waters Initiative Research Centre, UNSW Sydney, Sydney, Australia
e-mail: darren.sinclair@canberra.edu.au

C. Holley
University of Cape Town, Cape Town, South Africa

D. Sinclair
University of Canberra, Canberra, Australia

© Springer Nature Singapore Pte Ltd. 2018
C. Holley and D. Sinclair (eds.), *Reforming Water Law and Governance*,
https://doi.org/10.1007/978-981-10-8977-0_7

Yet, the results also suggest cap and trade schemes are not functioning at the peak of their powers because of seven key flaws, namely a lack of robust regulatory underpinning; limited accuracy in water accounting; challenges in addressing universality of impact and source; queries over environmental benefits; lack of accounting for wider social impacts; windfall gains; and limited operation across Australia. Some of these flaws are correctable, and the chapter pinpoints relevant areas for market policy reform. However, a number of the identified flaws require water law and policy to look beyond markets. The chapter argues in these areas, such as groundwater, complementary regulatory tools are needed to ensure Australia's future water security and sustainability.

1 Introduction

After 20 years of water reform in Australia, the clear policy 'winner' has been the cap and trade market based instrument. But are Australia's water markets a 'good' regulatory instrument for managing Australian water or are they a failed experiment? Over the last two decades, water governance in Australia has become synonymous with the concept of the market. Water trading has been promoted by the international community (World Meteorological Organization 1992), and practiced in areas such as California, Chile, Spain and South Africa (Burdack et al. 2014; Garrick et al. 2013), but it is Australia that has led the world in implementing market systems for governing water use. Nevertheless, Australia's water market is not the deregulation poster child of free market economists as even free markets are constituted by regulation (Shearing 1993). Rather, it is a 'cap and trade' market based regulatory instrument. This is a regulatory tool that involves significant, ongoing government intervention to establish and facilitate tradable water-access and withdrawal rights separated from land rights (Garrick et al. 2013).

Like other cap and trade schemes that emerged in climate, pollution, biodiversity and fisheries contexts (Arnason 2012; Stavins 1998; Warwick and Wilcoxen 2002; Curnow and Fitzgerald 2006), Australia's water market scheme is underpinned by a belief that environmental degradation occurs because of a failure to properly value environmental resources (Coase 1960; Cutting and Cahoon 2005; Roma 2006). On this view, what is needed is the creation of market signals so as to place a value on and charge appositely for the use of scarce assets (or the release of pollutant into the environment), and render environmental externalities visible (Gunningham et al. 2011; Gómez-Baggethun and Muradian 2015). Under nationally agreed water reforms, state governments needed to establish new markets for water resources (Gunningham et al. 2011). Because of the cooperative federal arrangements and place-based nature of hydrological systems, a composite of separate capped markets was pursued in unique water systems (local and regional) (NWC 2013). These sought to protect the environment by establishing an overall cap on acceptable resource use levels (e.g. sustainable water extraction levels in a catchment), assigning rights to extractors (e.g. separating land and water rights and setting an entitlement to extract a certain percentage of available water per season), creating and administering trading rules for those rights (e.g. allowing people to buy and sell water rights, often subject to assessment of impacts of trading extraction from one point to another) and designing and administering a compliance monitoring system (e.g. empowering state regulators to ensure users comply with given extraction levels) (NWC 2011; Karkkainen et al. 2000). The governments then left the rest to Adam Smith's 'invisible hand', which is thought to guide rational, self-maximising individuals to promote 'public interests' by relocating natural resources to those who value them most highly in both the long and the short term (Karkkainen et al. 2000; NWC 2011).

This system of 'trading' within a 'cap' is believed to provide a mechanism to coordinate diffuse knowledge and capacities away from the centre and bring order to complex systems (Hayek 1945; Burris et al. 2005; Gómez-Baggethun and Muradian 2015). This is achieved because change is driven by local people making choices in a context (created by governments) rather than by centralised public servants making direct decisions (Kiem 2013). Indeed, the use of centralised state based water regulation during the early and mid 1900s proved particularly unsustainable and ineffective as the knowledge and capacities of state governments were confounded by the many uncertainties associated with changing hydrological and hydrogeological systems, the unique local characteristics of surface and groundwater catchments, and the diversity in rural water users and uses. In contrast, trading between different local users is thought more likely to enable water to be reallocated across crops, irrigators, regions and the various demands of other water users in response to seasonal conditions (NWC 2013). The price signal for water in a market should also incentivise users to make efficient use of inputs and invest in improving the efficiency of their on-farm water use, encouraging flexible and 'voluntary' adoption of targeted behavioural change (NWC 2013; Kiem 2013). Moreover, in a capped system, trading would facilitate investment and structural adjustment (allowing new irrigation developments to establish and existing users to retire or move on) in response to changing conditions (NWC 2013). The upshot is that the 'trade' aspect of the cap and trade scheme is thought to produce greater efficiency, innovation, and public outcomes than pure 'top down' administrative, command and control approaches (Gómez-Baggethun and Muradian 2015).

With now over 20 years of market inspired reforms under Australia's belt, and with substantial trade now occurring within the Murray Darling Basin (MDB), it is timely to examine the achievements and limitations of water markets in Australia. Fortunately, this is a topic about which governments, practitioners and academics have produced considerable work (Liverman 2004; Castree 2010; Godden and Ison 2010; Connell and Grafton 2011; Tan et al. 2012; Murray-Darling Basin Authority 2017; Wentworth Group 2017). These reviews have collectively identified many successes, but also many weaknesses in Australia's water markets, predominantly from the perspective of economics, environmental management and environmental geography. This paper seeks to build on and extend this work, particularly in areas where far less attention has been directed, namely to legal and governance issues of Australian water markets in practice.

To do this, the paper reflects on the success of Australia's cap and trade schemes by interrogating the literature and new empirical data collected in NSW from a survey of 4000 water users and interviews with 48 government and non-government water stakeholders. This data is supplemented by 17 targeted interviews with government and nongovernment stakeholders from Queensland and Victoria.

The chapter proceeds in four parts. Following the introduction, Part 7.2 sets the context by providing a brief history of Australia's legal and policy architecture that created and governs water markets. Part 7.3 then focuses on the achievements and challenges of water markets. Our primary focus here is water markets in the non-urban context (Godden 2008), where over 65% of Australia's total water consumption occurs for agricultural purposes (Australian Bureau of Statistics 2014). It is accordingly here where some of the biggest efficiency, environmental, and social gains from market reforms can be made. Our data and analysis of non-urban water markets primarily considers legal and regulatory issues relevant to efficient and effective trading (rather than cap setting or allocation/licensing process). These are of course not mutually exclusive issues, as each element impacts on the overall effectiveness of the cap and trade scheme. Further, as the experience in European carbon markets revealed, the cap setting and allocation of rights (with specific conditions) can be major stand-alone challenges that, if done poorly, can be terribly inefficient (Goulder and Schein 2013). While we offer some brief reflections on cap setting and right allocation issues regarding environmental, socio economic and Indigenous interests, our primary focus is the regulation and governance of trading because it has received far less attention in the legal literature (Carmody 2013).

Three key achievements of cap and trade schemes are identified, namely better and more flexible responses to past and future droughts; efficiencies that contribute to economic and environmental benefits; and increasing trade and market functionality. These strengths are counterbalanced against seven challenges, namely implementing an effective regulatory underpinning; achieving accuracy in water accounting; addressing universality of impact and source; delivering expected environmental benefits; accounting for wider social impacts; addressing windfall gains; and operating equally across Australia. Part 7.4 summarises the proceeding analysis and concludes that while Australia's water markets are not a failed experiment, they risk failing in a host of key areas. We argue that despite their substantial success, there is an emerging implementation failure of effective and efficient market trading and this threatens to undermine the achievement of sustainable water use. As we explain, some of these limits are correctable, and we recommend increased attention on enhancing regulatory underpinning, improving accuracy of information and complete unbundling or rights in states such as Western Australia and Northern Territory to extend the operation of the market. However, we also suggest there are areas, such as groundwater use, where simply retooling and continuing with water markets is unlikely to be sufficient. In these contexts, we argue the market has significant limitations, and complementary regulatory tools will be needed to ensure Australia's future water security and sustainability. Part 7.5 concludes by identifying future research areas.

2 Water Governance Reforms and Markets in Australia

The early history of Australia's water management has been discussed in detail in Chap. 1. Here we briefly recap some of the core elements of the water market. Beginning in 1994 with the Keating Government's national competition reform packages, intergovernmental action was taken to arrest widespread water resource degradation and address the economic, environmental and social implications of water (Council of Australian Governments (COAG) 1994). A strategic framework was introduced to guide state implementation of new market based reforms designed to achieve an efficient and sustainable water industry (COAG 1994).

Consistent with Australia's cooperative approach to federalism and water's disrespect for human defined boundaries, these initial reforms set out a high level and long term national vision of water trading (Commonwealth of Australia 2014). Under this agreement, states and territories were afforded significant discretion to determine how best to implement this vision. The resulting diversity in state approaches, when combined with distinct basin and catchment caps, have produced a composite of many separate markets in Australia, each defined by water system boundaries and administrative arrangements (NWC 2013). Further, within those markets are segments for different water products, such as access entitlements and allocations (often of varying levels of security), and different trading transactions (NWC 2013). However, as we will see below, the most connected markets occur in the surface water context within the MDB, predominantly because of the hydrological connectivity between systems within the Basin (Grafton and Horne 2014).

The foundation of Australia's approach was the creation of firm property rights to extract water, within extraction limits. Water entitlements were separated from land title, and these provided an on-going share of a consumptive pool of water within a given catchment. Water allocations were the volume of water assigned to a water entitlement in a given water season under a specified water resource plan (Grafton and Horne 2014). To ensure the health of river systems and groundwater basins, allocations were also created for the environment (COAG 1994).

Trading arrangements for allocations and entitlements were instituted within each state (Wheeler et al. 2014a, b, c). However, in many cases actions were slower than expected, and the Millennium Drought soon threatened available water supplies, whilst demand for entitlements and allocations continued to increase. In response, the Council of Australian Governments (COAG) consolidated the 1994 reforms through an intergovernmental agreement on a National Water Initiative (NWI).

Led by the Howard and Anderson Government, and state and territory leaders, the NWI was signed in 2004. It aimed to embed a nationally-compatible water market, a cooperative planning process for managing surface and groundwater resources in each catchment and a government backed compliance and enforcement system (NWI 2004). To facilitate these paths, the states were required to take a number of actions, including establishing a transparent, statutory-based water planning process; making statutory provision for environmental and other public benefit outcomes; returning over-allocated or overused systems to environmentally-sustainable levels of extraction; progressively removing barriers to water-trading on an open market; developing water accounting which is able to meet the information needs of different water systems in respect to planning, monitoring, trading, environmental management and on-farm management; facilitating efficient water use and addressing adjustment issues; and recognising the connectivity between surface and groundwater resources and connected systems managed as a single resource (NWI 2004).

The NWI agreement also created a new independent statutory body, the National Water Commission (NWC), charged with a key role in information provision and monitoring of national and state performance, including undertaking biennial assessments of state progress in implementing markets and the NWI.

States progressively implemented the NWI (albeit unevenly), but they faced a number of flaws (discussed further below), including court challenges to 'caps' established as entitlement reductions to secure sustainability (Millar 2005; *ICM Agriculture Pty Ltd & Ors v The Commonwealth of Australia & Ors* (2009) HCA 51). Among other things, these challenges saw new initiatives being introduced, often with a particular focus on the MDB (e.g. the National Plan for Water Security). The most significant of these initiatives was the *Water Act* 2007 (Cth) which saw a new independent MDB Authority become responsible for MDB-wide planning and management (*Water Act* 2007 (Cth), Pt 9; Commonwealth of Australia 2014). As a part of this process, restrictions on trade were to be removed, especially across state boundaries through a new common trade framework and new rules for enforcing water market rules and water charge rules were introduced (see e.g. Australian Government 2009; *Water Market Rules* 2009 (Cth); *Water Charge (Termination Fees) Rules* 2009 (Cth)). A Commonwealth Environmental Water Holder (CEWH) was also created to enable government purchases of entitlements and investment in efficient water infrastructure so as to better protect the MDB's environmental assets (Commonwealth of Australia 2010, 2016).

Under a new Basin Plan agreed to in 2012 (Bowmer 2014; Gross 2011), caps were set to recover 2,750 gigalitres of water via investment in infrastructure efficiency and water buybacks (Murray Darling Basin Authority 2014a, b). Following the passing of the *Basin Plan* 2012 (Cth), Australian water reform has seen a raft of new developments and challenges, including the intensification of coal and other unconventional sources of extraction (Whitehead 2014; Parsons et al. 2014), the abolition of the NWC (National Water Commission (Abolition) Act 2015 (Cth)) and proposals to develop water resources and markets in northern Australia (Commonwealth of Australia 2015).

In summary, Australian legal and policy architecture for water markets has been on a significant journey since 1994. In the next section, we provide a brief overview of our empirical research methods, followed by identifying the key achievements and limits of these market based schemes.

3 Achievements and Limits of Water Markets

Cap and trade schemes offer a unique approach to governing water quantity in Australia. This paper examines the Australian schemes by using a mixed methods approach to triangulate data from quantitative and qualitative empirical research conducted in the MDB (Creswell and Clark 2011), as well as a review of available studies in the literature. The data includes opinions, attitudes and views collected from 65 interviews (New South Wales, Victoria and Queensland) and a survey of 4,000 NSW water users. As with most social research, the ethical and confidentiality requirements of the study require the chapter to preserve the anonymity of specific interviewees. The survey design and the mail-out process employed a modified Dillman (2007) approach. The survey was presented as a distinctive booklet and mailed with a cover letter and postage-paid return envelope. Two reminder/thankyou notices were posted to respondents and non-respondents. All non-respondents were then sent a new mail package (for further on the methods, see Holley and Sinclair 2015).

The survey was conducted from September to December 2012 across three regions in NSW, namely the Central West (CW), Murray and Murrumbidgee (MM), and North Coast (NC). The survey further refined the above three regions by focusing on two to six local government areas in each region that captured a diversity of water sources (regulated rivers, unregulated rivers and groundwater). The survey questions were designed to capture water users' views on, experiences with, and knowledge of water regulation, compliance and enforcement. Although the survey contained over 100 questions covering a range of topics, this chapter focuses only on those questions relevant to compliance and enforcement experiences of water markets.

Taking resource and practical constraints into consideration, the survey began with a raw list of 4,500 licence and approval holders (approximately 1,500 from each of three regions, and representing a full range of water users, from large entitlement holders extracting water for commercial use to people extracting water for stock and domestic purposes). This list was refined to create a more targeted mailing list, including ensuring multiple works/licence holders would only receive one survey and removing any repeat or incomplete addresses, as well as entries pertaining to local/state governments and commercial companies outside of NSW (who were unlikely to have the desired knowledge and experience with on-property water use). A final survey list was sent to 1,381 CW, 1,258 MM and 1,339 NC properties (totalling 3,978). The response rate was 22%.

The survey data was supplemented and tested through 48 follow-up interviews, with approximately one third in each NSW region. They also included five stakeholders from land care and catchment management authorities. Interviewees were selected to capture diversity, and included participants from both large and small farms, different industry types (e.g. cotton, cane, rice), different water types (groundwater, regulated and unregulated rivers), and a mix of those who had completed the survey and those who had not. The follow-up interviews provided more in-depth understanding. For example, in the survey farmers were asked whether they knew where to find the current water allocation for their water source and whether the regulator should take action against those who break the rules. The interviews then enabled the researchers to explore the particualrs of this knowledge (e.g. why knoweldge was or was not limited) and how reguatory actions should be taken (e.g. responsively versus strict punishment).

Given the NSW-focused survey and interviews, the chapter supplements this detailed data by drawing on 17 targeted interviews with key government agencies, peak bodies, farmers, NGOs, and other water holders from Queensland and Victoria. Each interview took approximately 30–60 min. Eleven interviews have been conducted in Victoria and six interviews have been conducted in Queensland and were a part of an ongoing study into water governance in Australia exploring the experience and attitudes of key stakeholders on a range of issues (e.g. aims and market rules; extent of implementation, monitoring and evaluation; and the nature of interaction between water planning and markets and regulation).

The following sections use this data to argue that market design features have delivered a range of achievements, but also confront fundamental limitations in practice.

3.1 Achievements of Water Markets

Better and More Flexible Responses to Past and Future Droughts As discussed above, one of the primary benefits of markets over top down command and control approaches is that they can coordinate local knowledge in ways that can flexibly react to changing circumstances and risks (Karkkainen et al. 2000). A number of interviewees recognised these benefits, confirming that the decentralised market mechanism had enhanced the capacity of individuals to have a say over water management: '[markets] actually give community and public interest constituencies direct control over water resources to get public interest outcomes' (Interview V0, NGO). This in turn was reported to have contributed to greater flexibility in the water communities' response to drought: 'when we started that drought, the irrigator didn't have a lot of say in how they could affect their own security. They do now with the carryover rules and as the market's got more developed, its expanded the tools you can use...I think we're better placed now than we've ever been to face a 2002 again' (Interview V8, Farmer).

Similar points have been drawn from economic modelling and research on water markets, which have found that the ability to trade water has been critical in maintaining irrigation sector incomes during drought (Wheeler et al. 2014a, b, c). For example, Wei et al.'s study suggests water trading played a key role in responding to the 2006–08 dry periods, including rates of around 20–31% of farmers either buying or selling water during these periods (Wei et al. 2011). Available modelling also suggests that the market provides significant adaptation opportunities for farmers under mild to moderate future climate change scenarios

(Wheeler et al. 2014a, b, c). Government interviewees confirmed this belief, pointing out that a market (as opposed to central administration) was the best mechanism for responding to times of scarcity: 'If there is less available…it gets harder to adapt and take into account new values and new benefits if it is not done through the market' (Interview V9, Government).

Market Efficiencies Produce Economic and, to a Lesser Extent, Environmental Benefits for Individuals and Regions A number of respondents argued putting a price on water and enabling its trade had ensured much greater efficiencies in the use of water: 'It's the ability of water to be able to go from the low value use to high value use and that's really changed practices and made the biggest success' (Interview V8, Farmer). As another respondent explained: 'the main success of it all has been the water tradability which really allows water to find its highest value. We can trade high flow now…We can now, with difficulty, trade inter State. But what it allows people to do is find the highest value for their water. Instead of wasting water on running cows or something, it's all heading to the highest value. So each ML is producing the most earning' (Interview Q3, Farmer).

These efficiencies produced substantial economic benefits, both for agricultural communities and individual farmers. At the community level, markets reportedly contributed to the resilience of agriculture and its transformation in the face of global shifts in demand and supply for commodities (Grafton and Horne 2014). Indeed, markets have facilitated water entitlement buyers and sellers to adjust their balance sheets or sell excess water in response to changing production (Grafton and Horne 2014). The result has been substantial economic returns to irrigators (both buyers and sellers) and their farming communities. One review of water trading in the southern MDB found that water trading increased the gross regional product of the southern MDB by some $370 million relative to a scenario where no trading was available (Grafton and Horne 2014; National Water Commission 2010). Further, the estimated value of entitlements on issue in the MDB, using the average prices per megalitre calculated for 2012–13, was approximately $13 billion, with an overall turnover in Australia's water markets of around $1.4 billion (NWC 2013).

At the individual level, Wheeler et al. modelling suggests that despite the sale of water entitlements potentially reducing on farm production, such negative impacts have been offset by many irrigators using water sale proceeds to reduce debt (and hence interest payments), as well as restructure and reinvest on farm (Wheeler et al. 2014a, b, c). In particular, water pricing has allowed farmers to improve their security and commercial certainty as water titles are viewed and treated as a financial asset. Financial institutions now commonly use water title (particularly high security water) as security for loans. A recent government survey found that 35 per cent of 1,200 respondents (irrigators) used water title as security over loans in 2010 (NWC 2014a, b). Interview respondents similarly confirmed these benefits. As one noted: 'people now have more security. I think that goes back to the fact of the change in the water licence now having a property right. That was the fundamental change in water in this State and in the country, that's been a huge thing.

Especially for security purposes and banks and all that sort of thing, that's made a massive difference' (Interview Q4, Industry Body).

In contrast to economic benefits, it is arguably more difficult to determine the extent to which water markets have supported a healthy environment. Outcomes take many years to materialise, environmental water programs have not been fully implemented in all states, and monitoring and assessment of environmental objectives in plans have often been inadequate, or not well targeted (NWC 2014a, b). Some studies have suggested that impacts of trading on the environment were small and largely positive (Wheeler et al. 2014a, b, c), however significant concerns remain about increased salinity, reduced end-of-system outflows, and degrading water quality as a result of detrimental change to the volume, location and/or timing of water use (Wheeler et al. 2014a, b, c). Even so, the 2007 commitment to the purchase of water entitlement from willing sellers in the Basin reportedly had significant environmental benefits (Grafton and Horne 2014). Since 2007, entitlements purchased through the buyback process and those recovered through the Sustainable Rural Water Use and Infrastructure Program have ranged between 22 and 426 GL annually, with a total cumulative volume registered at the end of 2012–13 being 1,599 GL (NWC 2014a, b). Up to one-fifth of all irrigators in the MDB sold water entitlements to the Commonwealth during this period, and these purchases provided substantial progress towards meeting the planned-for environmental target in the Basin Plan to reallocate 2,750 GL of water for environmental purposes (Wheeler et al. 2014a, b, c; Grafton and Horne 2014).

Although concerns are raised below about the impact of environmental flows, at least some respondents believed environmental flow management through the market also had significant environmental benefits. As one interviewee discussed, the market system has been an effective and efficient system for facilitating the provision of environmental water, particularly during the drought: 'We were able to sell some allocation in the north and use some of the money … purchase water that had previously historically been used for irrigation…and put the environmental flow over…and get a really good fish response and outcome in the estuary. So, you can see there, no water physically crossed the divide, but water was converted to money, the money was converted back to water, and an environmental watering event happened' (Interview V9, Government).

Increasing Trade and Market Functionality The key to securing the above benefits is ensuring 'the market' functions optimally and efficiently and trade is facilitated. Perhaps the greatest success then in Australia's water reforms has been the ongoing commitment to improving trade in water allocations and water entitlements (Wheeler et al. 2014a, b, c). Initially only water-right trading within one irrigation system was allowed, however over the course of time, trading rules permitted trade between connected irrigation systems within a single state and then between states within the Basin (Burdack et al. 2014). While progress on de-bundling water and land rights and facilitating trade has been slower in some states and territories (e.g. Western Australia & Northern Territory) than others, the majority of the 172 water plans across Australia facilitate trade ($\sim$ 100 water plans have been unbundled to facilitate trade, ~40 have been unbundled to some extent and less than ~10 have not been unbundled at all) (NWC 2014a, b).

Certainly, there are barriers to trade, including transaction costs as a result of regulatory restrictions, as well as hydrological and infrastructure constraints (discussed below) (Brooks and Haris 2014). Such barriers have meant trading tends to be concentrated in surface water (as opposed to groundwater) (Burdack et al. 2014; NWC 2013) and within the MDB. For example, in 2012-13, 78% of entitlement trade and 98% of allocation trade occurred in the MDB (NWC 2013). Collectively, the markets outside the MDB accounted for 300 GL of entitlement trade and 126 GL of water allocation trade in 2012–13 (or 22% and 2% of the total Australian markets for those products, respectively) (NWC 2013).

These barriers notwithstanding, progressive reforms in the MDB have improved confidence in the market (and made trade easier) by improving trading rules, developing a central online access point for comparing different water products and rules and the development of a Basin compliance strategy (Murray-Darling Basin Authority 2014b; c.f. Murray-Darling Basin Authority 2017). While public access to jurisdictional registers and 'searchability' is in need of improvement, and the full implementation of the trading rules will not happen until 2019, reductions in transaction costs and increased functionality (e.g. online access) of many state trade registers have occurred since 2004 (NWC 2010; Grafton and Horne 2014). These improvements were confirmed in surveys of irrigators asking whether trading of water allocation, entitlements and overall water trade had become easier in the last 5 years. Most agreed that trade had become easier (~40%) with strongest agreement found in Southern MDB (although over 50% disagree outside the MDB) (NWC 2014a, b). The ACCC has observed that compliance to both water market and water charge rules has also improved, and costs in some areas have declined (Grafton and Horne 2014).

Evidence of growing confidence in markets and trading appear to be borne out in available market data, both in terms of the level of trade and prices (NWC 2010). Although entitlement trades remain relatively minimal (around 4% by volume of the total entitlements on issue in 2012–13), the number of allocation trades are substantial (around 5 times the volume of entitlements traded—amounting to 27,136 intra-state allocation trades and 1,022 inter-state allocation trades, accounting for 6,184 GL of allocation traded in 2012–13) (NWC 2014a, b; NWC 2013). Of this volume traded, 6,054 GL occurred in the Murray–Darling Basin. The large majority of this trade occurs in NSW and Victoria. There is also a strong market in interstate water allocations accounting for 21% of trading volumes (NWC 2013).

In terms of price, although there have been declines in the price of entitlement trades (NWC 2013), reviews have found the seasonal water market is 'healthy', in so far as price clustering suggests that the trading activities of the water market produce characteristics that are developing in a similar fashion to more sophisticated, liquid and efficient financial markets (Brooks et al. 2013). Indeed, during 2012–13, the average price of approximately $44/ML in the southern connected MDB evidenced an increase of about 144% on the average price in 2011–12 and a reduction of about 70% on the average price in 2010–11. Total sales of water allocations in 2012–13 were valued at $287 million, which was more than three and a half times the estimated value of the water allocation market in 2011–12 (NWC 2013). Further, the introduction of the CEWH and its allocation trades do not appear to have substantially distorted market prices (NWC 2010). To date the literature is also yet to identify serious impacts on farmers from speculation (Skurray et al. 2013).

3.2 Limits of Water Markets

Water Markets Require a Regulatory Underpinning All markets are artificial constructs in that they require, even in the most unfettered of circumstances, institutional and regulatory underpinnings (Shearing 1993). At a minimum, this entails basic property rights and a regulatory, legal and judicial system to enforce those property rights and address disputes if and when they arise. In this respect, water trading under the NWI is no exception. The unbundling of water rights from land ownership has created a new form of right that can be traded, albeit with some important constraints, on a market. In order for this market to operate efficiently and effectively, however, there are several needs that have to be satisfied.

First, there needs to be confidence that the allocations of water rights within a given market are adhered to. To the extent that this is not the case, it has the potential to undermine the incentive to trade. In particular, those needing to purchase additional property rights may simply choose to obtain water additional to their existing allocations through illegal extractions, and thus avoid the cost of purchasing water rights. A crucial component of water trading, then, is an effective compliance and enforcement regime.

In this respect, the role of state-based regulatory agencies in having a robust compliance and enforcement regime in place is crucial to the success or otherwise of the national water reform enterprise, particularly in relation to the MDB. Indeed, compliance and enforcement is where the 'rubber hits the road' for national water policy implementation (Holley and Sinclair 2013a, b). And yet, there are reasons to conclude that this, despite some improvements, remains less than ideal.

In part, this is because environmental regulation in the agricultural sector has lagged behind other industries (Holley and Sinclair 2013a, b). Briefly, agriculture has often accorded a distinctive approach to governance and regulation. Political power, geographic isolation and the primacy of private property rights, in particular, have ensured that agriculture has not been subjected to the same degree of regulatory intrusion as manufacturing and mining, for example (Barr and Cary 2000; Holley and Sinclair 2013a, b). To the extent that regulation is imposed, 'soft' approaches that favour information, education, self-regulation and voluntarism have historically been used over traditional command and control (Gunningham et al. 1998; Holley and Sinclair 2013a, b). As such, there has been a reticence to impose robust compliance and enforcement (Gunningham and Sinclair 2004).

The case of DPI Water, the primary state water regulator in NSW (up until recent reforms, see e.g. Matthews 2017), is largely consistent with these phenomena, with a combination of reasons why it has historically struggled to fulfil its regulatory responsibilities. These include, in particular: an institutional and political culture that favoured voluntarism over deterrence; a lack of enforcement resources and a lack of relevant enforcement experience and training among staff; a lack of recognition across DPI Water of the role of compliance and enforcement, combined with an historical legacy of providing and encouraging farmers to take out water licences; and the outsourcing of meter reading to Water NSW which emphasises 'customer service' over compliance and enforcement.

In recognition of these past limitations, DPI Water put in place a series of policy, institutional and staffing initiatives since 2009 to enhance its compliance and enforcement activities. These have included a legal prosecution team, a compliance policy document, additional compliance and enforcement staff, including a special investigation and monitoring unit (separate from licensing officers), a risk-based compliance strategy targeting areas of high potential non-compliance, more sophisticated databases and risk assessment processes, and enhanced compliance technology, in particular, the installation of metering on licensed water extraction sites (Holley and Sinclair 2013a, b).

These reforms have been extended by DPI Water (and other state regulators) under the National Framework for Compliance and Enforcement Systems for Water Resource Management since 2012. The National Framework saw significant federal government investment designed to improve state water regulation across Australia (Australian Government Department of Sustainability Environment Water Population and Communities 2012). While NSW and a number of other states have demonstrated progress in meeting many milestones, completion rates across a number of areas (e.g. stakeholder education and monitoring) lag across a number of states (NWC 2014a, b). Indeed, recent inquiries into water compliance activities in New South Wales (Matthews 2017) and the Murray-Darling Basin (Murray Darling Basin 2017) suggest there is still significant work to be done, including completing ongoing institutional restructuring.

Given compliance and enforcement's crucial role in underpinning the water market reforms, together with the slow but increasingly positive DPI Water initiatives, it is pertinent to consider the views of water users towards the success or failures of compliance and enforcement in NSW.

In the first instance, the survey findings indicate that there is a significant degree of uncertainty about the levels of compliance and illegal activity in water extraction. While around 49% of respondents (n 604) were confident water users in their region complied with their licence conditions, an almost equal amount (45%) were unsure. Around 60% of respondents (n 583) were also unsure whether illegal water extraction was a big problem in their region, although the next largest group of respondents disagreed that illegal water extraction was a problem (34%). This uncertainty is a potential cause for concern, as it suggests that general perceptions of levels of compliance are less than optimal. At worst, it risks undermining norms of compliance as people who are regulated are less likely to comply with rules where norms of non-compliance are widespread in practice.

Beyond levels of compliance, the vast majority of non-urban water users have limited knowledge of DPI Water's compliance and enforcement activities, with

51% (n 504) wanting more information. Very few respondents claim to have direct interactions with, or received assistance from, regulatory officials (35%, n 430). Furthermore, only 25–34% agreed that DPI Water was fair and equitable (n 491), exercised good judgment (n 505), was able to detect illegal extractions (n 618) and effectively targeted its enforcement activities (n 613). More respondents agreed that DPI Water can ensure compliance (42%, n 615) and investigate illegal activities (45%, n 611), however, of those that had been subject to enforcement action (n 24–28), most thought that the action taken against them was not appropriate (65%) and that they had not been treated fairly (47%). This risks reducing support for laws, and at worst increases chances of recidivism.

Looking beyond DPI Water's current activities, there is strong support (nearly two-thirds, n 616) for DPI Water to undertake tougher enforcement action and more prosecutions against those who break the rules (that is, engage in serious breaches as opposed to minor/technical breaches). Overall, the survey findings from NSW indicate that although there is some success and support for compliance and enforcement in the water markets, there appears to be significant room to improve the regulatory underpinning of water markets, including levels of compliance, enforcement and prosecutions and perceptions of equality and fairness regarding state regulator interactions with water users.

Water Markets Require Accuracy A second crucial complement to regulatory underpinning and efficient operation of markets and water allocations is the accurate measurement of water extractions. In order for this to occur, extractions require effective metering. And yet, although various metering technologies have been implemented in non-urban contexts, their application has been patchy and uneven (Lavau 2013; Grafton and Peterson 2007; Holley and Sinclair 2013a, b). While surface water use has often been metered, the monitoring of groundwater extraction remains weak (or completely absent) (Holley and Sinclair 2013a, b). The accuracy of many current water meters is also said to be 'not high due to their age, lack of maintenance and improper installation' (Holley and Sinclair 2013a, b; Department of the Environment Water Heritage and the Arts (Commonwealth of Australia) 2009; Hamblin 2009). Although there is little data available, reports of existing meter recording errors range from +20% to −30% and +3% to −18%, and suggest 'worn or faulty meters tend to record less water than is actually extracted' (Holley and Sinclair 2013a, b; Department of the Environment Water Heritage and the Arts (Commonwealth of Australia) 2009). Further, the current lack of available real time data collection and on-line access via telemetry has the potential to limit water extraction accuracy and transparency (Holley and Sinclair 2013a, b). While recent government reforms have begun to address these deficiencies (Holley and Sinclair 2013a, b; Australian Government 2009; NSW Office of Water 2015b; NSW Government 2010), the net effect of these weaknesses in metering is that the volume of water diverted may exceed entitlement volumes, and undermine overarching goals of fair and efficient water use (Holley and Sinclair 2013a, b; Commonwealth of Australia 2009). More generally, the lack of accurate meters (whether over or under recording) is a significant impediment to the operation of water markets and their ability to guide water to the highest value uses (Raft and Hillis 2010). Indeed, recent studies into 120 metered sites in the Murray showed that only 20% were recording within ±5%, and one third were misreading by more than 20% with an overall average of a 2.2% under-reading of volume extracted (NSW Office of Water 2015a; Holley and Sinclair 2013a, b). If the 2.2% under-read was reflective of current metering across all rivers in the NSW MDB, it is estimated that around 140,000ML more would be extracted than is assumed, amounting to $21 m per year on the water market (NSW Office of Water 2015a; Holley and Sinclair 2013a, b).

Turning to the NSW survey, there was broad acknowledgement from water users of the positive role of metering in managing water sustainably, as well as of the importance of having accurate and well-maintained equipment (66%, n 608). There was substantial support for telemetry raised in interviews, in particular, so long as water users (not just government) were afforded access to real time data. Many respondents were of the view that the combination of improved metering and the use of real time water extraction data could improve the detection of illegal water extractions, discourage meter tampering and thereby ensure better compliance and equitable water extraction (Raft and Hillis 2010; Holley and Sinclair 2013a, b).

The overall tenor of survey and interview responses suggest that there is considerable scope for improving metering and thus accuracy in the market (Holley and Sinclair 2013a, b).

Markets Prefer Universality of Impact and Source The use of market trading as an environmental tool works best when there is universality of the environmental impact and access to the traded property right. In the case of the former, this means that where traded property is used or released the environmental impact is equivalent. For example, in the case of carbon emissions, the location of release is of no consequence for resultant environmental impact. In the context of water, however, this is not the case. Different regions and catchments have different ecologies and water requirements. Arguably, then, taking water from one region and using it in another, as market trading allows (certainly in relation to surface water), may produce less than optimal environmental outcomes. For example, in locations that are particularly vulnerable to salinity, trading water out of the area may reduce the capacity to rinse out the salt from the soil when the water table is shallow (Burdack et al. 2014). Such outcomes arise because allocations are determined by economic impact over environmental impact. In part, this may be compensated by government agencies reviewing trades, as well as entering the market to prioritise water for environmental flows that in turn can be used to target regions of greatest environmental need.

There is another twist in that the environmental impact of water extraction and use may vary over time (usually within a financial year). This is pertinent in that there may be significant delays between theoretical extractions and the ultimate use of the traded water, depending on seasonal economic requirements. Again, this has the potential to compromise the environmental benefits of water markets. Temporal issues may also impact on farm productivity. For example, farmers may sell some of their water allocation during the year, but if an unexpected drought arrives later in the same year, they can get caught short.

In addition to these issues in the trading of surface water, groundwater presents further challenges to the principle of universality. In the case of discrete groundwater aquifers, it is not possible to effectively trade between different aquifers. This is obviously inconsistent with the universality of impact and source principle. In such circumstances, rather than one market for water trades across the MDB, or even within state borders, trades may be confined to farmers in separate, discrete aquifers. In an ideal market, reducing the number of participants will inevitably reduce the overall efficiency of the trading system. Indeed, despite making up around 21% by volume of entitlements on issue in Australia, trading of groundwater entitlements is relatively limited in most jurisdictions, accounting for around 12% of total entitlement trading and only around 1% of groundwater allocation trading in NSW and Victoria (NWC 2013). In light of these challenges, the trading of groundwater may need more flexibility to encourage more participants in the market. Alternatively, as we discuss below, other policy approaches may be more appropriate.

Markets May Not Deliver the Expected Environmental Benefits Closely related to the issues of universality (in particular, its limitations in the context of water and diverse catchments), is uncertainty over the expected environmental benefits (e.g. greater water use efficiency, better environmental flows, more sustainable yields). Such benefits take many years to exhibit themselves in a measurable fashion. Further adding doubt is the fact that many environmental water programs are not yet fully implemented. For example, 'monitoring and assessment has often been inadequate, or not well targeted, to determine whether environmental objectives established under water planning arrangements have been met' (NWC 2010). As such, unintended (and deleterious) consequences may go unnoticed in many areas. In particular, the NWC has highlighted that water quality should be better integrated in planning processes. Gaps include understanding the water quality needs of environmental assets, as well as a lack of attention to water quality management and monitoring, particularly for environmental water which has tended to focus on surface water flows and timing rather than achieving groundwater and surface water quality objectives beyond specific salinity components (NWC 2010).

Such concerns were confirmed by interviewee respondents regarding the unintended environmental impacts of environmental flows, including degrading banks, increasing numbers of pest fish, and degrading tree species. As two respondents explained: 'If you asked anybody who lived in the valley or on the river, if they could wave a magic wand and fix the river up and improve the health of the river, the first thing they would do would be to get rid of the carp. You know just running more water down the river does nothing, it actually makes the carp worse because they create more breeding opportunities for them. So you know, if you're serious about environmental outcomes, go and do that stuff' (Interview Q4, Industry Body); and 'in the last two years, with huge prolonged flows for the environment, the bank has washed away two metres...we're getting more trees falling in the river through these high flows, they're just not natural. The environmental water holder said, "Well, we've got this water and that where it's earmarked to go and that's where it's going." Total disregard for local knowledge and all it's doing is killing heaps of trees' (Interview N1, Local Government).

Markets May Not Take Account of Wider Social Impacts A potential shortcoming of water markets is that they are subject solely to economic drivers, and, as such, may discount or ignore social impacts. For example, when water moves from one community to the next this can have an adverse knock-on effect by reducing employment and losing population, with a consequent reduction in local services such as schools, where there is reduction in water for productive use (Alston and Whittenbury 2011). This can also result in the so-called 'Swiss cheese' effect where there are increasingly larger distances between the remaining active irrigators, who, as a result of the other socio-economic impacts, become progressively more marginalised (Goesch et al. 2011). One difficulty here is that there will not always be a correlation between water trade patterns and key socio-economic indicators (e.g. employment in agriculture can fall in all regions regardless of whether the region was a net purchaser or seller of water) suggesting other factors besides water or drought (e.g. commodity prices) can have a greater impact on influencing social and economic change (Kiem 2013). Even so, as recent and ongoing reviews into the MDB demonstrate, there is a growing call for greater consideration of social and economic factors in the distribution of water (Commonwealth of Australia 2014). In this respect, water buybacks by government have been subject to significant criticism, not least because having government as the main buyer may undermine the water market as a vehicle for structural adjustment (Bell and Quiggan 2008). One respondent explained: 'the buy-backs have been absolutely catastrophic. There was no rhyme or reason, it was all ad hoc, it's absolutely destroyed some of the irrigation areas, bearing in mind that irrigation channels that have been there for forever and a day and have got an ecosystem established on them were suddenly shut off, so, they were destructive in what they did. But what they did was they bought here, there and everywhere and sometimes there was a 20 kilometre channel with one person left at the end, or two people left at the end, all the intermediaries had gone, so to get water to them was, you know, came at a big cost…. so it's been catastrophic…and took a heap of employment out of the community and contractors' (Interview N1, Local Government).

While these types of negative economic and social impacts were raised by a minority of interviewees, the majority confirmed a lack of broader social input, including from Indigenous communities, into the cap, rule setting and allocation process via water plans (Jackson et al. 2012). This lack of input was seen to augment the risk of adverse social impacts, as was an absence of purchasing power and resources of marginalised groups to engage in the market. Certainly, initiatives such as the Aboriginal Water Initiative in NSW and the Murray Darling Basin Authority's work with the National Cultural Flows Planning and Research Committee evidence positive steps, yet there is still significant progress needed. As one respondent reflected on the interests that lack voice in planning and the market:

'environmental interests and indigenous interests and the interests of other excluded groups that probably have never been brought into this equation, they are like marginalised communities' (Interview, V0, NGO). Such views were confirmed outside the Basin in recent studies, including a survey of respondents who were asked if water management was 'equitable', such that all parties were being treated equally and fairly under the water management regime. Some 67% of Indigenous respondents disagreed; compared to 40% of non-Indigenous respondents (Nikolakis et al. 2013). 73% of Indigenous respondents also disagreed that water management policies reflected Indigenous interests (of these, 29% strongly disagreed) (Nikolakis et al. 2013).

Markets May Not Account for Windfall Gains In conjunction with the broad policy implementation of water markets operating within the MDB, governments have also engaged in series of investments in both on-farm and off-farm infrastructure, designed to bring about improvements in water efficiency. While this may be a desirable outcome in and of itself, including as a means of addressing problems of over allocation (Cooper et al. 2014), it may also produce perverse outcomes in the operation of water markets. In particular, those private farmers who benefit directly from such public investments will be advantaged in that they may be able to sell any resulting water savings into the water market. In effect, they will have received a windfall gain at the expense of the public purse and other farmers who are not the recipients of equivalent infrastructure investments. This creates a major contradiction in the embrace of water markets by both farmers and politicians, as cost-reflective water prices for farmers are resisted and subsidised infrastructure remains a core component of the policy for dealing with over-allocation (Cooper et al. 2014).

Markets Do Not Operate Equally Across Australia Although it may appear self-evident, it is nevertheless important to acknowledge that water markets do not operate equally in all jurisdictions across Australia. For example, Western Australia and the Northern Territory still have yet to completely unbundle rights from land ownership. Even within the MDB, the capacity to trade is not uniform, with Victoria arguably lagging behind NSW in terms of un-bundled water rights. As noted above, certain groundwater systems also restrict trading. Such heterogeneity raises questions about the equality of treatment under compromised or absent markets, especially when a crucial underlying policy rationale of market trading systems is to avoid the economic distortions and/or inefficiencies of traditional public policy interventions.

To the extent that trading is compromised, through any or all of the above challenges, there are two key consequences. First, it undermines the economic efficiency of the cap and trade approach. Ideally, trading should facilitate the movement of water to those agricultural (or other industry) activities with the greatest economic dividend per unit of water. A compromised market will inevitably fail to deliver on this potential, and leave the water dependent economy poorer as a whole. Second, the lower the level of trading, the more the national water policy approach resembles traditional command and control regulation, with all the limitations often associated with such an approach. That is, the overall cap effectively becomes a conventional performance-based regulatory standard at the level of individual water users. Apart from necessitating robust and comprehensive compliance and enforcement regimes (which is far from evident to date), a reliance on a narrow performance based standard can crowd out other potentially valuable policy options that seek to provide greater participation, ownership, cooperation and flexibility.

4 Discussion

Market based instruments (MBI) arose to prominence in the environmental context in the 1980s and early 1990s as a promising response to complex and dynamic 'second wave' problems like climate change, which had proved to be incapable of resolution by first wave command and control regulatory practice and theory (United Nations Environment Program 2012; Steffen 2015).

Of these second wave wicked problems (Rittel and Webber 1973), few have proved more intractable than non-urban water use, which involves multiple diffuse points of extraction, dispersed over large geographic areas, impacting on numerous locally variable and dynamic surface and groundwater systems, and involve many different water users who are often resistant to government intervention, let alone to regulation, particularly by traditional regulatory techniques (Rubenstein et al. 2010). It of little surprise then that Australia's water reform journey would come to rest on a second wave instrument like a cap and trade scheme to try and achieve more efficient and effective water management across the nation (NWI 2004; Hamstead 2011).

Over the last 20 years, Australian governments have implemented this cap and trade system (Godden and Foerster 2011) and its defining features of property-based water rights (untied to land), setting caps through water allocation plans, and creating a regulated water market across (most of) Australia. Drawing on interviews, surveys and available secondary data, this chapter has identified the main features of Australia's cap and trade scheme, mapping the evolution of the water reforms and markets, and their successes and limitations in practice.

As we have seen, Australia's water markets appear to deliver benefits, at least in the MDB where significant surface water trading is occurring. According to interviewees and available data, this trading facilitates flexible responses to droughts. Market efficiencies also reportedly produce economic benefits for individual farmers and regional communities (albeit selectively), by providing new sources of income, securities for loans and reinvestment on farms. Moreover, these benefits appear likely to intensify, as there are encouraging signs that Australia's water market is increasingly functioning at a more optimal and efficient level, as a result of ongoing facilitation of trade and diffusion of information (Wheeler et al. 2014a, b, c).

Despite these benefits, Australia's water markets, like many MBIs (Howes et al. 1997), have proved challenging for centralised governments that face a variety of practical and contextual difficulties in seeking to develop and rely on the market. These difficulties include a limited regulatory underpinning and weak water accounting accuracy (McKay and Gardner 2013). While governments have taken some recent steps to address these shortcomings, they remain a significant risk. Without effective compliance and enforcement and accounting it will be difficult to optimise the economic, social and environmental outcomes for Australia's non-urban water resources. Indeed, if caps are exceeded due to illegal water extraction (in collaborative water allocation plans), if the various licences, approvals and tradable water rights (essential to efficient markets) are not adhered to, and if stakeholders lack confidence that there is an equitable sharing of water resources (particularly in periods of drought), then the entire edifice of the market can be undermined (Holley and Sinclair 2013a; NSW Office of Water 2015b; see also Matthews 2017; Murray-Darling Basin Authority 2017).

The findings raise further concerns about barriers to the effective operation of the market. This includes the potential challenge of (non) universality of environmental impacts and access to the traded property right, particularly in the context of groundwater trades. There is also significant uncertainty over the expected environmental benefits of markets, in part because of limited information and some concerns over the emphasis on flows at the expense of related environmental benefits (e.g. water quality). The water market confronts other significant limitations, including distorting cost reflective water prices through the use of subsidised infrastructure and the ongoing challenge of using the market to account for triple bottom line goals. Perhaps most fundamentally, markets largely remain a creature of the MDB, with functioning markets compromised or absent in many other parts of Australia. Certainly, the MDB is far and away Australia's primary irrigation area. However, the lack of trading in other areas (particularly in the proposed development area of Northern Australia) may be a significant problem for the overall efficiency of the market.

Stepping back from the specific achievements and challenges, the chapter asks: are MBIs a 'good' regulatory instrument for managing Australian water or are they a failed experiment? One view is that markets are indeed good and far from failing. While it is still early days in the development of markets in Australia, it is clear that they exhibit a number of benefits for efficiency, dealing with drought and delivering economic outcomes. Further, while a number of challenges have been canvassed, many of these are 'fixable' and thus are not fatal to the future success of the market. Problems such as regulatory underpinning, improving accuracy of information, subsidies and complete unbundling of rights in states such as Western Australia and Northern Territory to extend the operation of the market are issues that can conceivably be addressed through future policy and law reform and public investment. Indeed, problems such as enforcement and accounting are largely a function of governments historically focusing on the higher-level NWI policy goals rather than addressing implementation on-ground. As discussed elsewhere, there are numerous steps that governments can take to address most of these challenges, including through better technology, and in some (but far from all) cases they are beginning to do so in ways that are likely to improve market operation (Holley and Sinclair 2013a, b; Matthews 2017; Murray Darling Basin Authority 2017).

Despite such a positive outlook, and even if these issues could be overcome, there are a number of reasons to argue that markets may be suffering from more fundamental limitations, which are less readily correctable. One such limitation is the need to deal with (non) universality of impact and source, particularly for groundwater. Although predicated on free trade across geographical regions, this is not always possible or necessarily desirable given the different environmental circumstances of individual catchments and aquifers. For example, in the case of relatively discrete and autonomous groundwater systems, even given arguments for increasing trade outside of discrete groundwater aquifers, it may be both undesirable to trade water in or out, and virtually physically impossible. To the extent that such situations prevail, the policy of water trading may be compromised.

Further, accounting for wider social impacts of markets remains a significant challenge. It is true that new rights can be introduced (e.g. cultural flows), deeper engagement in water planning pursued, and other non-water policy instruments employed to counterbalance the market's social and economic impacts. However, it is a fundamental conundrum that while a strength of a cap and trade scheme is that it can facilitate flexible adoption of targeted behavioural change, this can also be a weakness as the market instrument leads to social, environmental and economic changes that are unplanned or not in line with the intended strategic direction (Kiem 2013). A related concern arises regarding the environmental benefits of markets, particularly around quality. While there is growing national focus on water quality, including under the Basin Plan and Water Resource Plans (Murray-Darling Basin Authority 2014a; NWC 2010; Carmody 2013), the inclusion of various contextual, interacting and multiple quality variables may prove to be particularly challenging. Of course, markets can be established to deal with quality issues (such as the Hunter Valley Salinity Trading Scheme) (Krogh et al. 2013). However, integrating these concerns into existing water markets could pose considerable policy and implementation obstacles, not least ensuring the various environmental resources and concerns are fully fungible.

What, then, do we make of these more fundamental challenges? One could argue that these more confounding problems outweigh the benefits, and are sufficiently grave as to abandon market based approaches along with other 'failed' policy experiments. However, such a move would need to overlook the extensive sunk costs already invested in establishing Australia's cap and trade schemes, as well as the likely political, revenue and public compensation impacts of such a decision. Further, abandoning the market altogether overlooks the relative immature nature of the market reforms themselves. As we saw in Chap. 1, water reforms have been a long and arduous journey, that have often involved trial and error and changes in policy that have meant the market system has been slow to evolve. There are also signs that this evolution is continuing, as evidenced by the Productivity Commission's inquiry into national water reform (as well as various other recent inquiries, Matthews 2017; Murray-Darling Basin Authority 2017; Parliament of

Australia 2012). Given this evolving context and relative youth of the market, it is plausible that more time is needed to truly test these more fundamental concerns and see if responses can be developed within the market system.

A more pragmatic response, and the one that we favour, is to address the areas of 'failed' policy by adopting alternative but complementary policies. For example, there may be other policy instruments that can plug the gaps in market operation for groundwater by providing flexibility and decentralised management that integrates quality and quantity and draws on local knowledge and expertise to account for social, economic and environmental impacts. Some of these options have been discussed elsewhere, including the model of Audited Self-Management (ASM) (Holley and Sinclair 2014). We have argued elsewhere that the ASM model, and its use of prescription, performance and process standards, offers an innovative way to re-engage with the agricultural community, while harnessing the benefits of flexibility (inherent to markets) and securing efficient and effective compliance and enforcement. Although ASM may take on a variety of forms, we have argued there are at least six core elements that are fundamental to its successful operation in the context of the regulation of water extraction, namely: (i) participating water users are willing and able to form a legal entity or collective capable of managing the ASM program; (ii) this entity is allocated, and therefore has legal responsibility for, a collective water right (in effect, a bubble licence) covering all the ASM participating members; (iii) within the collective licence, participants are able to determine individual annual extractions as they see fit (effectively trading within the bubble licence); (iv) all members have in place accurate metering that uses telemetry to generate real-time water extraction data; (v) the extraction data is made available to all participants (disaggregated to the individual level) and the government regulator (aggregated to the collective level); and (vi) the ASM program has in place appropriate integrity (e.g. auditor) and enforcement mechanisms to ensure compliance, including, if necessary, the capacity to draw on the support of the external government regulator (Holley and Sinclair 2014).

5 Conclusion

Australia's response to the challenge of achieving the sustainable management of water has been the introduction of a national cap and trade scheme. As we have seen, this approach led to more flexible responses to past and future droughts; efficiencies that contribute to economic and environmental benefits; and increasing trade and market functionality. However, it is far from clear that the cap and trade scheme is functioning at the peak of its powers, not least because of a lack of robust regulatory underpinning; limited accuracy in water accounting; challenges in addressing universality of impact and source; queries over environmental benefits; lack of accounting for wider social impacts; windfall gains; and limited operation across Australia. On balance, these successes and limits suggest water markets are not a failed experiment, but that without substantial reform they will fall short in a host of key areas.

We have recommended some 'low hanging fruit' policy options to address key inadequacies of the current cap and trade approach, including enhancing regulatory underpinning, improving accuracy of information, and complete unbundling of rights in states such as Western Australia and Northern Territory to extend the operation of the market. A failure to address these issues risks compromising the overall effectiveness of Australia's national cap and trade reforms.

We have also argued that there are crucial areas, in particular groundwater, where the market has major limitations, and complementary tools such as Audited Self-Management may be needed to secure Australia's future water sustainability. Pursuing such an instrument, however, may be difficult. The nature of economic instruments, like Australia's cap and trade scheme, is that they tend to 'crowd out' other regulatory approaches that might otherwise be employed to deliver flexibility to regulated entities (Gunningham et al. 1998).

Given this inherent exclusivity of the cap and trade approach to water policy reform, it may be that water regulators are effectively 'locked in' to this policy. This suggests that further inquiry by legal and governance scholars is needed to better examine the issue in more detail and to explore the scope for the existing water policy frameworks in Australia to accommodate a greater mix of complementary 'regulatory alternatives', particularly where water trading is difficult, limited and/or has unintended and unwelcome outcomes. One area where such exploration for alternatives could occur is northern Australia. Should plans proceed to develop northern Australia, it may be that a greater range of non-market alternatives can and should be explored, acknowledging the lack of well-developed markets or trading and that any potential trading across ecologically distinct geographic areas (northern and southern Australia) would appear unlikely.

Beyond this, there are at least three other legal and policy issues that have been beyond the scope of this chapter to address, but are closely related and warrant further attention. First is the mechanics of trades. The costs, administration, transparency, length of time, licensing, unintended consequences, water user decision-making and irrational behaviour associated with water trade and transfers have the potential to impact on the overall effectiveness of the cap and trade approach. Each warrants much more detailed investigation so as to further unpack and analyse the overall success of the cap and trade instrument. Second is the role of the Commonwealth Government (and other state water holders) as active participants in the water market. Examining their impacts on the behaviour of the market and individual water users is an important area for further exploration. Third and finally is the capacity of the existing policy paradigm to accommodate a large and growing entrant in the water space, namely mining and unconventional gas, which, to date, has only been partially accommodated within the water policy framework. Each of these additional issues requires detailed research and analysis in considering the future direction of Australia's national water policy approach.

To conclude, the analysis above sheds light on the implementation of water markets, an issue that is often overlooked in the context of water law and policy. It reveals progress made, particularly in MDB surface water trading, while highlighting past and emerging issues that impact on effective operation of Australia's cap and trade schemes. While policy steps can be directed to enhance implementation in many areas, there is little doubt that there are still fundamental challenges remaining for Australia's water markets and their future governance.

References

Alston, M., & Whittenbury, K. (2011). Climate change and water policy in Australia's irrigation areas: A lost opportunity for a partnership model of governance. *Environmental Politics, 20,* 899–917.
Arnason, R. (2012). Property rights in fisheries: How much can individual transferable quotas accomplish? *Review of Environmental Economics and Policy, 6,* 217–236.
Australian Bureau of Statistics. (2014). 5610.0–Water Account, Australia (2012–2013). Retrieved June 18, 2017, from http://www.abs.gov.au/AUSSTATS/abs@.nsf/Lookup/4610.0Main+Features12012-13?OpenDocument.
Basin Plan 2012 (Cth).
Barr, N., & Cary, J. (2000). *Influencing improved natural resource management on farms: A guide to factors influencing the adoption of sustainable natural resource management practices.* Kingston: Bureau of Rural Sciences.
Bell, S., & Quiggan, J. (2008). The limits of markets: The politics of water management in rural Australia. *Environmental Politics, 17,* 712–729.
Bowmer, K. (2014). Water resources in Australia: Deliberation on options for protection and management. *Australian Journal of Environmental Management, 21,* 228–240.
Brooks, R., & Harris, E. (2014). Price leadership and information transmission in Australian water allocation markets. *Agricultural Water Management, 145,* 83–91.
Brooks, R., Harris, E., & Joymungul, Y. (2013). Price clustering in Australian water markets. *Applied Economics, 45,* 677–685.
Burdack, D., Biewald, A., & Lotze-Campen, H. (2014). Cap-and-trade of water rights: A sustainable way out of Australia's rural water problems? *GAIA Ecological Perspectives for Science and Society, 23,* 318–326.
Burris, S., Drahos, P., & Shearing, C. (2005). Nodal governance. *Australian Journal of Legal Philosophy, 30,* 30–58.
Carmody, E. (2013). The silence of the plan: Will the convention on biological diversity and the Ramsar Convention be implemented in the Murray-Darling Basin? *Environmental Planning and Law Journal, 30,* 56–73.
Castree, N. (2010). Neoliberalism and the biophysical environment 2: Theorising the neoliberalisation of nature. *Geography Compass, 4,* 1734–1746.
Coase, R. (1960). The problem of social cost. *Journal of Law and Economics, 3,* 1–44.
Commonwealth of Australia. (2010). About Commonwealth environmental water. Retrieved June 18, 2017, from https://www.environment.gov.au/water/cewo/about-commonwealth-environmental-water.
Commonwealth of Australia. (2014). *Report of the independent review of the Water Act 2007.* Canberra: Australian Government.
Commonwealth of Australia. (2015). *Our North, Our Future: White paper on developing Northern Australia.* Canberra: Commonwealth of Australia.
Commonwealth of Australia. (2016). *About commonwealth environmental water.* Retrieved June 18, 2017, from https://www.environment.gov.au/water/cewo/about-commonwealth-environmental-water.
Connell, D., & Quentin Grafton, R. (Eds.). (2011). *Basin futures.* Acton: ANU E Press.
Convention on the Law of Non-Navigational Uses of International Watercourses, opened for signature 21 May 1997, UN Doc A/RES/51/229.

Cooper, B., Crase, L., & Pawsey, N. (2014). Best practice pricing principles and the politics of water pricing. *Agricultural Water Management, 145,* 92–97.
Council of Australian Governments (COAG). (1994). Council of Australian Governments' Communiqué, 25 February 1994, Hobart. http://ncp.ncc.gov.au/docs/Council%20of%20Australian%20Governments%27%20Communique%20-%2025%20February%201994.pdf. Accessed 10 Oct 2017.
Council of Australian Governments. (2004). Intergovernmental agreement on a National Water Initiative signed 25 June 2004. Retrieved June 18, 2017, from http://www.nwc.gov.au/__data/assets/pdf_file/0008/24749/Intergovernmental-Agreement-on-a-national-water-initiative.pdf.
Cresswell, J., & Clark, V. P. (2011). *Designing and conducting mixed methods research*. Los Angeles: Sage.
Curnow, P., & Fitzgerald, L. (2006). Biobanking in New South Wales: Legal issues in design and implementation of a biodiversity offsets and banking scheme. *Environmental and Planning Law Journal, 23,* 298–308.
Cutting, R., & Cahoon, L. (2005). Thinking outside the box: Property rights as a key to environmental protection. *Pace Environmental Law Review, 22,* 55–90.
Department of the Environment, Water Heritage and the Arts. (2009). *National framework for non-urban water metering regulatory impact statement*. Canberra: Commonwealth of Australia.
Department of Sustainability Environment Water Population and Communities. (2012). *National framework for compliance and enforcement systems for water resources management*. Canberra: Commonwealth of Australia.
Dillman, D. (2007). *Mail and internet surveys: The tailored design method*. New York: Wiley.
Gardner, A., Barltett, R., & Gray, J. (2009). *Water resources law*. Chatswood: LexisNexis Butterworths.
Garrick, D., Whitten, S., & Coggan, A. (2013). Understanding the evolution and performance of water markets and allocation policy: A transaction costs analysis framework. *Ecological Economics, 88,* 195–205.
Godden, L. (2008). Property in urban water: Private rights and public governance. In P. Troy (Ed.), *Troubled waters: Confronting the water crisis in Australia's cities* (pp. 157–185). Acton: ANU E Press.
Godden, L., & Foerster, A. (2011). Introduction: Institutional transitions and water law governance. *Journal of Water Law, 22,* 53–57.
Godden, L., & Ison, R. (2010). From water supply to water governance. In M. Davis & M. Lyons (Eds.), *More than luck: Ideas Australia needs now* (pp. 177–184). Sydney: Centre for Policy Development.
Goesch, T., et al. (2011). *The economic and social effects of the Murray-Darling Basin Plan: Recent research and next steps*. Paper presented at the 41st Australian Bureau of Agricultural and Resource Economics and Sciences (ABARE) Outlook Conference, 1–2 March 2011.
Goulder, L., & Schein, A. (2013). Carbon taxes versus cap and trade: A critical review. *Climate Change Economics, 4,* 1–28.
Gómez-Baggethun, E., & Muradian, R. (2015). In markets we trust? Setting the boundaries of market-based instruments in ecosystem services governance. *Ecological Economics, 117,* 217–224.
Grafton, Q. R., & Horne, J. (2014). Water markets in the Murray-Darling Basin. *Agricultural Water Management, 145,* 61–71.
Grafton, Q. R., & Peterson, D. (2007). Water trading and pricing. In K. Hussey & S. Dovers (Eds.), *Managing water for Australia: The social and institutional challenges* (pp. 73–84). Collingwood: CSIRO Publishing.
Gross, C. (2011). Why justice is important. In D. Connell & R. Q. Grafton (Eds.), *Basin futures* (149–162). Acton: ANU E Press.
Gunningham, N., Holley, C., & Shearing, C. (2011). *The new environmental governance*. New York: Earthscan.

Gunningham, N., Grabosky, P., & Sinclair, D. (1998). *Smart regulation: Designing environmental policy*. New York: Oxford University Press.

Gunningham, N., & Sinclair, D. (2004). Non-point pollution, voluntarism and policy failure: Lessons from the swan-canning. *Environmental and Planning Law Journal, 21,* 93–104.

Hamblin, A. (2009). Policy directions for agricultural land use in Australia and other post-industrial economies. *Land Use Policy, 26,* 1195–1204.

Hamstead, M. (2011). Improving water planning processes: Priorities for the next five years. In D. Connell & R. Q. Grafton (Eds.), *Basin futures* (pp. 339–350). Canberra: ANU EPress.

Hayek, F. (1945). The use of knowledge in society. *American Economic Review, 35,* 519–530.

Holley, C., & Sinclair, D. (2013a). Compliance and enforcement of water licences in NSW: Limitations in law, policy and institutions. *Australasian Journal of National Resources, 15,* 149–189.

Holley, C., & Sinclair, D. (2013b). Non-urban water metering policy: Water users' views on metering and metering upgrades in NSW. *Australasian Journal of Natural Resources, 16,* 101–131.

Holley, C., & Sinclair, D. (2014). A new water policy option for Australia? Collaborative water governance, compliance and enforcement and audited self-management. *Australasian Journal of National Resources Law and Policy, 17,* 189–216.

Holley, C., & Sinclair, D. (2015). *Water Extraction in NSW: Stakeholder Views and Experience of Compliance and Enforcement* (UNSW Connected Waters Initiative Research Centre, Feb 2015), http://www.connectedwaters.unsw.edu.au/sites/all/files/Water-extraction-in-NSW-stakeholder-views-of-compliance-and-enforcement-survey-report.pdf.

Howes, R., Skea, J., & Whelan, B. (1997). *Clean and competitive: Motivating environmental performance in industry*. New York: Earthscan.

ICM Agriculture Pty Ltd & Ors v The Commonwealth Of Australia & Ors (2009) HCA 51.

Jackson, S., Tan, P.-L., Mooney, C., Hoverman, S., & White, I. (2012). Principles and guidelines for good practice in Indigenous engagement in water planning. *Journal of Hydrology, 474,* 57–65.

Karkkainen, B., Fung, A., & Sabel, C. (2000). After backyard environmentalism. *American Behavioural Scientist, 44,* 692–711.

Kiem, A. (2013). Drought and water policy in Australia: Challenges for the future illustrated by the issues associated with water trading and climate change adaptation in the Murray-Darling Basin. *Global Environmental Change, 23,* 1615–1626.

Krogh, M., Dorani, F., Foulsham, E., McSorley, A., & Hoey, D. (2013). *Hunter catchment salinity assessment*. Sydney: Environmental Protection Authority.

Lavau, S. (2013). Going with the flow: Sustainable water management as ontological cleaving. *Environmental Planning D Society and Space, 31,* 416–433.

Liverman, D. (2004). Who governs, at what scale and at what price? Geography, environmental governance and the commodification of nature. *Annals of the Association of American Geographers, 94,* 734–738.

Matthews, K. (2017). Independent investigation into NSW water management and compliance—final report. Retrieved December 3, 2017, from https://www.industry.nsw.gov.au/__data/assets/pdf_file/0019/131905/Matthews-final-report-NSW-water-management-and-compliance.pdf.

McKay, C., & Gardner, A. (2013). Water accounting information and confidentiality in Australia. *Federal Law Review, 41,* 127–162.

McKibbin, W., & Wilcoxen, P. (2002). The role of economics in climate change policy. *The Journal of Economic Perspectives, 16,* 107–129.

Millar, I. (2005). *Testing the waters: Legal challenges to water sharing plans in NSW*. Paper presented at the 'Water Law in Western Australia' Conference 8 July 2005.

Murray-Darling Basin Authority. (2014a). *Handbook for practitioners: Water resource plan requirements*. Canberra: Murray-Darling Basin Authority.

Murray-Darling Basin Authority. (2014b). Murray-Darling Basin Authority annual report 2013–2014. Retrieved December 1, 2017, from https://www.mdba.gov.au/sites/default/files/pubs/Annual-Report-2013-14-MDBA.pdf.

Murray-Darling Basin Authority. (2017). The Murray-Darling Basin water compliance review. Retrieved December 1, 2017, from https://www.mdba.gov.au/sites/default/files/pubs/MDB-Compliance-Review-Final-Report.pdf.

National Water Commission (Abolition) Act 2015 (Cth).

National Water Commission. (2011). *Water markets in Australia: A short history*. Canberra: National Water Commission.

National Water Commission. (2010). The impacts of water trading in the southern Murray–Darling Basin: An economic, social and environmental assessment. Retrieved December 1, 2017, from http://webarchive.nla.gov.au/gov/20160615080824/; http://archive.nwc.gov.au/library/topic/rural/impacts-of-water-trading-in-the-southern-murray-darling-basin.

National Water Commission. (2013). Australian water markets: Trends and drivers 2007–08 to 2012–13. Retrieved March 27, 2017, from http://webarchive.nla.gov.au/gov/20140604112129/; http://www.nwc.gov.au/publications/topic/water-industry/trends-and-drivers-2012-13/section-2.

National Water Commission. (2014a). *Australia's water blueprint: National reform assessment 2014–Part one*. Canberra: National Water Commission.

National Water Commission. (2014b). *Water for mining and unconventional gas under the National Water Initiative*. Canberra: National Water Commission.

Nikolakis, W., Quentin Grafton, R., & To, H. (2013). Indigenous values and water markets: Survey insights from northern Australia. *Journal of Hydrology, 500,* 12–20.

NSW Environment Defenders Office. (2010). Water management–Submissions. Retrieved June 18, 2017, from http://www.edonsw.org.au/water_management_policy.

NSW Irrigators' Council. Submissions. (2017). Retrieved June 18, 2017, from http://www.nswic.org.au/news_submissions.shtml.

NSW Office of Water. (2010). *NSW sustaining the Basin Program: Metering project socio economic assessment*. Sydney: NSW Government.

NSW Office of Water. (2015a). *NSW water take measurement Strategy–Water take measurement in NSW: A way forward*. Discussion Paper, NSW Department of Primary Industries, July 2015.

NSW Office of Water. (2015b). Compliance policy. NSW Department of Primary Industries Office of Water. Retrieved June 18, 2017, from http://www.water.nsw.gov.au/__data/assets/pdf_file/0005/560192/compliance_policy_2015.pdf.

Parliament of Australia. (2012). The Murray Darling Basin Plan–Role of the Committee. Retrieved June 18, 2017, from http://www.aph.gov.au/Parliamentary_Business/Committees/Senate/Murray_Darling_Basin_Plan/Role_of_the_Committee.

Parsons, R., Lacey, J., & Moffatt, K. (2014). Maintaining legitimacy of a contested practice: How the minerals industry understands its 'social licence to operate'. *Resource Policy, 41,* 83–90.

Raft, S., & Hillis, G. (2010). *NSW sustaining the Basin Program: NSW metering project business case*. Sydney: NSW Office of Water.

Rittel, H., & Webber, M. (1973). Dilemmas in a general theory of planning. *Policy Sciences, 4,* 155–169.

Roma, A. (2006). Energy, money and pollution. *Ecological Economics, 56,* 534–545.

Rubenstein, N., Wallis, P., Ison, R., & Godden, L. (2010). *Strengthening water governance in Australia*. Briefing Paper National Climate Change Adaptation Research Facility.

Shearing, C. (1993). A constitutive conception of regulation. In P. Grabosky & J. Braithwaite (Eds.), *Business regulation and Australia's future* (pp. 67–80). Canberra: Australian Institute of Criminology.

Skurray, J., Pandit, R., & Pannell, D. (2013). Institutional impediments to groundwater trading: The case of the Gnangara groundwater system of Western Australia. *Journal of Environmental Planning and Management, 56,* 1046–1072.

Stavins, R. (1998). What can we learn from the grand policy experiment? Lessons from SO2 Allowance Trading. *Journal of Economic Perspectives, 12,* 69–88.

Steffen, Will. (2015). *Thirsty country: Climate change and drought in Australia*. Climate Council of Australia.

Tan, P.-L., Bowmer, K., & Baldwin, C. (2012). Continued challenges in the policy and legal framework for collaborative water planning. *Journal of Hydrology, 474,* 84–91.

United Nations Environment Program. (2012). Global environment outlook 5. Retrieved June 18, 2017, from http://www.unep.org/geo/pdfs/geo5/GEO5-Global_PR_EN.pdf.

Water Act 2007 (Cth).

Water Charge (Termination Fees) Rules 2009 (Cth).

Water Market Rules 2009 (Cth).

Wei, Y., Langford, J., Willett, I. R., Barlow, S., & Lyle, C. (2011). Is irrigated agriculture in the Murray Darling Basin well prepared to deal with reductions in water availability? *Global Environmental Change, 21,* 906–916.

Wentworth Group of Concerned Scientists. (2017). Review of water reform in the Murray-Darling Basin. Retrieved December 1, 2017, from http://wentworthgroup.org/wp-content/uploads/2017/11/Wentworth-Group-Review-of-water-reform-in-MDB-Nov-2017.pdf.

Wheeler, S., Zuo, A., & Hughes, N. (2014a). The impact of water ownership and water market trade strategy on Australian irrigators' farm viability. *Agricultural Systems, 129,* 81–92.

Wheeler, S., Loch, A., Zuo, A., & Bjornlund, H. (2014b). Reviewing the adoption and impact of water markets in the Murray-Darling Basin, Australia. *Journal of Hydrology, 518,* 28–41.

Wheeler, S., Zuo, A., & Bjornlund, H. (2014c). Investigating the delayed on-farm consequences of selling water entitlements in the Murray-Darling Basin. *Agriculture Water Management, 145,* 72–82.

Whitehead, I. (2014). Better protection or pure politics? Evaluating the 'water trigger' amendment to the EPBC Act. *Australian Environmental Law Digest, 1,* 23–36.

World Meteorological Organization, 'Development Issues for the 21st Century: the Dublin Statement and Report of the Conference' (Report of the International Conference on Water and Development, Dublin Ireland, 26–31 January 1992).

Part III
Collaboration and Participation—Litigation, Coordination and Water Rights

Public Participation in Water Resources Management in Australia: Procedure and Possibilities

Bruce Lindsay

Abstract Broad public participation in water resources management is necessitated by strong public good qualities. Nevertheless, it is only a relatively recent phenomenon in Australia. Government and commercial/industrial hegemony over water resources is predominant. The procedural vehicles for public participation in Australian water governance have been consultation and litigation. Both exhibit limitations. Consultation represents an administrative and discretionary approach to involvement; flexible, widely used, but lacking in policy direction and sophistication. This chapter discusses consultation as a method of participation generally as well as in the specific circumstances of Aboriginal engagement. Adjudicative approaches benefit from an established model of judicial practice, including reasoned outcomes and methods of inquiry, but are infrequently used in Australia, whether in civil litigation or in administrative decision-making. Each procedural framework presents possible pathways to widen community engagement in management of water as a crucial public resource. Lessons from the Australia experience of public participation in water governance are identified across both consultative and adjudicative modes.

1 Introduction

The management of water resources is a task of great public interest and concern. Water is a resource with strong 'public good' qualities, characterised by high degrees of variability including scarcity, and subject to complex and contested uses and values. By extension, public involvement represents an important factor in the water management task, whether to inform that task, mediate disputes and outcomes, or legitimate uses and practices.

Water law and policy in Australia is currently guided by the 2004 *Intergovernmental Agreement on the National Water Initiative* (COAG 2004:

B. Lindsay (✉)
Environmental Justice Australia, Melbourne, VIC, Australia
e-mail: bruce.lindsay@envirojustice.org.au

'NWI'), a political agreement between Commonwealth, state and territory governments. Profound changes in directions to water management have included institution or development of water markets, establishment of extraction limits, frameworks and tools of environmental management, improved information management and scrutiny, and recognition of Aboriginal interests in water systems (see e.g. in this volume, Holley and Sinclair; Owens; McPherson et al.). These developments are most advanced in the Murray-Darling Basin in south-eastern Australia, the most productive agricultural region in the country and a site of significant tensions between economic and environmental water uses. Discrete and highly evolved governance arrangements now operate in that Basin. Recent assessments have identified the waning of water reform in the public imagination (National Water Commission 2014), despite the need for water reform to be an ongoing project (National Water Commission 2011, 2014).

Public participation in the management of water resources has not historically been a prominent feature of governance, other than participation as 'users' and through representative organisations such as irrigator bodies. The 'public' character of water management has been associated with state ownership and control of water and strong governmental direction. Constrained models of public participation are reflected in the NWI (e.g. COAG 2004, cl 93).

2 Water Management in Australia

The origins of water law in Australia lie in the reception of the English common law, based on the riparian doctrine. By the late nineteenth and early twentieth century, Australian colonial (then state) governments, faced with distinct climatic and hydrological conditions and developmental needs uniformly adopted a different model: public ownership and control of water resources. Access to, use of and interference with those resources could then be allocated to public and private actors, through systems of licences, entitlements and allocations. For the most part, purposes toward which water management was directed were urban water supply, irrigated agriculture, and other industrial activities such as mining and power supply. Water was viewed as an important resource within industrial and developmental supply chains, governed under state direction. Common law rights to water were abolished in most jurisdictions (Stoeckel et al. 2012, [2.45]), although some residual forms of riparian rights do remain (see e.g. Stoeckel et al. 2012, [4.250]).

Within Australia's federal constitutional system, power to legislate in respect of water remains primarily a responsibility of state (and territory) governments. Each sub-national jurisdiction has water resources legislation, as well as interacting catchment management, land-use planning and environmental legislation. Nevertheless, since the 1990s, Commonwealth involvement in water law and policy has become significant, a fact arising from overallocation of water rights, crises in water-dependent ecosystems, climatic changes, and the inter-jurisdictional character of water basin management. Conflicts and deterioration of environmental conditions in the Murray-Darling Basin, encompassing parts of Queensland, New South Wales (NSW), Victoria, the Australian Capital Territory (ACT) and South Australia, was the most glaring of these issues. The Commonwealth entered into the field of water legislation with passage of the *Water Act 2007* (Cth), through exercise of powers to legislate with respect to external affairs (environmental treaties) and through limited referral of state legislative powers to the Commonwealth. The Commonwealth law is also restricted to management of the Murray-Darling Basin.

Water law and policy also features intergovernmental cooperation. A Council of Australian Governments (COAG) *Water Reform Framework* was adopted in 1994. A *National Action Plan for Salinity and Water Quality* was endorsed by COAG in 2000, and the NWI adopted in 2004. These documents have established the main policy bases of water reform.

A particular model of water governance has evolved in Australia, albeit implemented in only one geographic part of the continent, responding to the unique hydroecological factors, competing water needs and uses, and the federal system of government. It is a form of 'nested' and scaled governance, in which the Commonwealth has established a detailed framework for sustainability and planning at an overarching (Basin) scale and the states assume a subsidiary planning and management function, maintaining responsibility for the system of statutory water rights and allocations. Moreover, the approach exhibits characteristics of integrated management of water resources, for economic, social and environmental purposes.

3 Public Participation in Water Management: Modes of Procedure

The manner and ordering of public participation in water governance will be influenced by procedure. Procedure concerns processes by which decisions are made and governance occurs. Forms and purposes of procedure may vary widely, but it is typically a key variable in participation and thereby in 'fair treatment' (Galligan 1996). Fair treatment is crucial to sentiments of inclusion, justice, legitimacy, know-how, and reasonableness.

Three elements of participation are reflected in the NWI framework. The first of these is participation in water management as economic actors holding access rights to water. This is 'participation as user', through market-based procedures, and title-based, usufructuary rights. Such rights may have the character of property (Gray and Lee 2016). Participation of this type occurs through commercial use of rights or allocations, such as for agricultural purposes, and it is generally of a private, rather than public interest, character, although the public-private distinction can be difficult to discern where private interests function within a clearly defined collective structure (see Holley and Sinclair 2014). For current purposes, the focus is not on market-based modes of participation. One qualification should be made, however. There is a nascent trend to use private holdings of water rights for public interest purposes, notably environmental purposes, such as the restoration of wetlands, waterways or aquatic habitat (see e.g. Murray-Darling Wetlands Working Group discussed in Owens, this volume). This 'water trust' approach may prove to be a significant innovation in nongovernmental involvement in water management in Australia for public interest outcomes (Owens 2016). This model of participation in water resources management, while holding great potential, is presently limited.

A second dimension of public participation foreshadowed in the NWI is consultation. This approach is important at the water planning stage, but also in implementation of plans or policies and in monitoring or evaluation of actions and decisions (NWI, cl 93–97). National water planning principles include requirements to ensure consultation (NWI, Sch E [6(i)]). Consultation has been the subject of scrutiny and review, and this has disclosed limits and shortcomings, as well as opportunities for improvement (Holley and Sinclair 2013; Tan et al. 2012a, b).

Third, Aboriginal participation in water governance is a discrete category. The NWI provides for inclusion of Aboriginal representation in water planning processes and accommodation of native title outcomes impacting on water (COAG 2004, cl 52–54). The *Basin Plan 2012* provides for specific Aboriginal ('Indigenous') consultation requirements (*Basin Plan 2012* (Cth), Chap. 10, Part 14). Notwithstanding these special measures, Aboriginal participation in water management has met, at best, with partial success and a high degree of variability across Australia (Jackson et al. 2009, 2012).

A fourth category framing public participation not expressly identified in the NWI is the availability and exercise of rights to contest water uses and practices by way of litigation in courts, tribunals or quasi-judicial panels. These bodies provide scope for involvement through adjudicative procedure.

Distinct modes of procedure function through these approaches to participation. Participation occurs by way of access to, and exercise of, rights under entitlements in market systems. Consultation involves procedures of public administration. Litigation or adjudicative procedure requires access to and use of judicial procedure and seek the aid of court or administrative bodies to affect decisions.

4 Modes of Public Participation: Consultation

4.1 Consultation Generally

Consultation has application to water planning (Stoeckel et al. 2012, p. 33; Gardner, et al. (2009), p. 273; Tan 2008), as well as management practice more broadly (e.g. monitoring, implementation, review). The concept of consultation has received extensive consideration in literature on models and hierarchies of public participation (see e.g. Arnstein 1969; Cornwall 2008; Reed 2008; Fung 2006; Bruns 2003). Consultation is situated between mere information, at one end of a spectrum ('ladder', Arnstein 1969) of participation, and co-management, citizen power or delegation of decision-making, at the other end. Judicial consideration of 'consultation' (e.g. *TVW Enterprises Ltd v Duffy (No 3)* (1985) 8 FCR 93; *Communications Electrical Electronic Energy Information Postal Plumbing and Allied Services Union of Australia v QR Ltd* [2010] FCA 591; *Yallingup Residents Association (Inc) v State Administrative Tribunal* [2006] WASC 162; *Vanmeld Pty Ltd v Fairfield City Council* [1999] NSWCA 6) reinforces a sense that consultation is not a term of art. It has been said that consultation 'has about it an inherent flexibility' (*Communications Electrical Electronic Energy Information Postal Plumbing and Allied Services Union of Australia v QR Ltd* (2010) [2010] FCA 591, [44]). Consultative procedures often function as informative or advisory techniques, supply no essential right of veto, and do not of themselves lead to binding outcomes (*Communications Electrical Electronic Energy Information Postal Plumbing and Allied Services Union of Australia v QR Ltd* (2010) [2010] FCA 591, [44]). As distinct from adjudicative procedure, consultation ordinarily does not require or imply a hearing procedure, application of authoritative standards, or models of reasoned outcomes. Consultative modes align with what Galligan refers to as 'proceduralism' (Galligan 1996, pp. 29–31, 47). As distinct from negotiation, consultation does not establish the participants formally as actors within a framework of bargaining, intending to lead to agreement. Whether, in substance, consultations affect dealings *analogous* to negotiation and bargaining is likely to be a matter of rules, interests, practices, and actual power dynamics. Greater equality of bargaining power between governmental and other actors, including via legal tools to disperse and exercise power (such as title, resource allocations, recognised interests, or enforceable rights), tends to shift procedure in practice toward partnership, co-management (Bruns 2003), or delegation of decision-making (Holley and Sinclair 2014). In an ideal sense, consultation is an administrative tool, attached to the exercise of discretion, to inform its exercise and contribute to its legitimacy.

The 'flexible' nature of consultation infers a diversity of forms. Duties or powers to consult abound in water legislation (Tan 2008). For example, the *Water Act 1989* (Vic) contains more than 30 circumstances in which a form of public consultation can or must occur. Depending on the circumstances, consultation may be with rights holders, with statutory agencies, or with the general public. It may be by notice and comment, invited submissions, targeted consultation, or the establishment of consultation bodies. Other examples of formalised consultation bodies include the Basin Community Committee constituted under the *Water Act 2007* (Cth), s 202. 'Notice and comment' procedures might be said to represent a minimal level of procedural fairness and good faith (see e.g. *Communications Electrical Electronic Energy Information Postal Plumbing and Allied Services Union of Australia v QR Ltd* (2010) 198 IR 382; [2010] FCA 591, [44]; *TVW Enterprises Ltd v Duffy (No 3)* (1985) 8 FCR 93, 101). Stakeholder or community bodies are commonplace 'engagement' mechanisms, albeit often 'formulaic' in practice or effect (Tan 2008, p. 131).

Consultation can occur through more elaborated and deliberative techniques of participation, arising from legal requirements practices outside of procedures with legal status. Deliberative approaches to public participation have become more commonplace in water governance, although often operating absent a coherent policy framework or legal support structure. Deliberative techniques are prompted by the need to deal with challenges in the complexity of water management, problems with legitimacy and representativeness, and shortcomings in historic managerialist, 'top-down' and regulatory approaches to governance (Tan et al. 2012a, b; Holley 2010; Holley and Sinclair 2013). It has been argued that deliberative approaches, aimed (in principle, at least) at dispersal of decision-making power, are best deployed to complex, intractable or 'wicked' problems, that are more contested and/or resource-intensive (Holley et al. 2012).

Deliberative trends are consistent with models and norms of 'new environmental governance' (Holley et al. 2012), attempts at greater democratisation of decision-making and collaborative approaches. Concepts of *collaboration* in governance represent joint or cooperative, if not always successful (see Holley and Sinclair 2013), approaches to participation, including governmental and non-governmental actors, often aiming at consensus or agreement. In this respect, 'collaborative governance' can signify models of procedure tending toward negotiation and agreement, beyond mere consultation (Gunton and Day 2003).

To date, Australian attempts at 'collaborative governance' in water management have been modest in comparison with participatory projects elsewhere (see e.g. Holley 2015; Holley and Sinclair 2013). Large-scale deliberative processes have been employed in major water planning exercises—for example, the Northern Victoria Sustainable Water Strategy ('NVSWS') (Beckingsdale and Hind 2010). The NVSWS is a major water planning instrument for northern Victoria, an area heavily reliant on water resources for industry, especially agriculture. The preparation of the NVSWS is typical of the nature and degree of deliberative technique functioning in water management. The legislative framework of the Victorian Water Act provided a basis ('default', Holley 2015) for the deliberative approach. The process was referred to in terms of community 'engagement'. Procedurally, the NVSWS process established a basis of dialogue and consensus-building, including scope for adoption of proposals, although not a negotiation or agreement-making process (Beckingsdale and Hind 2010, pp 5–9). At the same time, there appears not to have been an overt policy basis to the approach adopted.

The NVSWS experience, as with most other water planning exercises in Australia, tends at most to view deliberative participation as methods of influencing, informing, and setting a normative basis conducive to governmental decision-making. It may include a *simulation* of cooperative venture. It may be limited to confidence-building measures (Tan 2008). Notably, deliberative participation tends not to upset, disrupt or even qualify governmental monopoly on decision- or policy-making. To this extent, it may be more appropriate to view many Australian experiences of deliberative water governance as experiments in, or extended forms of, consultation rather than collaboration or 'collaborative governance.'

4.2 Consultation with Aboriginal Peoples

Participation in water governance takes on special significance in relation to Aboriginal peoples in Australia. There are two primary bases for this status. The first of these is the unique and ancient connections of Aboriginal communities to water resources, including rivers, lakes, wetlands, wells and springs, and water-dependent ecosystems. Those connections are ancient and contained in narratives, law, custom, and inter-relationships of people and place (now often given the term 'Country'). They are ongoing connections. The second basis is the profound rupture to Aboriginal societies and connections to Country affected by colonisation. This rupture occurred through frontier wars and massacres, clearance, forced resettlement and suppression of culture, racialist policies, and attempted destruction of Aboriginal societies. Public resurgence of Aboriginal identity and self-determination, since the 1970s, commenced through land rights struggles and establishment of Aboriginal-run organisations and services, and includes assertion of sovereignty in relation to water (e.g. MLDRIN 2009).

While framed in the NWI in terms of 'consultation', Aboriginal participation in water governance across Australia is complex and variable (Jackson et al. 2012; Tan 2008, pp. 131–138; Behrendt and Thompson 2004; Hemming and Rigney 2014; Son 2012). Aboriginal peoples exercise deep moral and normative claims and attachments to water, grounded in preferences for self-determination (Morgan et al. 2004). Moreover, legally-cognisable rights and interests may be held by Aboriginal communities, individuals or organisations in respect of particular water resources. The particular configuration and force of those rights and interests can be highly variable across the continent. They may include interests in access and use, such as by native title, agreements under statute or at common law, special water rights and allocation mechanisms (e.g. Cultural Access Licences in NSW, or Aboriginal Water Reserve in Queensland), or cultural heritage. They may include management rights and interest, such as protected areas co-management. The availability and exercise of rights and interest in water resources establishes practical, normative and substantive legal foundations to procedural, participatory rights and practices. The character of those rights may be largely procedural, as in the 'right to negotiate' under native title law or participation rights in planning, or they may be substantive legal interests, such as the holding of water entitlements. These mechanisms provide a base of Aboriginal involvement in water governance. However, this is a base that, while manifest with potential, remains limited in terms of outcomes and effects, and generally fails to accommodate Aboriginal governance systems and perspectives in the domain of water governance (Jackson and Barber 2013; Tan 2008; Son 2012; Jackson et al. 2012; Durrette 2008).

Consultation provisions of the NWI inform water legislation on the broadest scale under the *Water Act 2007* (Cth) and, specifically, the *Basin Plan 2012* (Cth). The *Basin Plan* is a Federal legislative instrument made pursuant to the *Water Act*. A principal focus of these laws is limiting water extraction from the Murray-Darling system and achieving a sustainable balance between environmental, economic and social outcomes (see Carmody, this volume). Within a cascading regulatory framework, the Basin Plan establishes a binding framework for water resources planning ('WRPs') at regional level, to be prepared and administered principally by relevant state and territory Governments ('Basin States'). WRPs must be prepared subject to consultation with 'relevant indigenous organisations' (*Basin Plan 2012*, Chap. 10, Part 14). Other matters arising from the Basin Plan, such as environmental watering and risk management, are also functions in and through which Indigenous consultation can occur.

The general approach to Indigenous consultation is as follows. A WRP must identify and be prepared with regard to the views of Indigenous Peoples as to objectives and outcomes from water resources management (based on identified uses and values), relevant legal rights and interests (such as registered cultural heritage or native title), procedural obligations (such as Indigenous representation, encouragement of informed participation, and management of risks to Indigenous uses), and cultural flows. 'Cultural flows' are understood to be water rights legally or beneficially owned by Aboriginal people (MLDRIN 2009). A WRP may identify opportunities to strengthen protection of Indigenous values and uses. A WRP must also retain as much protection for Indigenous uses and values as exists under prior, interim or transitional arrangements.

The distinguishing features of Indigenous consultation provisions under the *Basin Plan* are obligations to 'identify' and 'have regard to' Aboriginal views regarding a range of matters. While it is intended that 'regard' requires 'genuine, proper and realistic consideration' (MDBA 2013, p 109), no obligation to adopt or defer to Aboriginal views, either wholly, partly or in preference to other considerations, is established under these consultation provisions. Aboriginal communities, via 'relevant' organisations, are obligatory 'voices' and actors within the *Basin Plan* governance framework and the scope of matters on which consultation is required is broad, including flows and quality of water, as well as water-dependent ecosystems, and aspects of administration (such as risk management). However, while broad, this model of participation may be described equally as 'thin' – of relatively weak force and effect.

Comparisons can be made to models of Indigenous participation elsewhere. Obligations on the Canadian Crown to consult with Indigenous Peoples in relation to matters affecting them include 'deep consultation', along a spectrum of engagement ranging from disclosure and dialogue through to negotiation and consent over legally protected rights and interests (*Haida Nation v British Columbia*, 2004 SCC 73, [43]–[48]). The deep consultation approach requires a process analogous to negotiation, absent an obligation to agree, including as appropriate testing and amending proposals, accounting for impacts on Indigenous communities, and accommodation (*Haida Nation v British Columbia*, 2004 SCC 73, [42]–[51]; *Gitxaala Nation v Canada*, 2016 FCA 187; Sossin 2010). Dynamics of collaboration, even partnership, are readily apparent in the model of 'deep consultation'.

For Canada, the approach is grounded in specific constitutional conditions, requiring maintenance of the honour of the Crown, the fiduciary relationship of the Crown to Aboriginal peoples, and the objective of reconciliation (*Haida Nation v British Columbia*, 2004 SCC 73). The constitutional relation of Aboriginal Peoples to the Australian state is distinct from the Canadian. No such duties of honour and reconciliation apply. The Crown is not a fiduciary. There are no treaties. The nature and content of 'deep consultation' is powerful and instructive for Aboriginal consultation in respect of water resources in Australia. But the latter proceeds from a much more rudimentary legal basis, a much barer form of title and a refusal of 'territorial rights' (McNeil 2013, 2015; Durrette 2008), hampering insistence on a richer form of dialogue.

In international law, the emerging concept of 'free, prior and informed consent' ('FPIC') provides another point of comparison. It is a concept adopted in international instruments, such as the *UN Declaration on the Rights of Indigenous Peoples* ('UNDRIP'), as well as applied in international jurisprudence (e.g. *Saramaka People v Suriname*, Inter American Court of Human Rights, Series C 172 (2007)). Various formulae of the concept have developed. Article 32 of UNDRIP, for instance, frames FPIC as a requirement to 'consult and cooperate in good faith… in order to obtain… free and informed consent…' This is a model of consultation, strictly controlled for the purposes of achieving consent. It is arguable that this formulation, or 'consent' provisions generally, do not afford Indigenous Peoples an absolute right to veto in respect of actions affecting them, but more in the nature of a qualified right for consent to be obtained, having regard to the significance, nature and extent of impacts, contributing 'more effective control [by Indigenous communities] over the broader consultation process' (Barelli 2012, p 24). The Australian Government has endorsed the UNDRIP, but no consultation principles framed by free, prior and informed consent with respect to land or natural resources have been incorporated into domestic Australian laws.

In respect of water resources, the model of consultation developed nationally in law thus far, in the text of the Basin Plan, is based on mandatory *consideration* of relevant Aboriginal organisations' views, not on a model necessarily intended to affect substantive outcomes in rights or interests. It appears purely procedural. While confining the overall nature of the obligation, the *Basin Plan* approach to consultation also pre-empts with whom the consultation is to occur ('organisations') and to that extent, how consultation is to occur.

The 'thin' consultation obligations of the *Basin Plan* are distinguishable, in the above examples, from both Canadian and UNDRIP approaches. The former requires no ultimate imperative or demonstration of passage toward consensus, effect on substantive outcomes, or, as in Canada, justice based on reconciliation (Sossin 2010). The UNDRIP model insists on participation conditioned by accord (consent) and the Canadian approach insists on the space of accommodation and bargaining even where final accord cannot be achieved. Aboriginal consultation under the *Basin Plan* does not require the former, and it is uncertain that 'genuine' regard of Aboriginal views compels any accommodation in the final analysis.

4.3 Lessons: Participation Through Consultation

Political and legal direction to incorporate public consultation into water governance is a phenomenon of the last two decades. Five instructive conclusions can be posed as to the role and character of this mode of public participation.

First, 'consultation' provides a ubiquitous and flexible rubric by which to frame public participation. It provides a normative framework directing public participation in water resources administration. The force and effect of those norms may vary, from perfunctory and manipulated involvement to a deeper engagement approaching collaboration, bargaining or agreement-making. Minimal content to consultation likely remains the norm, such as through notice and comment provisions and limited invitations to 'stakeholder' participation. Models and experiments in deliberative techniques, within the ambit of consultation, have emerged and represent important initiatives requiring ongoing development and support. The prevailing approach to 'consultation' or 'engagement' in water management has been to reinforce 'public confidence' (legitimacy) in decision-making but with 'restrained' effect (Tan 2008, 128–131).

Second, systemic weaknesses remain in the use of 'consultation' as the prevailing discourse of public participation. These include an absence of overarching policy concerning the meaning and exercise of public participation (Tan 2008, p 131). Remedying that gap is not assisted by the absence of wider, national narratives on public participation in governmental activities and conduct, such as is established by Articles 6–9 of the Aarhus Convention in Europe (UNECE 1998; see also Stec 2003, Chap. 2) or by the 'Open Government Initiative' of the Obama Administration in the United States (U.S.) (Orszag 2009; Lukensmeyer et al. 2011). There is also reluctance on the part of state actors to disturb established models of water governance including prioritisation of the power of established and historic interests, reproducing monopolistic or oligopolistic structures of governance and limiting dispersal of power in decision-making (Holley and Sinclair 2013). Consultation tends not to include or anticipate diffusion of power to broader sets of actors representing the public interest.

Third, policy approaches to public participation need to mature further, including in terms of the appropriate role and nature of consultation, its situation among the overall landscape of procedural possibilities (e.g. administrative, adjudicative, and market-based), and adaptation of procedure to the needs and circumstances of water governance. Those needs and circumstances traverse not only a wide of interests and issues, from urban and agro-industrial uses, ecological sustainability and restoration, and restorative justice to Aboriginal communities. They also encompass variabilities in scale and interactions with other spheres of natural resources management and the place of water and water systems in the social imagination. The maturation of policy in Australian water governance can draw on experience elsewhere (see e.g. U.S. EPA (2017)).

Fourth, the *Basin Plan* framework of consultation with Aboriginal peoples may be the most advanced example of this mode of participation, relating to water resources, established in law in Australia. Nevertheless, the form and the legal and policy bases of Aboriginal consultation are impoverished relative to international comparison. Absent constitutional or long-established recognition of title or jurisdictional authority over resources, Aboriginal participation in water governance is susceptible to political contingency (e.g. legislation) and/or legal increment (e.g. agreement-making, native title determination). Models of consultation and quasi-negotiation from other jurisdictions provide, however, rich comparison and lessons for Aboriginal participation in water governance in Australia. These include the Canadian doctrine of 'deep consultation' and the UNDRIP model of 'consultation… in order to obtain consent'.

Fifth, deeper, deliberative, more robust and sustained, models of public participation are likely to be influenced, if not underpinned, by the existence and operation of *legally-cognisable* rights and interests in water resources or in matters affecting the management of water resources. Those means can provide participants a 'seat at the table.' Aside from agro-industrial and governmental interests, this dynamic appears in the now broad-based, although still weak and 'inchoate' (Tan 2008, p 129), approaches to consultation with Aboriginal peoples. This occurs via means such as rights or allocation mechanisms, native title claims or determinations, agreement-making, cultural heritage interests, or co-management arrangements. The circumstances of Aboriginal communities in respect of water resources is unique. Legal or quasi-legal means, such as statutory recognition of interests, agreements on participation or plans mandating practices of participation, could be used to expand the base and efficacy of actors within deliberative modes of governance. Those mechanisms could be used to strengthen the role of environmental or community actors more generally within water governance.

5 Participation Through Adjudicative Procedure

Public participation in water resources management is also conceivable by way of litigation or, more generally, through the use of adjudicative procedure, whether in the exercise of judicial power or administrative power. Adjudicative procedures operate through a model of contest, including judgment according authoritative standards and a mode of reasoning based on evidence and marshalling of facts, to which standards are applied (Galligan 1996, pp. 241–242). Participation is central to adjudication (Galligan 1996, pp. 243–246), albeit a 'peculiar form of participation' through proofs and reasoned arguments (Fuller 1978, p. 364). The central role for reasoned argument and authoritative standards provides weight to the idea that adjudication also requires a degree of independence in decision-making, which is vested in a third party. Adversarial features of the formal trial-type hearing are not intrinsic to adjudication (see e.g. Flick 1984, pp. 9–22).

5.1 Litigation Under Water Resources Law

The use of water resources law as a vehicle for public action through Australian courts and tribunals has been scant. Until the 1980s and 1990s, the conservation and management of water resources concerned regulation for industrial (e.g. irrigated agriculture) and urban supply purposes, not for any imperative of environmental sustainability. Since the 1990s, water resources law has been driven by this latter imperative (Fisher 2000, p. 196; *Alanvale Pty Ltd v Southern Rural Water* [2010] VCAT 480, [24]). It conveys a clear public interest character. Given the common pool character of water resources, it is arguable that water law and management has always had a strong public interest character (*ICM Agriculture Pty Ltd v Commonwealth* [2009] HCA 51, [90]). This has been little reflected in public resort to courts and tribunals to regulate the resource. Despite the rise of environmental priorities, water resources management has remained largely impervious to wider public use of litigation to affect governance, its directions and reorientation toward ecological and Aboriginal priorities.

Much of the limited volume of litigation over water resources management emerging from the 2000s onwards has arisen out of the regulatory regime in which sustainability and limits on extraction prevail in water management. Those actions occur in state jurisdictions, as they typically involve matters of water access rights operating at the state level. Sources of this litigation have included "caps" on water extraction, water planning controls, whether controls are a form of expropriation of property, and whether or how sustainability should be implemented in relation to water resources. A prevailing theme is dispute and tension over limits to water extraction and, by extension, the practical meaning and effects of "sustainability" in water management. For the most part, litigation has been undertaken by agricultural irrigators (e.g. *Murrumbidgee Groundwater Preservation Association v Minister for Natural Resources* [2003] NSWLEC 322; *Arnold v Minister Administering the Water Management Act 2000* [2007] NSWLEC 531; *Harvey v Minister Administering the Water Management Act 2000* [2008] NSWLEC 165; *Hutchins Pastoral Co Pty Ltd v Minister Administering the Water Management Act 2000* [2014] NSWSC 46). Therefore a central feature is the commercial nature of interests at their heart and, to that extent, private interests (*Harvey v Minister Administering the Water Management Act 2000 (No 2)* [2008] NSWLEC 213, [7]).

A few cases in NSW and Victoria prioritise public interest factors, typically ecological sustainability and protection of environmental assets. The applicant in *Nature Conservation Council (NSW) Inc v Minister Administering the Water Management Act 2000* [2004] NSWLEC 33 sought (unsuccessfully) to establish that obligations operate under relevant water resources law to incorporate mechanisms within a water plan to benefit the environment, specifically performance indicators regarding environmental watering rules of a particular construction on the Gwydir River. Several Victorian cases (e.g. *Alanvale Pty Ltd v Southern Rural Water* [2010] VCAT 480; *Castle v Southern Rural Water* [2008] VCAT 2440; *Paul v Goulburn Murray Water Corporation* [2010] VCAT 1755; *Rigoni v Goulburn*

Murray Water Corporation [2011] VCAT 201) concerned permission to extract groundwater for agricultural uses in circumstances where impacts both on other landowners, as well as on the local environment, may have been unacceptable risks. Uncertainty and potentially wide-ranging impacts of extraction from groundwater systems drew in wider public interest issues into litigation of this particular resource. This includes the impact of extractions on other (non-urban) water users, as well as on groundwater dependent or influenced ecosystems. The Victorian line of cases, heard 'on the merits' by administrative review, established a tendency for the adjudicating Tribunal (Victorian Civil and Administrative Tribunal) to scrutinise forensically management of groundwater extraction according to sustainability principles, especially precaution. In the majority of those cases, decisions were made by the Tribunal to constrain or prohibit extraction of water resources, whether by affirming, varying or setting aside a decision of the original decision-maker.

5.2 Constraining Factors in the Use of Litigation as a Participation Vehicle

There are a range of reasons why litigation and the use of adjudicative procedures have achieved only very limited use as means of public participation in water resources management.

Standing Accessing courts or tribunals in order to resolve a matter in water management may confront the initial procedural issue of standing. Standing is rarely an issue where private rights are at stake. Access to these avenues of justice are more likely to arise as an issue where the public interest and third party involvement is at issue. Formulae for standing in relation to water resources litigation vary by jurisdiction and by the nature of the matter, for instance, whether judicial review, administrative review, or civil or criminal enforcement. Water resources legislation in jurisdictions such as NSW (*Water Management Act 2000* (NSW), s 336) and Queensland (*Water Act 2000* (Q), s 784) contain 'open standing' or 'citizen suit' provisions, allowing 'any person' to commence and maintain proceedings to restrain breach of the primary Act or regulations. In Tasmania and in South Australia, relevant legislation allows the relevant court or tribunal to grant leave to commence proceedings and public interest considerations are relevant to the exercise of discretion. In other jurisdictions, standing will attach to some form of interests test, such as a person affected under the *Rights in Water and Irrigation Act 1914* (WA), s 5E. A person seeking judicial review of decisions under the Victorian Water Act needs to meet the common law test demonstrating they have a 'special interest' in the matter. A person seeking administrative review of a decision under that Act typically needs to demonstrate their interests are affected (e.g. *Water Act 1989* (Vic), s 64 in relation to 'take and use' licences).

Rules regarding standing to seek enforcement of legislative or regulatory provisions under water law varies. Although the trend is toward liberal access to courts and tribunals, this is not uniformly the case, nor without tests and qualifications.

Costs In practice, it is more often the issue of costs rather than formal issues of standing that deter litigation, especially public interest litigation. Costs include legal costs, costs of experts and other disbursements, as well as the risk of adverse orders for costs if the case is lost. Australian jurisdictions follow the general rule that 'costs follow the event'. This is modified in some tribunals and specialist jurisdictions, such as the NSW Land and Environment Court and the Victorian Civil and Administrative Tribunal, which hear water cases. These mechanisms can serve to mitigate the 'chilling effects' of legal action, although they may not avoid them altogether (Watters 2010; Pain 2014). Costs act as a powerful 'informal filter' (Aronson and Groves 2013, [11.50]; also Hawke 2009, [15.86]; Mansfield 2015) for litigation. The nature of water systems and water law frequently render them complex. They concern valuable and often location-specific resources on which the fate of enterprises, industries and entire communities may rest. Disputes over water can inherently involve competing and conflicting purposes, needs and preferences, including economic, environmental, cultural and social. In short, these types of cases can be expensive, often because they are difficult. There may be related deterrents, such as requirements for security for costs, or undertakings as to damages arising from interlocutory actions.

Information and Evidence Water litigation can be technically complex and rely extensively on expert knowledge. Information and expertise may be held by statutory instrumentalities and/or consultants retained by them, or by key commercial or industrial interests. Australian courts and tribunals have evolved rules to manage access to information and potential evidence in civil proceedings on foot, as well as the uses of expert knowledge in evidence. As valuable as such rules may be, they do not necessarily overcome imbalances or monopolies on expertise prior to or in anticipation of dispute. They do not necessarily resolve more fundamental and structural issues in the nature and weaknesses of informational tools and scientific techniques (Martin and Williams 2014). They do not deal with or ameliorate systemic problems in the collection, acquisition, holding, dissemination, scrutiny and analysis of water information, whether this concerns the character of water systems, or uses and extractions from those systems (Holley and Sinclair 2013). Access to relevant and probative information can be a key hurdle to public participation in decision-making, including litigation (Gunningham 1995). Water information is notoriously problematic (Holley and Sinclair 2012).

The problem of information is multifaceted. At the coarse scale, improvements have been made in water information resources, such as through establishment of water title registers, operation of water accounts by the Commonwealth's Bureau of Metereology, and sources of water reporting such as by the (now defunct) National Water Commission. In specific disputes, there can remain the issue as to whether relevant information exists at all. This goes to the issue of monitoring and data collection and judgments as to what is relevant, how information should be classified and how resources should be allocated to collect it. There is the matter of whether existing information is accessible, which can concern legal obligations to make information available (such as through 'right to know' laws and positive obligations to collect and disseminate information) (Wilson 2012) and practical tools and platforms through which information is made available (e.g. databases, registries, etc.). There is the form in which information exists, not only as scientific data and modelling but in other epistemic models, such as Aboriginal knowledge systems (Marshall 2014; Cranney and Tan 2011). Finally, there are governance and institutional issues, such as what bodies should hold information, what protocols govern it, and how are systems to be resourced.

Exceptional Use of Adjudicative Procedure While the resolution of contests over legal rights and duties in respect of water has occurred through judicial process in Australia for decades (Fisher 2000, pp. 103–131), an adjudicative model is not central to the management of water resources. The prevailing model of Crown ownership and public administration of water resources, modified through water planning and markets, is complemented by occasional availability of adjudicative procedure. This may be in judicial review of administrative decisions, or in selective administrative review of decisions such as the granting of water licences, as illustrated in the cases cited above. At formative stages of decision-making, there may be scope for establishment of advisory bodies, such as panels, to inform planning or other actions, but these are rarely adjudicative bodies.

An instructive comparison to be made here is to the U.S. experience. By comparison to Australia's modified public administration, U.S. water law commonly includes more central roles for adjudicative procedure and litigation. Like Australia, U.S. water law is primarily a state matter (see USDA Forest Service 2005). Adjudicative procedure can occur in the settlement of claims to water rights, notably where the prior appropriation doctrine is a feature of governance. These can be in the nature of 'general stream adjudications', where courts or tribunals are employed to resolve claims to water systems. These cases can be notoriously complex, unwieldy and lengthy (Macdonnell 2015). They are accompanied by a significant public interest thread in the form of underpinning 'public trust' principles (Mudd 2013), safeguarded by public authorities but also a key tool for non-governmental, public interest intervention, including through judicial processes.

The more interesting comparative consideration is public participation via adjudicative procedure in the administrative cycle. In U.S. jurisdictions, water management may be accompanied by administrative hearings, such as at the point of transfers of licences or allocations, planning and the informing of administrators (see e.g. Cosens, this volume). Under the *California Water Code*, for example, an application to appropriate water under permit is subject to public notice (*California Water Code* § 1300–1317), scope for 'protest' (*California Water Code* § 1330–1335), and public hearing before the State Water Resources Control Board (*California Water Code* § 1350–1353). The Board possesses a power at large to undertake investigation into streams and water systems, take testimony on water rights and uses (*California Water Code* § 1051), and undertake hearings on transfers of long-term uses (*California Water Code* § 1535–1537) and temporary transfers (*California Water Code* § 1726). There are other circumstances under the *California Water Code* when administrative hearing procedures are used in decision-making processes, such as establishment or variation of water districts (e.g. *California Water Code* § 23645–23651), the administration of flood protection (e.g. *California Water Code* § 8730–8742), and in administrative enforcement provisions (e.g. *California Water Code* § 13300–13321).

The Australian experience of water law and administration is far less sympathetic to the use of adjudicative procedure, outside of administrative review of certain decisions (such as licensing). Under Victoria's *Water Act*, for instance, there are circumstances, typically discretionary, where expert or advisory panels contribute to administration (e.g. *Water Act 1989* (Vic) ss 11(6), 22F, 22S, 39, 48E, 50, 66, 139E, 161H), but without any requirement for hearing procedures. While Ministerial approval is required for a transfer (dealing) in the ownership of a water share (*Water Act 1989* (Vic) s 33X), there are neither provisions for objection nor hearing on the transaction.

Weaknesses in Substantive Laws Use of adjudicative procedure as a vehicle for public participation can depend upon the nature and effectiveness of substantive provisions. The most liberal procedure, such as third party standing or public interest costs protection, will be of little value where substantive rules and standards to be applied are vague, inappropriate, discretionary or ineffectual.

The example of management of groundwater dependent ecosystems in South Australia is a case in point (Nelson 2007). Framed by comparatively recent legislation (*Natural Resources Management Act 2004* (SA)), ecologically and culturally sensitive groundwater systems in northern South Australia are managed within a statutory planning model that is generally well adapted and appropriate to those systems. Nevertheless, it is argued this regime is deficient 'at a more operational level' (Nelson 2007, p 128). Gaps include detailed and ongoing information provision, ecological and hydrological baselines, application of precaution, and management in favour of ecological sustainability, adaptive management, and legal enforceability (Nelson 2007, pp 106–109).

Similar limitations arose in *Nature Conservation Council (NSW) Inc v Minister for Sustainable Natural Resources* [2004] NSWLEC 33. That litigation involved the validity of a water sharing plan and environmental water rules made under NSW water legislation. The elaborate statutory scheme was intended to give effect to controls on water extraction and to water management principles reflective of ecologically sustainable development. Liberal standing rights and public interest costs protections were available to the applicant. The applicant sought, but failed, to have the water sharing plan declared invalid on the grounds it did not establish and implement specific environmental performance measures and correct rule-based commitments of environmental water flows. Despite the relative strength of the legislation in providing for 'a scheme whereby the water source and its dependant ecosystems are protected', key features remained general, and retained discretion and a gap between commitment and specific actions (*Nature Conservation Council (NSW) Inc v Minister for Sustainable Natural Resources* [2004] NSWLEC 33, [83]).

These examples reprise certain important limitations in substantive water law. These include a lack of prescription—information, factual baselines, qualitative and quantitative standards or thresholds—as well as the difficult task of integrating general principles or commitments and discrete, measurable controls and actions in specific circumstances.

5.3 Lessons: Participation Through Adjudicative Procedure

Adjudicative bodies and processes are modelled on judicial approaches and institutions. Within water governance, adjudication on matters of water law in the courts plays a strategically important role, such as in supervision of the conduct of governmental decision-making. In other respects, adjudicative procedure is not prominent. Six insights and propositions may be gleaned from this condition.

First, adjudicative procedure can provide decision-making according to a generally known approach of judgment according to authoritative standards, norms of rationality and reasoned outcomes, with a greater or lesser degree of public participation. Adjudication can allow for the development and interpretation of authoritative standards. It can facilitate improvements in policy and practice on the part of administering agencies, as in the case of statutory tribunals (Administrative Review Council 1995, Chap. 6).

Second, adjudicative modes of procedure necessarily retain a foothold in water resources management as a result of judicial supervision of the extensive powers of Executive Government. The use of adjudicative approach as a vehicle for wider public participation is under-developed and could be widened. This is especially the case in the conduct of administration. Use of public hearings at formative stages, such as water planning, can provide a focus and structure for public participation. This approach is commonplace in land-use planning. Similar approaches could be employed to expand adaptive capacity at the point of water transfers (Cosens 2016). Adjudicative approaches need not be strictly adversarial.

Third, caution as to use of adjudicative methods in water resources management can be warranted, if complexity or excessive reliance on judicial procedure to resolve claims and issues leads to inefficiency without corresponding public participation and benefit.

Fourth, wider use of adjudicative procedure as a participatory mechanism could be undermined if 'access to justice' issues are not dealt with. This qualification applies to matters of formal procedural, such as standing or management of costs risks. It also applies to practical matters such as the availability of information and support to public interest actors to participate in water resources management.

Fifth, the role of specialist jurisdictions is likely to be important in the effective, efficient and responsive handling of contest over, or inquiry into, water management. This is indicated by the experience of the NSW Land and Environment Court and the Victorian Civil and Administrative Tribunal in adjudicating on water matters and by the acknowledged importance of specialist jurisdictions in handling environmental matters (Preston 2006).

Sixth, the design of substantive water laws, and the controls and conditions contained therein, are likely to have an impact on the effectiveness and value of adjudicative measures in identifying and enforcing public interest values, such as sustainability.

6 Conclusions

Water governance is a matter of significant public concern. Historically, its public character was signified by state management and control. Public participation is now a much broader concept, involving community and sectoral interests, affecting legitimacy and effectiveness in the management of water resources. Two important modes of public participation are the focus of this chapter: consultation and adjudication. Consultation is the favoured approach in Australian water governance, but it exhibits practical weaknesses and policy uncertainties. For Aboriginal people, with special attachments to water, legal foundations for consultation are relatively more developed, although remain circumscribed, including by limited recognition of legal rights. Adjudicative procedures are uncommon, especially outside of justiciable controversies. The chapter has identified 11 core lessons on participation that can be taken from the Australian context. These include, in relation to consultative and administrative approaches:

- The general value and ubiquity of consultative methods;
- The need for policy to guide consultation, its uses, methods and purposes;
- Consultation in water management remains relatively under-developed;
- Consultative methods are somewhat more advanced in the context of Aboriginal participation;
- Deliberative and 'deep' consultation will likely be effective where participants can avail themselves of legally-cognisable rights and interests in the governance process.

In respect of adjudicative approaches, core lessons include:

- Adjudicative methods have a role to play in decision-making processes;
- They are under-developed in Australian water governance, including in administrative decision-making;
- They should be used appropriately, balancing the benefits of this type of participation with inefficiencies;
- Access to justice issues, such as rights of standing and costs, must be tackled in order for adjudicative processes to be effective;
- Specialist jurisdictions are likely to be important to effective disposition of matters;

- Design of substantive water laws will influence the effectiveness of adjucative procedures.

Scope, conditions and the value of the wider use of public participation in water governance were discussed. While water will continue to be a site of conflict over preferred uses, values and outcomes, Australian approaches to integration of multiple 'voices' in governance will need to build from the initial steps of recent decades. This will likely occur in an iterative fashion, at different scales. Taking account of procedural possibilities and lessons will also be necessary. Conflicts may not be averted in this way but they can contribute to meaningful public confidence and innovation in a challenging legal and policy context.

References

Administrative Review Council. (1995). *Better decisions: Review of commonwealth merits review tribunals*. Report no. 39. Australian Government Publishing Service: Canberra.

Arnstein, S. (1969). A ladder of participation. *Journal of the American Institute of Planners, 35*, 216–224.

Aronson, M., & Groves, M. (2013). Judicial Review of Administrative Action, 5th ed. Lawbook Company.

Barelli, M. (2012). Free, prior and informed consent in the aftermath of the UN Declaration of the Rights of Indigenous People: development and challenges ahead. *International Journal of Human Rights, 16*, 1–24.

Beckingsdale, D., & Hind, J. (2010). Towards deliberation and dialectic: the community engagement process for the northern region sustainable water strategy: Final report. http://www.depi.vic.gov.au/__data/assets/pdf_file/0008/188774/NRSWS_engagement_process_sm.pdf. Accessed 5 Aug 2016.

Behrendt, J., & Thompson, P. (2004). The recognition and protection of Aboriginal interests in NSW rivers. *Journal of Indigenous Policy, 3*, 37–140.

Bruns, B. (2003). *Water tenure reform: developing an extended ladder of participation*. Paper presented at Politics of the Commons: Articulating Development and Strengthening Local Practices RCSD Conference, July 11–14, 2003, Chiang Mai, Thailand.

COAG (2004). *Intergovernmental Agreement on a National Water Initiative*.

Cornwall, A. (2008). Unpacking 'participation: models, meanings and practices. *Community Development Journal, 43*, 269–283.

Cosens, B. (2016). Water law reform in the face of climate change: learning from drought in Australia and the Western United States. *Environmental and Planning Law Journal, 33*, 372–387.

Cranney, K., & Tan, P. (2011). Old knowledge in freshwater: why traditional ecological knowledge is essential for determining environmental flows in water plans. *Australasian Journal of Natural Resources Law and Policy, 14*, 71–113.

Durrette, M. (2008). Indigenous legal rights to fresh water: Australia in the international context. Working Paper 42, Centre for Aboriginal Economic Policy Research, ANU, Canberra.

Fisher, D. (2000). *Water Law*. Thompson Reuters.

Flick, G. (1984). *Natural Justice: Principles and Practical Application* (2nd ed). Butterworths.

Fuller, L. (1978). The forms and limits of adjudication. *Harvard Law Review, 92*, 353–409.

Fung, A. (2006). Varieties of participation in complex governance. *Public Administration Review, 66*, 66–75.

Galligan, D. (1996). *Due process and fair procedures: A study of administrative procedures*. Oxford: Clarendon Press.

Gardner, A., Bartlett, R., & Gray, J. (2009). *Water Resources Law*. LexisNexis.

Gray, J., & Lee, L. (2016). National Water Initiative styled water entitlements as property: legal and practical perspectives. *Environmental and Planning Law Journal, 33*, 284–300.

Gunningham, N. (1995). Empowering the public: Information strategies and environment protection. In N. Gunningham, J. Norberry, & S. McKillop (Eds.), *Environmental Crime: Proceedings of a Conference Held 1–3 September 1993, Hobart*. Australian Institute of Criminology.

Gunton, T., & Day, J. C. (2003). The theory and practice of collaborative planning in resource and environmental management. *Environments, 31*, 5–19.

Hawke, A. (2009). *The australian environment act: Report of the independent review of the environment protection and biodiversity conservation act 1999*. Canberra, ACT: Commonwealth of Australia.

Hemming, S., & Rigney, D. (2014). *Indigenous engagement in environmental water planning, research and management: Innovations in South Australia's murray-darling basin region*. Goyder Institute for Water Research Technical Report Series No. 14/21, Adelaide, South Australia.

Holley, C. (2010). Public participation, environmental law and new governance: lessons for designing inclusive and representative participatory processes. *Environmental and Planning Law Journal, 27*, 360–391.

Holley, C. (2015). Crafting collaborative governance: Water resources, california's delta plan, and audited self-management in New Zealand. *Environmental Law Reporter, 45*, 10324–10337.

Holley, C., & Sinclair, D. (2012). Compliance and enforcement of water licences in NSW: Limitations in law, policy and institutions. *Australasian Journal of Natural Resources Law and Policy, 15*, 149–190.

Holley, C., & Sinclair, D. (2013). Deliberative democracy, environmental law and collaborative governance: Insights from surface and groundwater studies. *Environmental and Planning Law Journal*, 32–55.

Holley, C., & Sinclair, D. (2014). A new water policy options for Australia? Collaborative water governance, compliance and enforcement and audited self-management. *Australasian Journal of Natural Resources Law and Policy, 17*, 189–216.

Holley, C., Gunningham, N. & Shearing, C. (2012). *The New Environmental Governance*. Routledge.

Jackson, S., & Barber, M. (2013). Recognition of indigenous water values in Australia's northern territory: Current progress and ongoing challenges for social justice in water planning. *Planning Theory and Practice, 14*, 435–454.

Jackson, S., Tan, P.-L., & Altman, J. (2009). *Indigenous freshwater planning forum: Proceedings, outcomes and recommendations*. National Water Commission.

Jackson, S., Tan, P.-L., Mooney, C., Hoverman, S., & White, I. (2012). Principles and guidelines for good practice in indigenous engagement in water planning. *Journal of Hydrology, 474*, 57–65.

Lukensmeyer, C., Goldman, J., & Stern, D. (2011). Assessing public participation in an open government era: A review of federal agency plans. IBM Centre for the Business of Government. http://www.govexec.com/pdfs/082211jm1.pdf. Accessed 15 Jan 2011.

MacDonnell, L. (2015). Rethinking the use of general stream adjudications. *Wyoming Law Review, 15*, 347–381.

Mansfield, S. (2015). Enhancing access to the courts to improve western Australia's water resources. *Australian Environment Review, 30*, 136–140.

Marshall, V. (2014). A web of aboriginal water rights: examining the competing aboriginal claim for water property rights and interests in Australia. PhD dissertation, Macquarie University.

Martin, P., & Williams, J. (2014). Science hubris and insufficient legal safeguards. *Environmental and Planning Law Journal, 31*, 311.

McNeil, K. (2013). Aboriginal title in Canada: site-specific or territorial? Law on the Edge Conference, Canadian Law and Society Association/Law and Society Association of Australia and New Zealand, 1–4 July 2013, University of British Columbia, Vancouver. http://digitalcommons.osgoode.yorku.ca/all_papers/19.

McNeil, K. (2015). Indigenous territorial rights in the common law. Osgoode Legal Studies Research Paper Series. Paper 173. http://digitalcommons.osgoode.yorku.ca/olsrps/173.

MDBA. (2013). Handbook for practitioners: water resource plan requirements.

MLDRIN. (2009). Echuca Declaration. http://www.savanna.org.au/nailsma/publications/downloads/MLDRIN-NBAN-ECHUCA-DECLARATION-2009.pdf. Accessed 9 Jan 2017.

Morgan, M., Strelein, L., & Weir, J. (2004). *Indigenous rights to water in the murray darling basin: In support of the indigenous final report to the living murray initiative*. Australian Insitute of Aboriginal and Torres Strait Islander Studies, Canberra, ACT.

Mudd, M. B. (2013). Hitching our wagon to a dim star: Why outmoded water codes and 'public interest' review cannot protect the public trust in western water law. *Stanford Environmental Law Journal, 32*, 283–339.

National Water Commission. (2011). *NWI biennial assessment*.

National Water Commission. (2014). *Australia's water blueprint: National reform assessment*.

Nelson, R. (2007). Groundwater dependent ecosystems in law: troubled waters? *Asia Pacific Journal of Environmental Law, 10*, 99–130.

Orszag, P. (2009). Open government initiative. Memorandum. Office of Management and Budget, Dec 8, 2009, https://www.whitehouse.gov/open/documents/open-government-directive. Accessed 15 Jan 2017.

Preston, B. (2006). The role of public interest environmental litigation. *Environmental and Planning Law Journal, 23*, 337–350.

Reed, M. (2008). Stakeholder participation for environmental management: a literature review. *Biological Conservation, 141*, 2417–2431.

Son, C. (2012). Water reform and the right for indigenous Australians to be engaged. *Journal of Indigenous Policy, 12*, 3–23.

Sossin, L. (2010). The duty to consult and accommodate: procedural justice and as Aboriginal rights. *Canadian Journal of Administrative Law and Practice, 23*, 93–113.

Stec, S. (2003). *Handbook on access to justice under the aarhus convention*. Szentendre, Hungary: Regional Environmental Centre for Central and Eastern Europe.

Stoeckel, K., Webb, R., Woodward, L., & Hankinson, A. (2012). *Australian water law*. Thompson Reuters.

Tan, P.-L. (2008). *Collaborative water planning: legal and policy analysis* (Vol. 3). Tropical Rivers and Coastal Knowledge: Griffith University, Brisbane, Qld.

Tan, P.-L., Baldwin, C., White, I., & Burry, K. (2012a). Water planning in the condamine alluvium, queensland: Sharing information and eliciting views in a context of overallocation. *Journal of Hydrology, 474*, 38–46.

Tan, P.-L., Bowmer, K., & Mackenzie, J. (2012b). Deliberative tools for meeting the challenges of water planning in Australia. *Journal of Hydrology, 474*, 2–10.

Tan, P.-L., Bowmer, K., & Baldwin, C. (2012c). Continued challenges in the policy and legal framework for collaborative water planning. *Journal of Hydrology, 474*, 84–91.

US EPA. (2017). Public Participation Guide. https://www.epa.gov/international-cooperation/public-participation-guide. Accessed 15 Jan 2017.

USDA Forest Service. (2005). *USDA Forests Service Sourcebook of State Groundwater Laws in 2005*.

Watters, R. (2010). *Costing the Earth? The Case for Public Interest Costs Protection in Environmental Litigation*. Environment Defenders Office (Vic), Carlton, Victoria: The Case for Public Interests Costs Protection in Environmental Litigation.

Wilson, E. (2012). Australian freedom of information legislation v the aarhus convention: Is australia falling below international standards? *Australasian Journal of Natural Resources Law and Policy, 15*, 1–16.

Pain, N. (2014). Protective costs orders in Australia: Increasing access to courts by capping costs. *Environmental and Planning Law Journal, 31*, 450–458.

A Governance Solution to Australian Freshwater Law and Policy

Jennifer McKay

Abstract This chapter presents research outputs on three governance players: the law, the organisations and a collective look at civil society through epistemic communities and non-government organisations. All three players seem to be holding different shaped balls and rules and hence are poorly coordinated. The chapter points to ways to increase co-ordination and places that burden on the law. The conclusion reached is that the *Water Act 2007* should be applied to the entire nation and water supply organisations need a harmonised corporate form. Civil society has played crucial roles in the past but their eye is off the ball at present. There is a need for stronger epistemic communities and NGOs. The chapter compares urban water supply laws and organisations in Australia and China. China was selected as it shows a change in governance process. Driven by acute need, China accepted foreign investment in the water sector and developed individual cities. Of late, China has changed this process and adopted a catchment approach to integrate urban water supply in the context and created an innovative position of Chief of the river.

1 Introduction to the Major Policy Events and Cases in Australian Water Law and Policy by Type of Water, Rivers, Aquifers and Manufactured Water Since 2003

> There is a deep and understandable community anxiety about water … no other single substance has a greater impact on the human experience or on our environment … We need to make every drop count (John Howard, National Press Club on a National Plan for Water Security, January 25 2007).

J. McKay (✉)
School of Law, University of South Australia, City West Campus, Australia
e-mail: jennifer.mckay@unisa.edu.au

Australian freshwater law and policy is (still) 'monstrously long, tangled' and 'exhausting'.[1] Freshwater supply to urban areas and for agriculture in Australia is spatially variable, administratively complex and reliant on the three governance dimensions of law, organisational arrangements and civil society/epistemic communities. Fisher (2000) remarked:

> [t]here can be little doubt that a system of federal government ... not only fragments the management of water resources, but also renders their management a complex exercise.

Later in 2007, Malcolm Turnbull, the then Minister, when introducing the *Water Bill (Cth)* in 2007 said:

For more than a century our greatest system of rivers and aquifers, the MDB, has been managed between four states each of which has had competing interests with the others. Federal management of the Murray River was called for in the 1890s, but the vested interests of the states prevailed.

Research conducted with government officials has repeatedly found that the water planning process is highly political and that its processes do not engender trust (McKay et al. 2006). Water customers do not trust local politicians and desire more federal intervention in water governance (McKay et al. 2011).

In this chapter the definition of governance is adapted from Philipp Lange et al. (2013), governance is the process of—more or less institutionalised—interaction between public and/or private entities ultimately aiming at the realisation of collective goals. The particular collective interest in this context is sustainable use of water in rivers, aquifers and large scale manufactured water sources such as recycled water and desalination.

The Millennium Drought focused the minds of all governments federal, state and local on water issues as civil society reacted to urban water restrictions (Jacobs 2006). Governance has emerged as a concept in political science, sustainability science and other fields as a response to the growing awareness that governments are no longer the only relevant actors when it comes to the management of societal issues. Within the governance literature it is argued that over the past decades governing has increasingly become a shared responsibility of state, market and civil society (Kooiman 2003; Haque 2001)

The objectives of the Council of Australian Governments, Intergovernmental Agreement on a National Water Initiative (NWI) (National Water Commission 2004) were to provide for sustainable use of water, increasing the security of water access entitlements, and ensuring economically efficient use of water. These objectives were to be achieved principally by strengthening environmental flow provisions for water in rural catchments, removing barriers to markets in water, and providing for public benefit outcomes through water planning mechanisms (McKay 2011).

---

[1] La Nauze describes the water debates as 'monstrously long and tangled' and 'the most exhausting' of the entire Melbourne Convention (La Nauze 1972, pp. 153–210).

One instrument was the regional water plan and this ideally would be the overarching set of rules determining the consumptive pool in a regions and hence the amount of water available to be traded. The water plans, passed by the state governments under different laws, and the determination of the consumptive pool and the amount available for agriculture, causes many problems (McKay 2011). While water planning does take many forms under state laws (Hamstead et al. 2008; McKay 2011), in preparing surface and groundwater management plans for areas of concern, state and territory jurisdictions were meant to follow nationally consistent guidelines for undertaking transparent statutory planning relying on best available information (National Water Initiative, cl 23(ii)). Jurisdictions were expected to consult and involve communities, including Indigenous groups (National Water Initiative, cll 52, 95). Tradeoffs between competing outcomes for water systems are to be considered and settled using the best available science, social and economic analysis and community input. In all of these, disputes arouse between sectors of the community and the quantum of water for consumptive uses was one huge source of anger.

The Productivity Commission in 2011 was commissioned to report on the urban water sector and it recommended improvements in decision making processes of governments and increased transparency. According to the Productivity Commission, 2011:

Deficiencies in the institutional and governance arrangements are, in turn, leading to policies and water supply decisions that are costly to consumers of water, wastewater and stormwater services.

In this respect, the Productivity Commission (2011) identified a range of key problems in the urban water sector:

- **Conflicting and inappropriately assigned objectives and policies**: There is a lack of clarity and transparency about the way government objectives and policies are being applied in the urban water sector to service delivery, environmental, public health and social matters. Governments are assigning multiple objectives to their agencies, utilities and regulators, with inadequate guidance on how to make trade-offs among them.
- **Lack of clarity about roles, responsibilities and accountabilities**: Often governments are influencing or making decisions in non-transparent ways.
- **Too great a focus on water restrictions, water use efficiency and conservation**: water use is relatively unresponsive to changes in price, indicating that consumers place a high value on water consumption. Water restrictions are likely to have cost in excess of a billion dollars a year from the lost net value of consumption alone.
- **Constraints on efficient water resource allocation and supply augmentation**: Although some of the recent investment in desalination plants might have been appropriate in the circumstances to maintain security of supply, there is sufficient evidence available to conclude that many projects could have been: deferred for a number of years, smaller in scale, or replaced with investment in lower cost sources of water.

According to the Productivity Commission, the above issues in decisions regarding costly desalination projects stem from inadequate institutional arrangements that have led to 'increasing politicisation of supply augmentation processes' (Productivity Commission 2011). Additionally, shortcomings such as inadequate transparency, and the lack of high quality, publicly available evidence about the costs, benefits, and risks of the choices available for supply augmentation projects were seen to result in negative community perceptions of government's decisions and dissatisfaction over water prices (Productivity Commission 2011, Productivity Commission 2014). According to Dolnicar and Hurlimann (2010) little factual information was provided to the Australian public about water conservation, water augmentation projects and other water options currently being used in Australia.

The reports above and other research are reflected in Table 1 which provides a summary of the players, the water issues, rural or urban and the types of water, illustrating the complexity. The institutional complexity is illustrated in Fig. 1. This section illustrates the legal and institutional fragmentation which characterize Australian water governance and points out some particular legal issues from cases and statutory interpretation of state based laws. Part of the problem is that several epistemic communities participate in the water debate, the private sector, local governments who are purchasers from the private sector, state governments, often also purchasers, national government only in the MDBA area, non-government organizations and academics. The non-government organizations are particularly varied in their scope. Some are local like the *Lock the Gate Alliance* and others are branches of international groups such as World Wildlife Fund. An epistemic community is a transnational or national network of knowledge-based experts who help decision-makers to define the problems they face, identify various policy solutions and assess the policy outcomes. The definitive conceptual framework of an epistemic community is widely accepted as what Haas (1992) describes as:

> a network of professionals with recognized expertise and competence in a particular domain and an authoritative claim to policy relevant knowledge within that domain or issue-area.

Sources original research cases in italics discussed in text. Nb state law provides the rules and each state has amended its water allocation laws and other natural resources management laws to include sustainable development as on objective (McKay 2006).

One solution to the problem of how to achieve sustainable water use is to use several sources of water. This is the case in Australia and the US and several other countries. However, as Leroux and Martin (2016) state in relation to the US 'what constitutes an optimal mix of water supply assets (freshwater and manufactured water) has yet to be determined'. This is the universal problem made more acute by climate change in some regions (Fig. 2).

The epistemic communities involved in water management in Australia are hard to define and often remain nascent until there is a local issue. Although the

A Governance Solution to Australian Freshwater Law and Policy

Table 1 Cross cutting players in water governance in Australia 2004-present types of water in Australia, water planning process and other players in water Governance aiming to achieve sustainable development

	Civil society and Epistemic communities and NGOs	Rivers water plans mandated by NWI*	Aquifers water plans mandated by NWI	Manufactured water not in the water plans mandated by NWI
Urban supply	Private providers; Government regulators; User groups government sponsored some NGOS	City scale so often small areas covered *Tepko Pty Ltd v Water Board case* Governments make choices as to water supply types. Not justiciable	City scale so often small areas	Large scale desalination plants not included in water plans
Rural supply agriculture	Private providers Government regulators Several competing NGOs farmers, mining, indigenous	Large areas covered *Lee v Commonwealth*	Large scale *Arnold ICM Commonwealth* can fund states to revise water plans to reduce water consumption by 50%	Recycled water regulated by type of crop under Health laws Can be in water plan
Industrial/ commercial use often sign alliance contracts in a group which attempt to minimize litigation	Private sector lobbying groups *Graham Barclay Oysters Pty Ltd v Ryan* ('*Graham Barclay Oysters*')	Included in water plans (*Graham Barclay Oysters*)	Included on water plans (*Graham Barclay Oysters*)	Small scale plants heavily regulated Health laws

members of an epistemic community may originate from a variety of academic or professional backgrounds, they are linked by a set of unifying characteristics for the promotion of collective amelioration and not collective gain. This is termed their 'normative component'. Some groups however, are only interested in local gain and pushing a business as usual agenda. The introduction of water plans generated local conflicts in several states as detailed in McKay 2006. The introduction of the first Murray Darling basin plan also generated local problems as some groups thought that the Plan favored the environment too much (see below).

Fig. 1 Complexity in institutions involved in water management in Australia

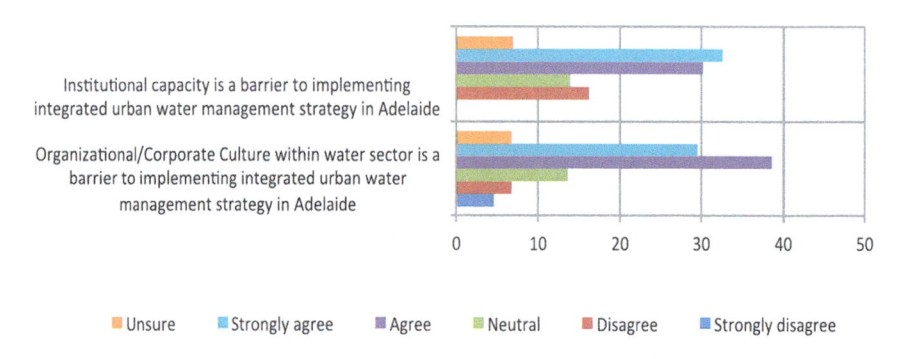

Fig. 2 Key stakeholder perceptions on organisational culture and institutional capacity as barriers

2 Australian Water Governance Institutional Aspects by Scale, Urban/Rural and Type of Water

This section provides some reflections on the overlapping governance issues on the three aspects of scale, urban rural and type of water. The Council of Australian governments reforms of 1994 (COAG 1994) required the states to include Ecologically Sustainable Development (ESD) in the state laws. (McKay 2005, 2011 and 2017) This they did in peril of not receiving funding under section 96 of the Constitution. Unfortunately, they used different formulations of the statement of ESD, which emerged out of the UN Conference on Environment and Development in 1992 Rio. They did include important aspects such as the precautionary principle, however, as Peel (2005) has demonstrated, the problems rested in the implementation and the implications of the precautionary principle for science-based decision making. The question, as Peel puts it, concerns 'to what extent are "science based" frameworks capable of operating in an environment of scientific uncertainty. When decision makers come to make decisions about environmental impacts'? Furthermore, what impact does uncertainty have on scientific rigour and objectivity of processes of health or environmental decision making? (Peel 2005).

The work below focusses on the broad ESD objectives in relation to water supply and asks the same questions as Peel.

2.1 The Law Statutory and Cases on the Broad ESD Rules

This legal requirement to include ESD was implemented in water supply legislation of the states from 1994, and the Commonwealth included this in the Water Act 2007. This is an example of the law leading the community, and in many ways, the law was a leader in shaping community views. Many sectors did not agree with the process at all (see McKay 2006 and further work below). Table 2 below summarise these issues.

2.2 Public Authority Decision Making and Liability and Private Sector Engagement in Water Supply Delivery

Several organisations, public and private, are at the hub of implementation of the laws above. The tension for the public authority is how much control they have over the resource and the real risk for them is in actions for negligence. This aspect is discussed by the High Court in *Graham Barclay Oysters Pty Ltd v Ryan* ('*Graham Barclay Oysters*') and it also deals with a private sector user (Table 3).

The rest of this section will summarise work done by McKay and Margaret 2017 interviewing water planners about the water planning processes on a couple of key dimensions such as trust in the process.

In *Graham Barclay Oysters Pty Ltd v Ryan* ('*Graham Barclay Oysters*'), the High Court considered in detail the circumstances in which a public authority may be liable under the tort of negligence for exercising or failing to exercise its statutory powers. As a result of the judgments handed down, the liability of public authorities in negligence actions has been significantly reduced (Stubbs 2003). Justices Gummow and Hayne (with whom Gaudron J agreed) began their analysis by observing that a duty of care is not established merely by showing that a public authority had knowledge of a risk of harm and a power to minimise that risk. Rather, the existence of a duty of care on the part of a public authority depends on

Table 2 Timeline and commentary on major water policy and law initiatives and cases since 2004

Water policy	Outcome
2004 the *National Water Initative* (NWI) and its objectives	The NWI provides for sustainable use of water, increasing the security of water access entitlements, and ensuring economically efficient use of water. These objectives are to be achieved principally by strengthening environmental flow provisions, removing barriers to markets in water, and providing for public benefit outcomes through water planning mechanisms. This included a requirement that water planning frameworks should recognise Indigenous needs in relation to access and management. Commentators such as Peoples Water Engagement Council 2012 and Jackson and Langdon (2012) have found that Indigenous engagement is lacking. In most water sharing plans for New South Wales' inland rivers (where competition over water is high and environmental degradation manifest), native title rights are currently not allocated any water. By contrast, in plans for coastal rivers where the competition is less intense, measures range from low impact ones that correlate to existing rights under native title legislation, to those that provide for new use such as Aboriginal commercial licences. Nevertheless, even these measures are limited in their scope and may contain onerous provisions, such as charging for water used for cultural/environmental purposes (Jackson and Langton 2012)
Water Act 2007 (Cth)	Williams and Kildea have said that '[t]he Water Act 2007 (Cth) is the most extensive Commonwealth intervention into water resource management in Australia since Federation'. After more than a century of the MDB being managed according to different, often competing, state-based water laws and policies, a Commonwealth water law had been passed with a coordinating framework based upon catchments, not state boundaries
ICM Agriculture Pty Ltd v The Commonwealth [2009] HCA 51	The question asked of the High Court was whether Commonwealth funding of action by NSW Government, brought the NSW Government actions under the just terms provisions of the Constitution. The water plans dramatically reduced the amount of water available in the consumptive pool to be allocated to farmers. The Court decided in the negative, as the funding of actions was not enough. The actors were the NSW Government. The relevant state laws had an ESD objective Statutory water licences are still unclear legally as to the type of property they amount to. However, it is clear that they can be altered by the Minister of the

(continued)

Table 2 (continued)

Water policy	Outcome
	state to achieve ESD. Each state had been able to do that through laws passed around Federation which rendered water common property. Contrary to this, in *ICM*, Justice Heydon stated that the licences were a form of property and the Minister could not reduce the licenses at will as they were stable. The powers of intervention in the public interest did not render the bore licences so slight or insubstantial or so inherently susceptible to modification or extinguishment, that they were incapable of being property. The breadth of the powers might affect the value of the property, but they were not so broad as to prevent it being categorised as property (per Heydon at 218) The remaining six justices declared water resources as common property hence all private interests derive from the public ownership and the Minster could reduce the volume of consumptive water to achieve ESD. The fact that the Commonwealth gave NSW the funds to do this did not mean that the just compensation clause in the Constitution applied
MDBA Plan 2010 and 2012	The controversial plan, which sparked wild protests in Griffith in 2010 and 2012, has sent shockwaves through irrigation towns, stripping confidence from business and real estate markets. The *Basin Plan* was prepared on a triple bottom line basis and, as a consequence, the SDLs in the *Basin Plan* provide significantly less water to the environment than was originally proposed. Williams and Kildea, along with the Australian Network of Environmental Defenders Offices both argue that this puts the *Basin Plan* at risk of being 'struck down by the High Court as being developed inconsistently with the terms of the Act'. The principal argument that the *Basin Plan* may be invalid is that the procedures that were required by law to be observed in connection with its making were not observed
2012 Victorian Government	The Victorian Government stated that it was concerned that Water Resource Plans (WRPs) and the Basin Plan undermined the security of Victorian water entitlements and so has not produced plans
Lee v Commonwealth [2014] FCA 432 surface water and section 100	An appeal to the Federal Court of Appeal in *Lee v Commonwealth* (Lee v Commonwealth of Australia [2014) 315 ALR 427 was dismissed by Middleton, Barker and Griffiths JJ. Special leave to appeal to the High Court was refused on 15 May 2015 in *Lee v Commonwealth* [2015] HCATrans 123 North J dismissed the claim, holding that section 100 of the Constitution only operated with

(continued)

Table 2 (continued)

Water policy	Outcome
	respect to laws based upon the trade and commerce power, and it did not operate with respect to water laws based upon the external affairs power, and that the *Water Act* was primarily based upon the latter. Supported by *Morgan v Commonwealth*, (1947) 74 CLR which held that the prohibition imposed by s 100 applied only to laws which were capable of being made under ss 51(i) and 98 of the Constitution. It did not apply, for example, to laws supported by the defence power, s 51(vi)
	Lee is particularly important as it supports the *Water Act* on the basis of the external affairs power. This means that it could apply to the entire nation and make it immune from a withdrawal of the referral of power by the states. (Williams et al. 2012)

Source Keremane et al. 2014

the 'terms, scope and purpose' of the relevant statutory provisions. The purpose of examining the statute is to determine whether the statutory regime 'erects or facilitates a relationship between the authority and a class of persons that, in all the circumstances, displays sufficient characteristics answering the criteria for intervention by the tort of negligence'. The judgment expresses the issues well and raises the ever present problem of justiciability:

1. Citizens blame governments for many kinds of misfortune. When they do so, the kind of responsibility they attribute, expressly or by implication, may be different in quality from the kind of responsibility attributed to a citizen who is said to be under a legal liability to pay damages in compensation for injury. Subject to any insurance arrangements that may apply, people who sue governments are seeking compensation from public funds. They are claiming against a body politic or other entity whose primary responsibilities are to the public. And, in the case of an action in negligence against a government of the Commonwealth or a state or territory, they are inviting the judicial arm of government to pass judgment upon the reasonableness of the conduct of the legislative or executive arms of government; conduct that may involve action or inaction on political grounds. Decisions as to raising revenue, and setting priorities in the allocation of public funds between competing claims on scarce resources, are essentially political. So are decisions about the extent of government regulation of private and commercial behaviour that is proper. At the centre of the law of negligence is the concept of reasonableness. When courts are invited to pass judgment on the reasonableness of governmental action or inaction, they may be confronted by issues that are inappropriate for judicial resolution, and that, in a representative democracy, are ordinarily decided through the political process. Especially is this so when criticism is addressed to legislative action or inaction. Many citizens may believe that, in various matters,

Table 3 Australian water governance—A preliminary overview of civil society organisations

Epistemic communities	Discipline: social science and humanities	Discipline: engineering and biological sciences	Lobby groups
Academics	Law, economics, political science and sociology: Law about 6 all over Australia not Economics about 20 all over Australia linked to professional bodies Political science about 5 Sociology a few	Engineering and biological sciences: Engineering many in all Universities and linked to Professional Bodies Biological sciences ditto	
Australian NGOS	River basin Management Society WaterAid Australia is an international NGO dedicated exclusively to the provision of safe domestic water, sanitation and hygiene education to the world's poorest Water for life|Oxfam Australia Act on Climate—Australia NGOs • Water Trust Alliance—Murray Darling Wetlands This first-of-its kind, national alliance of six key NGO's was formed to… Australian Conservation Foundation; Environmental Water Trust established by Nature Australian Conservation Foundation • Environmental Water Trust established by Nature Conservation Council NSW • Healthy Rivers Australia • Murray Darling Association • Murray Darling Wetlands Working Group • Nature Foundation of SA This first-of-its kind, national alliance of six key NGO's was formed to strengthen the role of communities in reviving wetlands and rivers and to bring the non-government sector of water management to the attention of the federal and state governments	Australian water association 700 corporate and 5000 individual members to help build a sustainable water future Australian Society for Limnology Australian Wetlands and Rivers (University of New South Wales) River Basin Management Society Waterkeepers (river care) Wetland Care Australia Wild Rivers—joint website by Queensland Conservation Council and The Wilderness Society (Qld)	Farmers Federations (each state) Lock the gate speak up irrigator lobby group

Source Submissions to National Water Commission and Productivity Commission 2011

there should be more extensive government regulation. Others may be of a different view, for any one of a number of reasons, perhaps including cost. Courts have long recognised the inappropriateness of judicial resolution of complaints about the reasonableness of governmental conduct where such complaints are political in nature. And

2. The proposition that the New South Wales government exercised substantial managerial control over the Wallis Lake oyster industry requires further analysis. If taken at face value, it virtually forecloses further debate. Control is a well-established basis for the existence of a duty of care in a public authority or a private citizen Managerial control, if it existed, would seem to equate the position of the state with that of the Barclay companies, which admittedly owed a duty of care. But what exactly does it mean to say, in a market economy, that the state has substantial managerial control over an industry conducted by private enterprise? Does it mean any more than that the government has governmental power? In the Federal Court, at the first instance, Wilcox J referred to the following aspects of control:

(a) the State owned, and had powers of control over, the lake;
(b) through the Department of Fisheries, it established, and supervised the operations of, a mosaic of oyster leases;
(c) through the Department of Fisheries, it issued, and enforced the provisions of, aquaculture permits;
(d) through the Department of Health, the State supervised the depuration process, including the nature and location of water intake points and the design, construction and maintenance of depuration tanks and ultra-violet facilities;
(e) through the EPA, the State had powers under the *Clean Waters Act* to remove, disperse, destroy or mitigate pollution of waters (s 27) and to carry out inspections and investigations of premises (s 29);
(f) through a number of agencies, the State was a participant in the Wallis Lake Estuary Management Committee, one of whose objectives was to prepare a management plan designed 'to sustain a healthy, productive and attractive estuary'; and, most importantly,
(g) through the Minister for Fisheries, it had the power—at any time, to prohibit the taking of oysters from the lake.

The practical way around this is for governments to use self-regulation plans.

In a study in metropolitan Adelaide of 55 key water planners, issues concerning governance were canvassed including policy and legal challenges, barriers and solutions. The results have been published in Keremane et al. (2014) and Cuddy et al. (2014).

Comments included:

- 'Unclear who is responsible or the driver for what. Near impossible to get diverse water supply projects being undertaken. State Gov. has no funding, staff or capacity to implement or administer/approve others to implement.'
- 'Too many BODIES trying to apply too many POLICIES for such a complex and life-critical resource.'

- 'Emphasis is still on traditional sources of water. Whilst excellent progress on using recycled water sources (inc. stormwater) has been made, this is not matched at a State Gov. level and so governance arrangements remain unclear and forgotten.'
- 'Highly fragmented with differing responsibilities with established cultures.'

There is growing support for Integrated Urban Water Management but the institutional arrangements are not always clear. The interviews showed that the impediments were socio- institutional such as too great a variety of regulatory regimes and lack of overall co-ordination, Lack of clarity about roles, responsibilities and accountabilities within the urban water sector, extreme and frequent levels of restructuring and institutional role separation within the public sector departments and conflicting agendas and/or differences in power among water agencies related to addressing water rights issues, and dealing with opponents to recycling or reuse.

There is no 'one solution' to address these issues but better governance models addressing these issue would be valuable. The governance models need also to promote long-term thinking (much longer that electoral cycles). The focus need to be on implementing institutional change through reform approaches that emphasise introducing developed coordinating mechanisms and improving intra- and inter-organisational relationships in a context of engaged community consultation.

This may require modifying existing legislation and policies to conform to a consistent framework based on the NWI principles.

2.3 Civil Society/Epistemic Communities and NGOs Engaged in Water Policy Development

In this domain of governance, the institutions are diverse and not very well coordinated, and hence make sporadic contributions to the debates on water policies. The table in this section shows this but this area is worthy of further research. In several studies members of civil society have suggested that they would prefer more federal government involvement in water management (see Keremane et al. 2014; Cuddy et al. 2014).

Society does not have one clear voice on water policy, even by rural or urban sector. Customer groups are often sponsored by the urban water authority in each state. Groups such as the Citizens against Drinking Sewage and Lock the Gate often focus on single issues, they are fragmented in space and may be short lived. Recently, however, some alliances are being formed between groups and this is a positive development.

3 China Urban Water Supply Institutional Models and Views of the Community About Foreign Players

This section was constructed after the research showed that China has used many foreign companies to supply water in their first phase of early development. Of late, the Chinese have withdrawn from the public private partnership model and instituted an integrated water resources management regime based on regions. The literature on this in English was limited so Dr Wu was assigned to answer the question what do civil society think of this process? A case study approach was adopted of Lanzhou Veolia water.

The Chinese public has never stopped debating on foreign investment's entry into China's water industry. The peak time often arises when there is water contamination or water tariffs are increased. The public have major concerns in three areas:

The first is with regards to national security. Water is different from other commercial goods. Given its vital interests to humankind, the public are concerned with national security when water is under the management of foreign companies (Zhang 2014; Wang 2014; Zong 2014; Tianya bbs 2014; Tianya bbs 2014a).

The second is related to public health. As long as there is water contamination reported and the water is under the management of foreign companies, public opposing foreign investment into China's water will be debated (Zhang 2014; Wang 2014; Zong 2014; Tianya bbs 2014; Tianya bbs 2014a).

The third is related to water tariffs. The foreign companies often increase water tariffs significantly (about 50%) when they take over the management of water supply and sewage from the Chinese government. This is always the peak time that the public oppose foreign investment into China's water too. The public request information transparency and auditing but the foreign companies often refuse (Lu 2009; Lang 2010).

The public criticism has put great pressure on foreign investment, which explains why foreign investment in China's water is much less active in recent years (Ma 2014; Zhang 2016). Instead, local private companies have been more actively participating in China's water industries.

3.1 Case Study: Lanzhou Veolia Water

The portable water was found to be contaminated in April 2014 in Lanzhou, the capital city of Shannxi Province, with a population of more than 3.6 million. Drinking water was found containing dangerous levels of benzene (over 20 times the safe level) in Lanzhou. The water supply company, Lanzhou Veolia Water Co, is majority-owned by the City Government, with Veolia China, a unit of French firm Veolia Environment, holding a 45-percent stake (Lang 2010). As soon as the news was released local residents were sent into panic. Bottled water sold out immediately and local residents had to wait in long queues to access the safe water later provided by government.

One article about water contamination was posted on Tianya bbs, a very popular online social network (http://bbs.tianya.cn/post-free-4259644-1.shtml), on the next day and received 76,702 views and 889 comments in two weeks. The article was reposted by a few other online media sites, for example Qudunet, (http://117.126.243.8/qudunet/page_detail.php?id=90452&page=1; receiving 232 comments and a large number of views). Huanqiu Times (2014) released an editorial commentary (http://opinion.huanqiu.com/editorial/2014-04/4969219.html) on the following day, instigating a heated debate amongst the public. In total, 5,087 people participated in the debate.

Besides these, some people posted their opinions on blogs and received significant public attention. Mr Zhang (2014) posted an article in his blog and expressed his concerns for national security arising from the entry of foreign investment into China's portable water industry. Considering there is no technical difficulties for Chinese people in managing water by themselves, he questioned why the Chinese government needs Western countries to take control of China's water supply and sewage.

Mr Zhang (2014) argued that in Western countries, for example France, management of water by foreign companies is not permitted, whether it be tap or bottled water. In America, water cannot even be managed by American companies and is solely under the management of American army. Zhang (2014) strongly recommended that the Chinese government take back the control of water management from foreign companies and transfer control and responsibility to the Chinese Government. Zhang (2014)'s opinions received a large amount of support from public.

Hongge Net (2014), an online media site, posted one article responding to Zhang (2014) by expressing the concerns of national security due to foreign investment into China's water. The author further argued that the immediate increase in water tariff by Veolia was not acceptable without any spending on water infrastructure. It was reported that Veolia didn't spend money on water infrastructure in Lanzhou but they immediately increased water tariffs by a large proportion (49%) once they took over the water supply and sewage for the city.

Mr Wang (2014) posted an article on Sina China, one of the most famous online media sites, and reported that local residents were unhappy about Veolia increasing water tariffs but not putting any actual and extra effort into the city's water management. Veolia reported that they didn't make any changes to water supply and water sewage for the city, aside from the change of names from Lanzhou Water to Lanzhou Veolia Water. The article was followed and supported by many people.

Veolia was criticised because they didn't immediately release the information about water contamination to residents when it was found. They hid the information for about 10 h during which the local residents still drank the contaminated water. The local residents were shocked and thought it was unacceptable. They again questioned the utility of foreign companies managing China's water: 'the foreign companies as businesses only care of their investment returns. They have no care about the local residents' health'.

Mr Zhang (2014) argued that this was not the first time that Veolia was found to be responsible for water contamination. Veolia was found connecting a water sewage pipe to portable water supply pipes in 2007. Veolia was also found to discharge polluted water (treated but not meeting the requisite standards) to sea in Qingdao.

According to Hsien (2010), who is a well-known Hong Kong-based economist, commentator, author and TV host in China, Veolia is the largest foreign investor into China's water, taking over 60% of China's water business (supply and sewage). It has projects in 34 provinces in China. The reason that Veolia could have such a large share of China's water was because of the very high purchase price they offered in the bids, almost three times of the actual assets possessed (Lang 2010). The CEO of one private water company commented: 'this is crazy. It is impossible for Veolia to get its returns back through appropriate running because the price was unbelievably high. The local governments are crazy too.' The Chinese economists doubted the authenticity of Veolia's financial statements because they had increased assets but did not have investment and profit reported. It was assumed the reason was hidden in the contract between Veolia and Lanzhou government. The contracts specified that when the 30 year contract expires, Lanzhou government needs to purchase back the Lanzhou Water from Veolia.

In addition to the criticism of foreign investment in this case mainly on Veolia, there were also different opinions. Ms Wang (2014a) posted an article on Wangyi China, another famous online media site in China, arguing the statement that foreign investment threatens China's water security was overstated. The author considered the Chinese government should be responsible for the water issues rather than the foreign companies. She argued:

1. The statement that foreign companies have main control over the water industries is not true. The Chinese Government still keeps the control of the water industries;
2. The water supply and sewage infrastructure in Lanzhou needed to be upgraded urgently but the local governments did not put in the necessary spending. Rather they hoped the foreign companies (Veolia) would bear the cost;
3. The foreign companies had to meet the requirement of Chinese government as water has public interests. So those foreign companies experienced loss because of policies and regulations in China;
4. Water contamination had no relation with whether it foreign or local companies are in charge of water management. Even the local Chinese government-owned companies could have had the same issues;
5. Water tariff was still controlled by local governments. The foreign companies had no rights to increase water price by themselves.

However, the author did not provide enough evidence for her opinions. The article received 12,084 views and 1,516 comments on the same day. The majority of comments disagreed with the author's opinions and some of them had provided evidence in opposing the author's arguments. But there was an agreement amongst public that local governments need to take the roles and responsibilities of water management in securing public health and national security.

4 Research Agenda for Governance in Australian Water

There are several issues requiring more legal doctrinal research and socio-legal research.

- The corporatisation phenomena requires more research, particularly into how actors balance the public interest when making decisions. The work on *Graham Barclay Oysters* above showed several uncertainties in the relative powers and responsibilities of public and private sectors. The public-private partnerships approach needs some refinement and better definition. One key legal aspect worthy of attention are the so-called alliance contracts. These aim to inhibit the use of courts in dispute resolution and the legality of this and the changes this could impose on contract law need to be investigated.
- The role of civil society and epistemic communities and NGOs requires more work as well. In particular, research is required into how they interact, including some detailed case studies of the governance structures used to reconcile disputes.
- Legal doctrinal work on the issues arising from the *Barclays* case and the degree of control rendering a public authority liable for non-feasance in supervision. This is acute if manufactured water is to be used for purposes closer to human contact than at present.

5 Summary and Reflections on Questions

Water governance has been shown to be complex. The three stakeholders—private, public and civil society are not well integrated. The law is firmer thanks to *Lee*'s case and clearly provides the *Water Act* with a firm foundation based on the external affairs power of the Constitution. Hence the objects in Sect. 3 (see Table 2) could be applied to the entire nation and this would start to bring some coherence. The national interest concept would achieve some legal definition and this would be a positive outcome. In concert with this a standard corporate form would also improve cross-jurisdictional dialogue (see McKay 2005).

Australia's approach to designing water governance is not efficient as it is beset by too many jurisdictional differences, it is not equitable between members of different states. The inequities are not great but the focus is still on local issues, not national issues.

The implementation of water governance is very disparate between the states and the *Water Act 2007* only applying to the Murray-Darling Basin introduces tiers of objects and tests and promotes further inefficiency and inequitability. Resilience at a local level is high at present but climate change will likely change this.

What conditions have enabled or blocked its success, including environment, social and political and legal?

The partisan local interests have blocked successes and non-compliance with the *Water Act* and the drafting of plans in an issue. The law in each state has been robustly implemented by judges (see McKay (2011)) and *Lee*'s case opens up more power to the Commonwealth over water.

How does Australia's hybrid governance system of collaborative planning, markets and regulation map onto or contrast with different international water governance practices, particularly in China?

The case study shows that China adopted elements of public-private partnerships and that the community is unlikely to accept these in the future.

What does Australia's use of markets, regulation and collaborative planning (both separately and their interaction) suggest for our understanding of water and broader environmental governance theory, different governance modes and normative policy design?

Normative policy design needs agreed norms and this is lacking in Australia. Markets have been criticised by some and embraced by others. The implications for broader governance theory is that the three governance players need to be operating in a less diverse system in order to have it operate in a way to achieve ESD.

References

AAP. (2006). Toowoomba says no to recycled water. *Sydney Morning Herald* (online). Retrieved Mar 27, 2017, from http://www.smh.com.au/news/national/toowoomba-says-no-to-recycled-water/2006/07/29/1153816419568.html.

Australian Network of Environment Defenders Offices, Submission to Senate.

Council of Australian Governments. (2004). Intergovernmental Agreement on a National Water Initiative (National Water Commission).

Cuddy, S. M., Maheepala, S., Dandy, G., Thyer, M. A., MacDonald, D. H., McKay, J., Leonard, R., Bellette, K., Arbon, N. S., Marchi, A., Kandulu, J., Wu, Z., Keremane, G., Wu, Z., Mirza, F., Daly, R., Kotz, S., & Thomas, S. (2014). A study into the supply, demand, economic, social and institutional aspects of optimising water supply to metropolitan Adelaide—preliminary research findings: Summary report from Project U2.2. Goyder Institute for Water Research Technical Report Series No. 14/20, Adelaide, South Australia. ISSN:1839-2725.

Dolnicar, S., & Hurlimann, A. (2010). Australians' water conservation behaviours and attitudes. *Australian Journal of Water Resources, 14*(1), 43–53.

First Peoples Water Engagement Council. (2012). Advice to the National Water Commission. Retrieved Mar 22, 2017, from http://nwc.gov.au/__data/assets/pdf_file/0004/22576/FPWEC-Advice-to-NWC-May-2012.pdf.

Fisher, D. E. (2000). *Water law*. North Ryde: LBC Information Services.

Fisher, D. E. (2010). The High court and techniques of judicial reasoning. *Environmental and Planning Law Journal, 27*, 85–97.

Graham Barclay Oysters Pty Ltd v Ryan (2002) 211 CLR 540.

Hamstead, M., Baldwin, C., & O'Keefe, V. (2008). *Water allocation planning in Australia—Current practices and lessons learned*. Canberra: National Water Commission.

Haas, P. M. (1992). Introduction: epistemic communities and international policy coordination international organization. *International Organisation, 46*(1), 1–35.

Haque, M. S. (2001). The diminishing publicness of public service under the current mode of governance. *Public Administration Review, 61*(1), 65–82.

Howard, J. (2007). Speech to the National Press Club on January 25 2007. Retrieved Mar 20, 2017, from http://parlinfo.aph.gov.au/parlInfo/download/media/pressrel/K81M6/upload_binary/k81m68.pdf;fileType=application%2Fpdf#search=%22media/pressrel/K81M6%22.

Jacobs, M. (2006). Securing Australia's urban water supply: Research notes for selected case studies prepared for research notes prepared for department of the prime minister and cabinet, unpublished.

Jackson, S., & Langton, M. (2012). Trends in the recognition of Indigenous water needs in Australian water reform: The limitations of 'cultural' entitlements in achieving water equality. *Journal of Water Law, 22*, 109–123.

Keremane, G., Wu, Z., & McKay, J. (2014). Institutional Arrangements for Implementing Diverse Water Supply Portfolio in metropolitan Adelaide—Scoping Study. Goyder Institute for Water Research Technical Report Series No. 14/14, Adelaide, South Australia. ISSN:1839-2725.

Kooiman, J. (2003). *Governing as governance*. London: Sage.

La Nauze, J. A. (1972). *The Making of the Australian Constitutio*, pp. 153–210. Melbourne University Press.

Lafferty, W. M. (2004). Introduction: Form and function in governance for sustainable development W. In W. M. Lafferty (Ed.), *Governance for sustainable development: The challenge of adopting form to function* (pp. 1–3). Cheltenham: Edward Elgar.

Lange, P., Driessen, P. P. J., Sauer, A., Bornemann, B., & Burger, P. (2013). Governing towards sustainability: Conceptualizing modes of governance. *Journal of Environmental Policy and Planning, 15*(3), 403–425.

Lee v Commonwealth of Australia [2014] FCA 432.

Leroux, A. D., & Martin, V. L. (2016). Hedging water supply risks: An optimal water portfolio. *American Journal of Agricultural Economics, 98*(1), 276–296.

McKay, J. M. (2005). Water institutional reforms in Australia. *Water Policy, 7*, 35–52.

McKay, J. M. (2011). Australian water allocation plans and the sustainability objective—conflicts and conflict resolution processes. *Hydrological Sciences Journal, 56*(4), 615–629.

McKay, J. M. (2017). Australian jurisprudence of justice in water management: Present limitations, future issues and law reform suggestions. In A. Lukasiewicz, S. Dovers, L. Robin, J. M (eds.) *Natural resources and environmental justice: Australian perspectives*.

McKay, J., Schilizzi, S., & Graham, S. Clayton: CSIRO Publishing, pp. 215–232.

Murray-Darling Basin Authority. (2013). Handbook for practitioners—water resource plan requirements (October 2013). Retrieved 20 Mar, 2017, from www.mdba.gov.au/sites/default/files/pubs/WRP-Handbook-for-Practitioners.docx.

Moon, E. (2017). Why does the Condamine coal mine need to use so much water? The Conversation. Retrieved 22 April, 2017, from https://theconversation.com/why-does-the-carmichael-coal-mine-need-to-use-so-much-water-75923?utm_medium=email&utm_campaign=Latest%20from%20The%20Conversation%20for%20April%202018%202017%20-%2072085459&utm_content=Latest%20from%20The%20Conversation%20for%20April%2018%202017%20-%2072085459+CID_522c4d597334031b2ae00cb2e2f7021f&utm_source=campaign_monitor&utm_term=Why%20does%20the%20Carmichael%20coal%20mine%20need%20to%20use%20so%20much%20water.

Peel, J. (2005). *The precautionary principle in practice: Environmental decision making and scientific uncertainty*. Annandale: Federation Press.

Peoples Water Engagement Council. (2012). Advice to the National Water Commission. Retrieved March 25, 2017, from http://nwc.gov.au/__data/assets/pdf_file/0004/22576/FPWEC-Advice-to-NWC-May-2012.pdf.

Pennie, S. (2016). Floods drown the MDBA plan. The Land. Retrieved March 22, 2017, from http://www.theland.com.au/story/4233133/from-burning-to-drowning-water-plan/.

Rothwell, D. (2012). International law and the murray-darling basin plan. *Environmental Planning and Law Journal, 29,* 268–280.

Stubbs, M. (2003). Prosper the government, suffer the practitioner: the graham barclay oysters litgation. *University of New South Wales Law Journal, 26*(3), 727–740.

Standing Committee on Legal and Constitutional Affairs, *Inquiry into Provisions of the Water Act 2007,* March 2011.

State Government Victoria. (2012). The Victorian Government Submission to the Proposed Basin Plan. Retrieved March 20, 2017, from http://www.depi.vic.gov.au/__data/assets/pdf_file/0007/176578/Basin-Plan-Proposal-April2012.pdf. *Water Act 2007* (Cth).

Williams, G., & Kildea, P. (2011). The water act and the murray-darling basin plan. *Public Law Review, 22,* 9–14.

References Chinese Section

Du, J. (2014). 兰州水务事件主角是又一个被媒体骄纵坏的私企. http://www.guancha.cn/dujian-guo/2014_04_12_221458.shtml. Received 489 commnets.

Hongge Net. (2014). 威立雅与兰州水厂的前世今生, 红歌会网. http://news.ifeng.com/a/20140613/40722213_0.shtml.

Huanqiu Times. (2014). 社评:兰州水污染,三月辟谣四月成真之悲. http://opinion.huanqiu.com/editorial/2014-04/4969219.html.

Lu, Y. (2009). 外资大规模进入中国水务市场 水价能否让百姓全部买单. 2009年08月06日 08:05时代周报. http://finance.ifeng.com/topic/news/sjsz/news/hgjj/20090806/1048270.shtml.

Lang, H. P. (2010). 郎咸平揭秘中国水价涨声一片的内幕2010/01/28. http://book.ifeng.com/shuzhai/detail_2010_08/07/1905317_0.shtml.

Ma, W. (2014). 外资水务在华扩张受阻 水务市场化改革面临考验, 2014年08月01日 23:30 华夏时报. http://finance.sina.com.cn/china/20140801/233019891212.shtml.

Tianya bbs. (2014). 威立雅集团.又一个闹出大事的大型私企 (转载). http://bbs.tianya.cn/post-worldlook-1082929-1.shtml.

Tianya bbs. (2014a). 兰州水污染:谣言总是成真的背后及国人可怜的底线2014-04-13. http://bbs.tianya.cn/post-free-4259644-1.shtml.

Wang, X. (2014). 兰州威立雅水务公司的罪与罚, 新浪网. http://www.wyzxwk.com/Article/shehui/2014/04/317776.html.

Wang, L. (2014a). "外资威胁中国水务安全"是言过其实, the other side, issue 1046, Wangyi. http://view.163.com/special/reviews/water0414.html.

Zhang, H. (2014). 震惊:美国自来水由军方管理,中国自来水由外资经营——由兰州自来水事件引发的对外资控制中国自来水的质疑. http://blog.kjsdhedn.net/cbym/19989.html.

Zhang, K. (2016). 洋水务中国受阻水务市场化改革面临考验. http://blog.hosjb.com/lsmd/18893.html.

Zong, H. (2014). 兰州自来水苯超标事件背后的真相, 红歌会网. www.szhgh.com/Article/news/politics/2014-04-13/49348.html.

Lessons from Australian Water Reforms: Indigenous and Environmental Values in Market-Based Water Regulation

Elizabeth Macpherson, Erin O'Donnell, Lee Godden and Lily O'Neill

Abstract The Australian model of water governance is considered one of the most effective, efficient and resilient approaches to designing and implementing water governance. In place since the early 1990s, the Australian approach is a hybrid governance system involving collaborative planning of water resources together with market mechanisms and statutory regulation. However, in implementing the model, successive reforms have yet to completely redress the historical exclusion of Aboriginal peoples from water law frameworks, and have struggled to account for the needs of a healthy and sustainable aquatic environment. In this chapter we examine the trajectory of water law and policy reform in Australia, including two of the most recent developments: the push to intensify water development in the northern Australian *White Paper* and the collaborative planning approach set in the *Water for Victoria* policy. Our study of the incremental and evolving Australian water law reforms highlights the difficulty of ensuring fairness in the operation of hybrid governance systems for water regulation, and reveals important lessons for international policy-makers embarking on and implementing water reforms in their

---

The authors thank Cameron Holley and Darren Sinclair for organisation of an excellent colloquium that stimulated the initial research. Authors acknowledge assistance provided by CCIF grant 501319GL, The University of Melbourne. The authors' views are their own.

E. Macpherson (✉)
School of Law, University of Canterbury, Christchurch, New Zealand
e-mail: elizabeth.macpherson@canterbury.ac.nz

E. O'Donnell · L. Godden · L. O'Neill
Centre for Resources, Energy and Environmental Law, Melbourne Law School (CREEL), Melbourne, Australia
e-mail: erin.odonnell@unimelb.edu.au

L. Godden
e-mail: lcgodden@unimelb.edu.au

L. O'Neill
e-mail: lily.oneill@unimelb.edu.au

© Springer Nature Singapore Pte Ltd. 2018
C. Holley and D. Sinclair (eds.), *Reforming Water Law and Governance*,
https://doi.org/10.1007/978-981-10-8977-0_10

own jurisdictions. From its inception, strategic planning for innovative water law reform must be supported by meaningful engagement with Indigenous peoples, and embed Indigenous and environmental values and rights in water planning and governance.

1 Introduction

The Australian approach to water law and governance has been characterised by innovative change that reflects extensive reforms to water law, policy and institutional practice. Initiated in 1994, expanded in the National Water Initiative in 2004 and culminating in the federal *Water Act 2007* (Cth), the reforms sought to integrate strategic management objectives through water planning and responded to the environmental effects of over-allocation in the Murray-Darling Basin. The result is a hybrid governance system of collaborative planning of water resources together with market mechanisms and statutory regulation that provides a robust model for potential application in many countries facing water over-allocation across multiple jurisdictions, periodic drought and water scarcity.

The model that developed in Australia is recognised globally as being effective, efficient and resilient in its design and implementation of water governance (Garrick and O'Donnell 2015; Grafton et al. 2011). Yet these far-reaching reforms around water planning and allocation also largely failed to address Indigenous peoples' interests until relatively recently. Sophisticated Indigenous water resource management systems were displaced by the colonisers who introduced the British common law system of riparian water rights, allowing the unregulated use of adjacent water sources by settler landholders. Australian jurisdictions are now experimenting with a range of policy and legal measures that are aimed at providing a more inclusive and equitable platform for Aboriginal peoples and Torres Strait Islanders to engage in water law and governance. Similar experimentation is occurring in other countries, including in Latin America and New Zealand (Ruru 2013; Macpherson 2017), as Indigenous communities seek a stronger voice in water management and assert claims to water rights based on traditional historical associations with water. There is also a strong impetus for water law and governance to respond to human rights imperatives.

This chapter provides an analysis of Australian water law reforms to inform the growing transnational interest in environmental water policy and Indigenous rights to water. It details the poor historical environmental management that lead to over-allocation, and then the environmental flow regimes that were devised in order to protect environmental values in river systems, some using market mechanisms in the interests of the environment (Ison and Wallis 2011). The chapter also details developments such as Indigenous strategic reserves and cultural flows provided for in water planning—water entitlements or shares which are set aside or held by Indigenous people to improve spiritual, cultural, social and environmental conditions, but usually fall short of commercial rights—intended to provide an 'equitable' share of water resources for indigenous use (Jackson and Barber 2013). There is evidence of similar mechanisms being adopted in other places: including in South American countries, like Chile (O'Donnell and Macpherson 2013). Yet, the Australian experience highlights the difficulty of ensuring fairness in the operation of hybrid governance systems for water regulation; seen particularly in the failure to redress historical and current exclusion of Aboriginal peoples' rights to water, and the continuing task of meeting the needs of a healthy and sustainable aquatic environment. Both of these pose challenges for the ongoing stability and integrity of the Australian model (Lindsay this volume).

The study of evolving Australian water law reforms in this chapter therefore reveals important lessons for international policy-makers embarking on and implementing water law reforms in their own jurisdictions; around the need to appreciate and plan for indigenous and environmental values in the design phase. We examine the trajectory of water law and policy reform in Australia, including the latest push for water development in the northern Australian *White Paper,* and the direction set in the recent *Water for Victoria* plan. We argue that strategic planning for innovative reform must be supported by meaningful and constructive engagement with Indigenous peoples[1] in planning and implementation, and embed environmental and Indigenous values in water governance.

2 Water Governance in the Driest Continent

Australia has an unenviable status as the world's driest continent supporting a permanent population (Cullen et al. 2002). Much of Australia's inland area receives less than 500 millimetres of rainfall annually, exacerbated by high evaporation rates, while the coastal fringe has moderate rainfall (Australian Government Bureau of Meteorology 2017). Many of Australia's 246 river basins do not have permanent flow regimes but are marked by periods of intermittent flow. Australia is heavily dependent upon groundwater with a large aquifer system, the Great Artesian Basin, supplying water for inland uses (Tan and Quiggin 2004). Australia also experiences cycles of drought and severe flooding and geographic and seasonal distribution of rainfall varies greatly (Australian Government Bureau of Meteorology 2017).

Indigenous exclusion from water law frameworks has its origins in Australia's colonisation from 1788. Until 1992, the law accepted that the British Crown assumed sovereignty over Australia because Aboriginal people were 'without laws, without a sovereign and primitive in their social organization' (*Mabo v Queensland [No 2]* (1992) 175 CLR 1, 36 ('*Mabo*')). Across the continent, British colonisation resulted in the decimation of Aboriginal populations through murder, disease and

---

[1]The term 'Indigenous people' refers to both Aboriginal and Torres Strait Islander communities. As this discussion is focused on mainland Australia, this Chapter uses "Aboriginal peoples" in relation to the relevant communities affected, and 'Indigenous' in relation to a broader discussion of water rights for Australian Indigenous people.

starvation (McHugh 2011, p. 1). The landmark Australian High Court decision in *Mabo* found that Aboriginal and Torres Strait Islander rights to land and water survived the acquisition of sovereignty by the British Crown, and, in certain circumstances, could be recognised by the Australian common law. 'Native title' under the *Native Title Act 1993* (Cth) recognises the rights as the communal, group or individual rights and interests in the land and waters of Aboriginal or Torres Strait Islander people. Proof of the rights requires that claimants demonstrate that they are a society that has continued its connection to those lands and waters since pre-sovereign times according to their traditional laws and customs (*Native Title Act 1993* (Cth) s 223; *Mabo* pp. 59–60; Tehan 2003; Strelein 2009).

Human settlement of the driest continent is now concentrated in the south-east (and parts of the south-west), including major urban areas and agricultural hinterlands (Australian Government Bureau of Statistics 2012). The north is less intensely settled, other than mining and scattered agricultural developments. Northern Australia includes a higher percentage of Aboriginal land; including tenures held pursuant to statutory land grants and native title (National Native Title Tribunal 2016).

The contrast in water conditions between different geographic regions in Australia has led to land characterisation as either part of the 'dry south' or the 'wet north.' Schemes have repeatedly been devised to develop the 'wasted' water resources of the north (Head 1999, p. 153), both for northern irrigated agricultural expansion and transfers via pipeline to the south, where agriculture was already firmly established (Gibbs 2009, p. 2968). There have been parallel debates about whether a lack of water in inland areas would pose an ultimate limit to settlement (Gibbs 2009, pp. 2964–2969).

By the mid-twentieth century commentators began to argue that irrigated agriculture in Australia, particularly in northern regions, was not economically 'efficient' (Davidson 1969). Davidson, for example, argued that the damming of the Ord River in Western Australia—first mooted in the early twentieth century, and achieved at astronomical cost in 1972 with the main dam creating Lake Argyle—was a waste of government resources. Agricultural expansion north of the Tropic of Capricorn was not economically rational, Davidson argued, due to harsh seasonal conditions, lack of labour, distance from markets, pests, and inappropriate soils. Instead, money should be channeled into dry-land agricultural innovation (Davidson 1965).

However, the need to develop the land, and avoid land and water going to waste has been a key theme in Australian policy debates around water governance (Head 1999, p. 153). Such an approach rested, Head argues, on colonial views of 'the empty landscape, the invisible Aborigine, and the idealization of agricultural land use' (Head 1999, p. 141). These ideas continue to inform plans to open up new areas of Australia to development, despite the increasing attention given to Aboriginal and environmental issues, and have been 'at least as powerful in driving the process as have any rational or quantitative assessments' (Head 1999, p. 142).

A commitment to irrigation has placed many Australian rivers under increasing pressure, especially in the southern parts of the continent. Irrigation dominates the consumptive use of water within Australia (Australian Government Bureau of Statistics 2013), a high proportion of which occurs in the Murray-Darling Basin. While only 10% of the Basin is formally categorised as 'over-allocated', the entire Murray-Darling Basin might be considered over-allocated where there is insufficient environmental water to deliver sustainability and protect ecological functions (Connell 2007). A large number of Australia's other catchments and ground water supplies have allocations exceeding available sustainable flow, and there are ongoing attempts to address these problems.

Water governance has inevitable social, as well as natural, implications, and 'uneven power relations' influence social access to water rights (Jackson and Barber 2016). The commitment to irrigated agriculture, dominant in public discourse on water governance since colonisation, and the impetus to open 'waste' lands up to development, has also ignored Indigenous peoples, and the uses they have made of water since long before the acquisition of sovereignty pursuant to sophisticated water access rights (Jackson and Langton 2011, p. 109). The way in which law has enabled and responded to these challenges of water governance in Australia is considered in the following sections.

3 Legal Frameworks Enabling Water Governance in Australia

3.1 Early Approaches: Common Law Riparian Rights and State-Based Statutes

The need for water to support productive land development pervaded early legal frameworks for water governance in Australia. Until the late nineteenth century the common law riparian doctrine provided for rights to water as derivate of property in land (Gardner et al. 2009). A landholder, by virtue of its title, held the right to take and use water on or adjacent to its land. Water that did not flow in a defined river channel was exploitable by a landowner without restriction. Thus, ground and surface waters—very significant parts of the total flow volume of many Australian drainage basins—could be harvested without limit. The common law favoured user rights (Tan 2002, p. 15), but the limitations of this model became apparent in the water-scarce southern and interior areas of the Australian mainland, prompting a move to a different regulatory framework (Tan 2002, p. 16). After severe droughts in the 1880s (Evans and Howsam 2005), Australians began to demand public 'responsibility' for water governance (Powell 1989, pp. 100–104). Towards the turn of the century, several Australian states vested the right to the use and flow of surface and ground water resources in the Crown, and implemented a system of water licences and concessions to authorise the 'consumptive use' of water, focused upon irrigated agriculture.

River basins became the key focus of the new state-based water law frameworks, to support irrigated agriculture via tightly managed hydrological regimes as a 'tenet of productivism' (Molle 2009, p. 486). Water infrastructure was seen as key to supporting the new approach (Musgrave 2008, p. 35), and extensive public funding was poured into irrigation schemes in the early-twentieth century, including those in the Murray-Darling Basin, to irrigate large areas of dry inland within a worldview of nation-building (Wooding 2009, p. 58). However, the environmental aspects of water attracted little emphasis, and by the early 1980s significant and sustained concerns were raised about impacts on water quality and quantity (Musgrave 2008, p. 39). The legal and institutional structures built around river basins, dams and irrigated agriculture began to rupture, triggering major water law reform.

Moreover, the productivist approach to early water governance in Australia completely ignored the relationships Indigenous Australians had with water resources since long before colonisation (Jackson and Barber 2016). Neither the British Crown nor subsequent Australian governments acknowledged that Indigenous Australians held state-recognised titles with respect to land or resources until the late twentieth century after the introduction of state-based land rights legislation from the 1970s and High Court decision in *Mabo* in 1992, although some areas of lands were specifically reserved for Aboriginal use. A nexus linking water use to land holding as a basis for agricultural development applied during the same period (Council of Australian Governments 1994), meaning that Indigenous groups, who typically did not hold Torrens (registered) land titles, did not enjoy access to statutory water entitlements, and therefore could not lawfully make use of water on or adjacent to their traditional territories.

3.2 Introducing the Hybrid Model: Major Water Reforms of the Late Twentieth Century

By the 1980s, there was increasing pressure on Australian governments to react decisively to intensifying environmental degradation in the Murray-Darling Basin, prompting nation-wide, comprehensive water reform (Department of the Environment and Heritage 2004). At the same time concepts of sustainability and integrated catchment management were introduced to state-based water legislation, including the *Water Act 1989* (Vic) (s 1(b)); *Water Management Act 2000* (NSW) (s 3); *Water Act 2000* (Qld) (s 10(1)) and *Natural Resources Management Act 2004* (SA) (s 7). In 1994 the Council of Australian Governments (COAG) agreed to cap water entitlements (finalised in the Murray-Darling Basin in 1997), improve transparency of water pricing, separate water from land and establish property rights to water, and allocate water to the environment, as part of a broader productivity-based policy reform agenda (Australian Government National Water Commission 2011).

As Australia has a federal system of government there is inevitable overlap between governments in managing water, with considerable tension between jurisdictions. Water regulation has, historically been the responsibility of Australian states and territories. However, a series of high-profile environmental cases established that the Commonwealth Government could use 'indirect' constitutional legislative heads of power to support environmental legislation with respect to natural resources (Crawford 1991). The National Water Initiative was agreed by COAG in 2004, formalising the new approach to water governance and linking state achievement of water reform goals to the receipt of Commonwealth funding. The Initiative committed Australian states to complete the separation of water rights from land; facilitate water trading; set aside legally protected water for the environment; and return over-allocated systems to an environmentally sustainable level (Council of Australian Governments 2004).

The Commonwealth Government progressively assumed a stronger role in water management at the end of the twentieth century under the COAG reforms and in response to major droughts in south-eastern Australia from 1997 to 2009 that significantly reduced water allocations for irrigation and urban water use (State of Victoria 2009, p. 16). The introduction of the *Water Act 2007* (Cth), together with a $10 billion water plan to save the Murray-Darling Basin, was premised upon the states referring constitutional powers to the Commonwealth, enabling it to pass its own legislation (confirmed in Sects. 9 and 9A of the *Water Act*). For the first time, the Commonwealth Government had specific water management responsibilities. The reforms introduced new institutions, in the form of the Murray-Darling Basin Authority and the Commonwealth Environmental Water Holder, a sustainable diversion limit on water extraction under the Murray-Darling Basin Plan 2012 (Cth), and significant investment in water recovery for the environment, via purchase programs and investment in water efficiency (Department of Sustainability, Environment, Water, Population and Communities 2011).

After these reforms, together with emblematic changes to the nature of water access entitlements and their unbundling from land titles and transferability in markets, the Australian approach to water governance was dominated by strategic water planning at the catchment level. 'Water planning' is an attempt to match water supplies and water demands, both present and future, so that water resources are managed in a reliable and environmentally sustainable manner, dependent on assessment of the current and future availability and use of water and community stakeholder consultation (Council of Australian Governments 2010). Under the Guidelines for the National Water Initiative environmental flows must have the same level of legislative protection as consumptive water use (Council of Australian Governments 2010, pp. 32–33), while recognising the 'significant' economic value of water to irrigated agriculture, among other water users (Council of Australian Governments 2010 p. 18).

While the COAG reform process was gaining momentum in the early 1990s, Australia was also beginning to recognise Indigenous land rights including, notably, with the recognition of native title in the *Mabo* decision and subsequent *Native Title Act*. However, because water rights were unbundled from landholding as part of the first stage of the COAG reforms, those Indigenous landholders that acquired land rights, for example in the native title process, would not, as a matter of water law, acquire the right to use water on those lands (Macpherson 2016). This can be contrasted to the situation prior to reforms where the grant of land would have entailed rights to water. Further, because after the COAG reforms rights to use water could be acquired independent of land ownership, third parties could obtain rights to use water on areas over which, for example, there may be recognised native title. Accordingly, if Indigenous water rights are to be provided for, they require specific protections.

The National Water Initiative did acknowledge Indigenous interests and required water plans to take account, at least, of native title rights to water (Council of Australian Governments 2004, [53]). However, it failed to make substantive provision for Indigenous water rights or to effectively engage Indigenous peoples. Most agitation for Indigenous water rights has occurred in subsequent years.

The non-binding, policy character of the Initiative meant that few measures emerged to provide specific water-holding entitlements for Indigenous Australians. It provided that Indigenous peoples' water needs should be taken into account in water planning processes, in the form of water allocations for 'traditional cultural purposes' (Council of Australian Governments 2004, [28]–[34], [59]); to maintain culturally important ecosystems. Thus emerged the concept of 'cultural flows' in the Murray-Darling Basin (Murray Lower Darling Rivers Indigenous Nations and Northern Murray–Darling Basin Aboriginal Nations 2007), for which water should be allocated under either environmental regimes or through a non-tradeable environmental and public benefit water access entitlement (Council of Australian Governments 2004, p. 32). Variations of this concept have been taken up by some state government and local water authorities who appear increasingly keen to incorporate Indigenous water values into water planning. For example, there is a requirement under the Murray Darling Basin plan for consultaion with Aboriginal groups to this effect. The effectiveness of these schemes is yet to be gauged in many instances.

The focus of Australian water planning on Indigenous cultural water values rather than commercial water rights implicitly relies on assigning responsibility to the native title model for attaining holistic Indigenous water rights. The native title process as noted has, to date, failed to recognise a specific right to use water for commercial purposes (Strelein 2009; Australian Law Reform Commission 2015, Chap. 8). Since the litigation culminating in the 2013 High Court decision of *Akiba v Commonwealth* (2013) 250 CLR 209 some commentators have predicted that Australian native title jurisprudence will evolve to recognise a right to use water for any (including commercial) purpose (O'Donnell 2011). However, *Akiba* has not been followed with a decision on water and subsequent cases frame 'commercial water rights' in a particular limited sense, contemplating trade in or sale of resources themselves rather that the use of water for associated commercial activities (*BP v Western Australia* [2014] FCA 715; *Western Australia v Willis* [2015] FCAFC 186).

Where Indigenous water users require water for commercial purposes, the Initiative contemplates that these entitlements may be acquired through market mechanisms, such as water trade and/or direct purchases of water entitlements and licences, the cost of which could be partially borne by governments if they so choose (Council of Australian Governments 2004, p. 32). However, little has been done to provide Indigenous people with commercial water rights until now and those mechanisms that have been introduced, such as Aboriginal access licences under the New South Wales legislation, have enjoyed little uptake (Altman and Arthur 2009, p. 9). Possibly the most ambitious Indigenous water initiatives are 'strategic reserves' of water entitlements (some models include reservations for future Indigenous economic development opportunities) contemplated under certain water plans in parts of northern Australia, although recent governments have shied away from the approach (Stewart 2013). The result is an ongoing discrepancy in Indigenous land and water holdings in Australia, by which Indigenous-held 'water access entitlements' are estimated at only 0.01% of total Australian water allocations (Jackson and Langton 2011), despite Indigenous land now making up more than 30% of the Australian continent (Altman and Markham 2015). In light of these stark statistics, and increasing calls from Indigenous groups for greater water equity, the recent developments in the Australian water reform story discussed below place more emphasis on the need to allocate water entitlements to Indigenous peoples from the consumptive pool, in addition to safeguarding environmental (and cultural) flows (Jackson and Barber 2013, 2016).

3.3 Indigenous and Environmental Values in Recent Water Policies

Water planning continues to be of immense importance in Australian water allocation and governance. Water policy frameworks have been revisited recently in both northern and southern regions. In northern Australia, where water resources have historically been viewed as 'underexploited', policy frameworks continue to emphasise productive land potential through intensifying irrigated agriculture, with less emphasis on environmental or cultural water values. Perhaps reflecting the more progressive political tendencies in the south, (but also in response to distinct perceived hydrological conditions), the southern approach places more importance on restoration of environmental water and indigenous rights, including the use of market mechanisms. However, whether the more progressive southern policy objectives are translated into statutory or policy change is yet to be seen.

3.3.1 Our North, Our Future

The importance of water for productive land use is emphasised in the latest iteration of the Commonwealth Government's plans to open up northern Australian regions of Queensland, Northern Territory and Western Australia to development: *Our*

North, Our Future: White Paper on Developing Northern Australia (Australian Government 2015). The *White Paper* encourages increasing population and bolstering the northern economy by promoting fisheries and agriculture, cutting red tape, increasing business links with Asia, and promoting tropical medicine. In relation to water infrastructure, it provides (Australian Government 2015, p. 40):

> Northern development depends on water. Up to 17 million hectares of land in the north have soils which are potentially suitable for agriculture, but there is only water sufficient to irrigate about one tenth of that area. Building the right water infrastructure in the right place will be crucial to realise the full potential of the north. Both surface and ground water in northern Australia serves a variety of functions, including cultural and spiritual use by Indigenous communities. River flows and groundwater are vital for supporting natural environments as well as other productive uses.

In order to extract maximum productive potential from northern water resources $200 million will be set aside under the plan for water infrastructure, identifying irrigation and damming potential in river systems across Queensland, the Northern Territory and Western Australia, as well as opportunities to use groundwater in the Western Australian Pilbara region (Australian Government 2015, pp. 43–44). The plan also provides for economic feasibility studies for dams and other potential water infrastructure across northern Australia (18). Thus far, water markets have not been as widely established in northern Australia as in the south-east. The northern approach continues the Australian preference for increased use of market tools, promising the development of 'secure and tradeable water rights as part of a new National Water Infrastructure Development Fund'.

This latest policy approach to developing northern Australia is a continuation of the early engineering approach to water regulation (Crase et al. 2009, p. 446), which has paid insufficient heed to environmental or cultural factors (Turville et al. 2015, p. 2). Turville, Cullen and Tan suggest that the original damming of the Ord River, earmarked for extension in the *White Paper*, involved 'scant regard' for either Aboriginal communities or the environment and was technically deficient (Turville et al. 2015, pp. 1–2). Jackson and Barber note how the original Ord development 'fundamentally altered the existing indigenous water scape' in the region, with floodwaters displacing indigenous people and cultural sites (Jackson and Barber). Concerns continue to be expressed about the environmental impacts of expanded water infrastructure in northern Australia (Feuvre et al. 2016, p. 176), although these attract little consideration in the *White Paper* (O'Donnell and Hart 2016).

Given the large Indigenous populations and landholdings in northern Australia, the *White Paper* would be expected to address Indigenous water values and consultation requirements. The paper does reference the need to consult traditional owners, as well as emphasising the importance of Aboriginal-run projects, including Indigenous ranger programs (Australian Government 2015). However, it continues, as has been the historical practice in northern Australia, to place Indigenous concerns within the broader quest for land development. Joe Morrison of the Northern Land Council, a significant northern Aboriginal land and political organisation, explains (Morrison 2015a):

Ever since the north was settled – by conquest, not by consent – there have been a cascade of reports which have purported to map various El Dorados, just waiting to be discovered and developed by men of vision… These are not empty lands… Aboriginal people are not afraid of development. We want development, but we want it to be ethical.

The *White Paper* proposes that Aboriginal landholders work their land through irrigated agriculture, in the interests of long-term economic independence (Australian Government 2015):

The Commonwealth Government supports northern jurisdictions taking actions that support Indigenous Australians to derive greater economic benefits from water on Indigenous land. Water can provide opportunities for Indigenous Australians in diverse areas such as aquaculture, nature based tourism and intensive horticulture. Access to water can also provide an opportunity to participate in water markets, where they exist.

This focus on productive land potential in the north, as opposed to resource conservation, builds on other Commonwealth discussion papers that underscore the need for commercial water rights for Aboriginal economic development (Australian Government 2012, pp. 1–2). Yet, a question remains as to whether this productivist approach represents the views of northern Indigenous people. The Northern Land Council, for example, suggests that it has not been adequately consulted in relation to the White Paper, and insufficiently engaged in water planning more generally (Morrison 2015b).

3.3.2 Water for Victoria

In the south-eastern state of Victoria, the demographic and geographic profile is much different, being densely settled and intensively cultivated. Victoria released Australia's newest strategic planning document for water, the *Water for Victoria* plan, in late 2016, as a 'long-term direction' for the governance of Victorian water resources. The themes of the plan prominently include both environmental and Indigenous aspects, including waterway and catchment health and climate change, and recognising and managing Aboriginal values. These aspects, however, are to be provided for within a framework that utilises the 'water grid' and markets as effective governance tools (State of Victoria 2016, p. 140):

Victoria's water markets allow farmers, the Victorian Environmental Water Holder and water corporations to buy and sell water entitlements and seasonal allocations, so they can manage their own risk according to their willingness to pay. The water markets allow us to share water security benefits in ways that are equitable, responsive and transparent.

Regulation also plays a key role in Victoria, setting rules and standards for the efficient, and equitable, functioning of water markets (State of Victoria 2016, p. 148). The plan encourages greater regulation to ensure accurate information and transparency for effective pricing.

In terms of environmental water values, the plan goes little further than more innovative reforms by previous governments, notably, the creation of the Victorian Environmental Water Holder in 2010 (O'Donnell and Macpherson 2013), which, similar to the Commonwealth Environmental Water Holder, manages a portfolio of water allocations and flows in the interests of the environment within the market framework (O'Donnell 2012). The plan signals further investment in waterway health, integrated catchment management and protections of beneficial uses (State of Victoria 2016, p. 50). Also, there is to be further investment in environmental watering, acknowledging the impact of climate change, and increased monitoring and research (State of Victoria 2016, p. 54).

In terms of Aboriginal water values, the *Water for Victoria* plan does appear to break new ground, probably in response to the historical dearth of Indigenous water protections in Victoria and increasing lobbying by Indigenous groups in the context of a commitment to treaty negotiations by the state government. The 2016–2017 Victorian Budget committed $4.7 million to embedding Aboriginal values and knowledge in Victorian water management. The plan highlights a need for greater Indigenous involvement and consultation in water planning, and incorporating traditional knowledge via an 'Aboriginal Water Reference Group', with representation from Traditional Owners with knowledge of water management (State of Victoria 2016, p. 98).

The plan even suggests that in the future Aboriginal Victorians may be provided with water access for economic development purposes, facilitating 'economic self-determination' via water-related Aboriginal enterprises (State of Victoria 2016, p. 98). This prospect has not been raised before in the Victorian context, including under Victoria's statutory regime for traditional owner settlements in lieu of native title, which apply only to 'traditional purposes' (*Traditional Owner Settlement Act 2010* (*Vic*) s 79). Remarkably, the plan refers to a potential to reallocate water access entitlements to traditional owners, including by acquiring water entitlements from water corporations in areas where this is allowable within sustainable limits, investing in water savings projects that create new entitlements, and buying water in water markets in areas where entitlements or use is capped at sustainable limits (State of Victoria 2016, p. 106). However, the aims of the plan are vague and non-committal; suggesting that any real reform is many years away, if to eventuate at all. The plan advocates working with traditional owners to develop a 'roadmap for access to water for economic development', to 'consider opportunities' for Aboriginal access to water for economic development purposes, and to 'notify traditional owners' when opportunities for access arise (State of Victoria 2016, p. 106). Beyond the rhetoric, there are no concrete requirements in *Water for Victoria* to reform water legislation or create specific legal water entitlements for Indigenous people in catchment-level plans.

The absence of effective policy directions reflects the competitiveness and political sensitivity of water distribution in Victoria, and the potential cost involved in water reallocation (Jackson and Langton 2011). The legal position is that the Victorian approach continues to 'compartmentalise' Indigenous water values as heritage values (Jackson 2010), principally via the concept of 'cultural flows'. Typically, such flows do not draw upon the pool of consumptive water allocations. Rather, the plan contemplates that Indigenous groups are to share benefits with environmental flows and programmes; although the groups are to be directly involved in water planning and the identification of cultural values in waterways. Thus, this latest iteration of Australian water policy, aside from highlighting a need for open and meaningful consultation with Indigenous groups, demonstrates the iterative and lengthy process of community-focused water reforms, and the trade-offs implicit in safeguarding Indigenous and environmental values in competitive water markets.

4 Lessons for an International Audience

4.1 Including Indigenous Peoples in Water Governance

The Australian experience with water reform highlights the importance of strategic planning to accommodate Indigenous interests within hybrid governance systems for water resource management that rely on a combination of government regulation and water markets. The initial Murray-Darling Basin water law reform process, in which Indigenous water rights were barely considered, demonstrates that detailed water resource assessments and plans, and the design of water entitlements, need to include not only technical assessments of water allocation based on hydrology and ecology, but also community values, especially the cultural values of Indigenous peoples. Water resource assessments and planning are tools widely used in many countries. Thus, measures designed to include Indigenous peoples' interests in water potentially have wide application.

Internationally, there is growing emphasis on free, prior and informed consent and the need for equitable and transparent approaches to engagement with Indigenous communities. These principles have pertinence for the water law reform process in many countries. Undoubtedly, the first step in adequately accommodating Indigenous values is meaningful consultation; in order for non-Indigenous water planners to better understand Indigenous water perspectives and, as emphasised by Lindsay in chapter 10, to seek the consent of Indigenous people to water developments that affect them (Lindsay this volume). The introduction of water reforms in Australia in the early 1990s largely overlooked Indigenous water rights and there was minimal consultation. Recent iterations of the Australian reform measures do emphasise free, prior and informed consent, and acknowledge the importance of incorporating indigenous values in water governance, although what may be encompassed by 'meaningful engagement' with Aboriginal peoples is still being tested.

The *Water for Victoria* plan is indicative of emerging strategies that seek to facilitate consultation with and the participation of Indigenous people in water governance, in an effort to better embed Aboriginal values and knowledge in water management. The Victorian plan encourages greater Indigenous involvement in water planning, and the incorporation of traditional knowledge via an 'Aboriginal Water Reference Group', with representation from traditional owners with knowledge of water management (State of Victoria 2016, p. 98). Whether this non-binding policy document does in fact lead to meaningful consultation, in which Aboriginal Victorians are actually able to influence the reform of water law and policy, is yet to be seen.

Moreover, until now most legal and policy debates about Indigenous water rights in Australia, in reliance on the native title model, have presented Indigenous interests as having a 'traditional-cultural' character (O'Bryan 2012; Lane 2000; Australian Human Rights Commission 2009). In situations where some form of 'entitlement' is contemplated, the cultural flow model has predominated (Godden and Gunther 2009, p. 244). The cultural flows model as a first step can play a useful role in promoting the inclusion of Indigenous peoples in water governance processes with governments and Indigenous representative groups recognising their significance in this regard (Murray Lower Darling Rivers 2007). Current legal mechanisms providing for Indigenous water rights in Australia, though, do not effectively support Aboriginal water use for commercial purposes, focusing instead on cultural and environmental interests. These important values merit protection, but a broader perspective that takes into account the history of exclusion from culturally appropriate economic opportunities is required (Macpherson 2016). Arguably, a limitation of the concept of cultural flows is that such interests have tended to be narrowly realised via in-stream protections, similar to the environmental flows model, although the concept itself does contemplatate commercial water use. The cultural flows model may not be sufficient to achieve substantive Aboriginal outcomes unless water law frameworks also set aside a share of the consumptive pool for Indigenous users or otherwise provide for water for cultural economy purposes.

Indeed, the protection of customary rights to water and/or cultural flows finds resonance other jurisdictions, such as New Zealand, which has moved much further along the spectrum in providing robust legal and institutional protections for indigenous peoples 'cultural' rights to water (see e.g. *Te Awa Tupua (Whanganui River Claims Settlement) Act 2017* (NZ)). The Australian experience affirms the importance of protecting traditional cultural flows but also acknowledges the need for more holistic approaches that can encompass commercial utilisation of water.

Despite significant legal barrriers and a lack of formal acceptance, Indigenous communities seek to exercise their management rights over water in accordance with cultural norms (Jackson and Barber 2016). To achieve equity, and consistent with international human rights, Aboriginal peoples' water rights should have the same characteristics as the rights available to non-Indigenous water users since Australia was colonised. Thus, Aboriginal people should be able to choose whether to exercise water rights for cultural or economic development purposes (Jackson and Langton 2011). Australian experience here lags behind international best practice as some jurisdictions have accorded indigenous communities the capacity to exercise water rights for commercial uses and in order to support economic development, notably in Latin America (Macpherson 2017). This much, at least, appears to be contemplated in the *White Paper* for developing northern Australia and the *Water for Victoria* policy, both of which reference the potential use of water by Indigenous people for commercial purposes. Importantly, if Indigenous interests in water are not factored into planning for water markets, then Aboriginal peoples will again be excluded from their economic benefits.

Furthermore, an important lesson from the Australian experience is to account for Indigenous water interests when designing wider water reforms that encompass market measures. Typically a prerequisite of water market implementation is the 'full' allocation of entitlements within the relevant water system, so some accounting for current and future Indigenous interests is vital. By contrast, in the Murray-Darling Basin in southern Australia there was no provision made in the consumptive pool for Indigenous water use when existing entitlements were 'grandfathered' as water markets were introduced. A critical resulting problem is the expense and difficulty of buying-back entitlements in a fully allocated system if Indigenous water rights are to be contemplated. For this reason, policy approaches that propose to set aside 'strategic indigenous reserves' in water catchments for future allocation to Indigenous peoples in areas where water is still plentiful are a welcome development (Jackson and Barber 2013). Unfortunately, as noted above, there is a tendency to move away from such policy approaches in northern Australia in recent years and the *Water for Victoria* plan remains little more than aspirational in this regard.

While there is a diversity of views, many Aboriginal communities recognise a need for economic development (including via the exploitation of water resources) as providing employment and long-term viability for their communities, even while, 'they are very determined to protect their country and sacred sites' (Australian Law Reform Commission 2015, p. 54). Collaborative water planning processes must ensure Aboriginal and multiparty confidence in holistic water planning and development outcomes (Jackson et al. 2012, p. 22).

In northern Australia, Indigenous people are an important stakeholder group with respect to any plans for water development (O'Neill et al. 2016). Aboriginal people have significant statutory landholdings together with a growing number of native title determinations. Aboriginal peoples and Torres Strait Islanders are a larger percentage of the population across northern Australia, meaning that their opinions, interests and aspirations may hold more political sway than in the south. Thus, culturally appropriate Aboriginal consultation and inclusion is all the more vital as Aboriginal people are the demographic group most likely to be impacted by water development (Greiner 2000). Notwithstanding, there is less guidance on the role of Indigenous people in developing water reform in the *White Paper*, which makes scant provision for future consultation. Taking into account economic, social and environmental factors is a stated principle behind any proposed new water infrastructure (Australian Government 2015, p. 51). Exactly how the detail of social impact will be measured or accounted for is uncertain. Previous northern development planning and project implementation failed to consult with Aboriginal people, or to effectively account for social and cultural impacts (Greiner 2000). Consultation and participation processes must be significantly different in any new phases of development if it proceeds.

In this respect, northern Australian Governments could find a useful model in the extensive consultation process developed for the proposed Kimberley Browse liquefied natural gas (LNG) development. The consultation with Aboriginal peoples in the Kimberley region went well beyond typical public participation processes. In 2006, then Western Australian Premier Alan Carpenter said that the Kimberley Browse development would only go ahead with the support of Kimberley traditional owners and it would be 'a dialogue, not an imposition or a demand' (Western Australian Government 2006, p. 8443c). The initial level of political commitment to obtaining Aboriginal consent was viewed in some quarters as giving Aboriginal people a de facto veto over the Browse development (O'Neill 2014). Whether further expansion of water infrastructure should take place in northern Australia also is likely to be strongly contested. An effective process for engaging Aboriginal peoples is vital, but the broader exclusion of Aboriginal interests from commercial uses of water, in the context of interplay of environmental priorities and Aboriginal interests, must also be addressed. As this overview has identified, there are principles and models in place in Australia that can be drawn upon to ensure more robust consultation and participation with Indigenous communities around their involvement in water governance. Australian water law and policy has long functioned as de facto social and economic policy—now there is a need to be explicit about the extent to which Australian water law and policy will actively provide culturally appropriate opportunities for Indigenous participation in water governance and economic development that is in line with emergent international practice.

4.2 Embedding Environmental Values

Australia has been more explicit in its attempts to embed environmental values and outcomes in water governance. Internationally, its models and approaches in the implementation of environmental flows are leading edge (Grafton et al. 2011), although in the face of fierce competition for water, protecting environmental water interests in many Australian jurisdictions remains challenging. Such challenges confirm the importance of strategic planning for environmental water at the early stages of policy design. The experience of the Murray-Darling Basin and the significant levels of effort and funding required to restore overallocated and degraded river systems shows that it is much easier to protect existing environmental flows than to restore these flows in the future (O'Neill et al. 2016). First, it is less complex (procedurally and legally) to implement flow protection mechanisms based on new rights than it is to change the water rights of existing users—and much less expensive (Horne et al. 2017). The protection of environmental flows as part of initial water planning and allocation, however, does require significant upfront investment in understanding water resources, ecology and social and cultural values of the affected area (Pahl-Wostl et al. 2013).

There are critical insights from the long, costly and as yet unfinished attempts to provide adequate environmental water in the Murray-Darling Basin that should be noted by other jurisdictions contemplating water reforms. Experience in Australia has demonstrated that establishing and maintaining adequate environmental flows to protect both the resource base and the ecological functions of water dependent ecosystems is complex (Connell and Grafton 2011). Environmental flows need to be identified and protected using a range of legal mechanisms with adequate legal security. As a minimum, these mechanisms should include: a cap on water extraction, conditions on dam location and operation, and a minimum flow or environmental water allocation to protect the environment during dry years (Horne et al. 2017).

Other jurisdictions should note that capping water extraction will protect the important high flow events on which the water dependent ecosystems rely. Not only will this facilitate trade, but setting a conservative cap will encourage high value, efficient water uses initially, while also enabling the potential for the cap to be raised in future (if warranted). This cap needs to be expressed as a proportion of the available flow rather than an absolute volume, so that it continues to protect environmental flows during dry years as well as wet years.

Further, dams need to be located where they minimise environmental impacts and, where possible, should be off-stream. One of the critical lessons from the Murray-Darling Basin on-stream dams is that reducing the effect of large on-stream dams on the timing and frequency of flow events is extremely challenging once the dams are in operation (US Army Corps of Engineers and The Nature Conservancy 2011; Richter and Thomas 2007). On-stream dams need to be operated to protect the seasonality, frequency and duration of important flow events.

Climate change and increased water extraction are likely to increase the frequency and duration of droughts and extended dry periods (Steffen 2015). By protecting critical drought refuges, ecosystems will be able to respond and recover when water availability improves once more. Establishing a minimum flow release from on-stream dams, and creating environmental water rights, is an extremely effective mechanism for ensuring that environmental managers have the capacity to respond flexibly in dry periods, and to meet ecological needs efficiently (Horne 2009). The crucial role of groundwater as a critical reserve must also be recognised (Martin 2016).

Finally, environmental water needs to be managed effectively and efficiently to deliver maximum benefit for the available water (Horne et al. 2009, p. 779). Environmental water management depends on interactive partnerships between catchment managers, water resource planners, river and storage operators and the community (Garrick et al. 2011, p. 167; Garrick and O'Donnell 2015). These relationships are complex and governance arrangements need to be addressed at the outset. At the federal level, the Commonwealth Environmental Water Holder operates as the single decision-maker for the use of environmental water rights each year (O'Donnell 2013). A single organisation responsible for implementation helps to streamline decision-making, achieve economies of scale in environmental water management, and enhance accountability by having an identifiable decision-maker (O'Donnell 2012, 2013).

Disappointingly, recent Australian water policy pays insufficient attention to environmental water. The northern Australian *White Paper* is almost completely silent on the issue of environmental flows, beyond the brief acknowledgment of the need for a sustainable limit on extractions. *Water for Victoria*, promises more investment in environmental water programmes and monitoring but specifies no improvements to existing regimes. The Australian experience shows that embedding environmental values in water governance is not a one-off event, but must be an ongoing process of reflective and responsive reform to learn from and adapt to prior mistakes (O'Donnell and Garrick 2017b).

5 Conclusion

Internationally, Australia is regarded as a highly innovative water manager—with much of this credential tied to the extensive law reforms initiated under the National Water Initiative in the Murray-Darling Basin. While internal debates continue about the level of effectiveness of these reforms in achieving long-term sustainability, many valuable lessons may be drawn from the Australian experience. These include that any process for water reform must embed environmental values in water planning, creating legal rights to environmental water to protect existing ecological values, and establishing a system of governance to promote the effective and efficient use of environmental water. Policy developments must be scrutinised for impacts upon Indigenous peoples (and indeed other cultural groups as relevant to the respective jurisdiction), in terms of substantive opportunities for involvement in water planning processes, and inclusion in water entitlement allocations and corresponding economic benefits. Reforms of water law and policy frameworks offer an opportunity for Indigenous peoples, if effectively supported to do so, to be an integral element of any strategic planning for hybrid water governance in the future.

References

Altman, J. C., & Arthur, W. S. (2009). Commercial water and indigenous Australians: A scoping study of licence allocations. CAEPR Working Paper No 57/2009, Centre for Aboriginal Economic Policy Research. http://apo.org.au.

Altman, J., & Markham, F. (2015). Burgeoning indigenous land ownership: Diverse values and strategic potentialities. In S. Brennan et al. (ed.), *Native title from mabo to akiba: A vehicle for change and empowerment*? The Federation Press.

Australian Government Bureau of Statistics. (2013). Australian Yearbook 2012. Geographic Distribution of the Population. http://www.abs.gov.au/ausstats/abs@.nsf/Lookup/by%20Subject/1301.0~2012~Main%20Features~Geographic%20distribution%20of%20the%20population~49.

Australian Government. (2012). Position statement: Indigenous access to water resources. National Water Commission. http://www.nwc.gov.au/__data/assets/pdf_file/0009/22869/Indigenous-Position-Statement-June-2012.pdf.

Australian Government. (2015). Our north, our future: White paper on developing Northern Australia. http://northernaustralia.gov.au/files/files/NAWP-FullReport.pdf.

Australian Human Rights Commission. (2009). 2008 Native Title Report. Human Rights and Equal Opportunity Commission. https://www.humanrights.gov.au/our-work/aboriginal-and-torres-strait-islander-social-justice/publications/native-title-report-2008.

Australian Law Reform Commission. (2015). Connection to country: Review of the native title act 1993 (Cth). Report No 126. http://www.alrc.gov.au/publications/alrc126.

Connell, D. (2007). *Water politics in the murray-darling basin*. Federation Press.

Connell, D., & Grafton, R. (2011). *Basin futures: Water reform in the murray-darling basin*. ANU E Press.

Council of Australian Governments. (1994). Report of the Working Group on Water Resource Policy to the Council of Australian Governments. Unpublished.

Council of Australian Governments. (2004). Intergovernmental agreement on a national water initiative. http://www.pc.gov.au/inquiries/current/water-reform/national-water-initiative-agreement-2004.pdf.

Council of Australian Governments. (2010). National water initiative policy and guidelines for water planning management.

Crase, L. R., O'Keefe, S. M., & Dollery, B. E. (2009). The fluctuating political appeal of water engineering in Australia. *Water Alternatives, 2*(3), 440–447.

Crawford, J. (1991). The constitution and the environment. *Sydney Law Review, 13*(1), 11–30.

Cullen, P., Flannery, T., Harding, R., Morton, S.R., Possingham, H.P., Saunders, D.A., Thom, B., Williams, J., Young, M.D., Cosier, P., & Boully, L. (2002). Blueprint for a living continent: A way forward from the wentworth group of concerned scientists. http://wentworthgroup.org/2002/11/blueprint-for-a-living-continent/2002/.

Davidson, B. (1969). *Australia wet or dry: The physical and economic limits to the expansion of irrigation*. Melbourne University Press.

Davidson, B. (1965). *The northern myth: A study of the physical and economic limits to agricultural and pastoral development in tropical Australia*. Melbourne University Press.

Department of Sustainability, Environment, Water, Population and Communities. (2011). Water For the Future. http://www.environment.gov.au.

Department of the Environment and Heritage. (2004). Integrated water resource management in Australia: Case studies—murray-darling basin initiative. http://www.environment.gov.au/node/24407.

Evans, B., & Howsam, P. (2005). A critical analysis of the riparian rights of water abstractors in England and wales. *Journal of Water Law, 16*(3), 90–94.

Feuvre, L., Matthew, C., Dempster, T., Shelley, J. J., & Swearer, S. E. (2016). Macroecological relationships reveal conservation hotspots and extinction-prone species in Australia's freshwater fishes. *Global Ecology and Biogeography, 25,* 176–186.

Gardner, A., Richard, B., & Gray, J. (2009). *Water resources law*. Chatswood: LexisNexis Butterworths.

Garrick, D., & O'Donnell, E. (2015). Exploring private roles in environmental watering in Australia and the US. In J. Bennett (ed.) *Protecting the environment, privately*. World Scientific.

Garrick, D., Lane-Miller, C., & McCoy, A. L. (2011). Institutional innovations to govern environmental water in the Western United States: Lessons for Australia's murray–darling basin. *Economic Papers 30*(2):167–184.

Gibbs, L. M. (2009). Just add water: Colonisation, water governance, and the Australian Inland. *Environment and Planning A. 41*:2964.

Godden, L., & Gunther, M. (2009). Realising capacity: Indigenous involvement in water law and policy reform in South-Eastern Australia. *Journal of Water Law, 20*(5/6), 243–253.

Grafton, R. Q., Libecap, G., McGlennon, S., Landry, C., & O'Brien, B. (2011). An integrated assessment of water markets: A cross-country comparison. *Review of Environmental Economics and Policy, 5*(2), 219–239.

Greiner, R. (2000). The northern myth revisited: A resource economics research response to renewed interest in the agricultural development of the kimberley region. In *Annual conference of the Australia agricultural and resource economics society, Sydney.*

Head, L. (1999). The northern myth revisited? Aborigines, environment and agriculture in the ord river irrigation scheme, stages one and two. *Australian Geographer, 30,* 141–158.

Horne, A., O'Donnell, E., & Tharme, R. (2017). Mechanisms for allocating environmental water. In A. Horne, M. Stewardson, A. Webb, M. Acreman, & B. Richter (Eds.), *Water for the environment: from policy and science to implementation and management.* Elsevier.

Horne, A., Stewardson, M., Freebairn, J., & McMahon, T. A. (2009). Using an economic framework to inform management of environmental entitlements. *River Research and Applications, 26,* 779–795.

Horne, A. (2009). An approach to efficiently managing environmental water allocations. PhD Thesis. University of Melbourne.

Ison, R., & Wallis, P. (2011). Institutional change in multi-scalar water governance regimes: A case from Victoria, Australia. *The Journal of Water Law, 22*(3), 85–94.

Jackson, S. (2010). Compartmentalising culture: The articulation and consideration of Indigenous values in water resource management. *Australian Geograper, 37*(1), 19–31.

Jackson, S., & Barber, M. (2013). Recognition of indigenous water values in Australia's Northern Territory: Current progress and ongoing challenges for social justice in water planning. *Planning Theory & Practice, 14*(4), 435–454.

Jackson, S., & Barber, M. (2016). Historical and contemporary waterscapes of North Australia: Indigenous attitudes to dams and water diversions. *Water History, 8*(4), 385–404.

Jackson, S., & Langton, M. (2011). Trends in the recognition of indigenous water needs in Australian water reform: The limitations of 'cultural' entitlements in achieving water equity. *Journal of Water Law, 22,* 109–123.

Jackson, S., Tan, P.-L., & Nolan, S. (2012). Tools to enhance public participation and confidence in the development of the howard east aquifer water plan, northern territory. *Journal of Hydrology, 474,* 22–28.

Lane, P. (2000). Native title and inland waters. *Indigenous Law Bulletin, 4*(29), 11.

Macpherson, E. (2016). Commercial indigenous water rights in Australian law: Lessons from Chile. Ph.D Thesis. University of Melbourne.

Macpherson, E. (2017). Beyond recognition: lessons from chile for allocating indigenous water rights in Australia. *University of New South Wales Law Journal 40*(3).

Martin, P. (2016). Creating the next generation of water governance. *Environmental and Planning Law Journal, 33*(4), 388–401.

McHugh, P. (2011). *Aboriginal societies and the common law: A history of sovereignty.* Status and Self-Determination: Oxford University Press.

Molle, F. (2009). River-basin planning and management: The social life of a concept. *Geoforum, 40,* 484–494.

Morrison, J. (2015). Keynote Speech at garma festival, gulkula. http://www.nlc.org.au/media-releases/article/keynote-speech-at-garma-festival.

Morrison, J. (2015) Resilient communities and sustainable prosperity—Northern indigenous development (Townsville, 22 July 2015).

Murray Lower Darling Rivers Indigenous Nations and Northern Murray–Darling Basin Aboriginal Nations. (2007). Agreed Definition of Cultural Flows. http://www.mdba.gov.au.

Musgrave, W. (2008). Historical development of water resources in Australia. In L. Crase (ed.), *Water policy in Australia: The impact of change and uncertainty, resources for the future.*

O'Bryan, K. (2012). The national water initiative and Victoria's legislative implementation of indigenous water rights. *Indigenous Law Bulletin, 7*(29), 24–27.

O'Donnell, M. (2011). Indigenous rights in water in Northern Australia. Project 6.2. Northern Australia indigenous land and sea management alliance—tropical rivers and coastal knowledge. Charles Darwin University.

O'Donnell, E. (2012). Institutional reform in environmental water management: The new Victorian Environmental Water Holder. *Journal of Water Law, 22,* 73–84.

O'Donnell, E., & Macpherson, E. (2012). Challenges and opportunities for environmental water management in Chile: An Australian perspective. *Journal of Water Law, 23(1),* 24–36.

O'Donnell, E. (2013). Australia's environmental water holders: Who is managing our environmental water? *Australian Environment Review, 28,* 508–513.

O'Donnell, E., & Hart, B. (2016). Damming Northern Australia: We need to learn hard lessons from the south. The Conversation. 10 February.

O'Donnell, E., & Garrick, D. (2017a). Environmental water organizations and institutional settings. In A. Horne, M. Stewardson, A. Webb, M. Acreman, & B. Richter (Eds.) *water for the environment: From policy and science to implementation and management.* Elsevier.

O'Donnell, E., & Garrick, D. (2017b). Defining success: A multi-criteria approach to guide evaluation and investment. In A. Horne, M. Stewardson, A. Webb, M. Acreman, & B. Richter (Eds.), *Water for the environment: From policy and science to implementation and management.* Elsevier.

O'Neill, L., Godden, L., Macpherson, E., & O'Donnell, E. (2016). Australia, Wet or Dry, North or South: addressing environmental impacts and the exclusion of Aboriginal peoples in northern water development. *Environmental and Planning Law Journal, 33*(4), 402–417.

O'Neill, L. (2014). The role of state governments in native title negotiations: A tale of two agreements. *Australian Indigenous Law Review, 18*(2), 29–42.

Pahl-Wostl, C., Arthington, A., Bogardi, J., Bunn, S. E., Hoff, H., Lebel, L., et al. (2013). Environmental flows and water governance: Managing sustainable water uses. *Current Opinion in Environmental Sustainability, 5,* 341–351.

Powell, J. (1989). Watering the garden state: Water, land and community in Victoria 1834–1988. Allen & Unwin.

Richter, B. D., & Thomas, G. A. (2007). Restoring environmental flows by modifying dam operations. *Ecology and Society [online] 12*(1). http://www.ecologyandsociety.org/vol12/iss1/art12/.

Ruru, J. (2013). Indigenous restitution in settling water claims: the developing cultural and commercial redress opportunities in Aotearoa, New Zealand. *Pacific Rim Law & Policy Journal, 2,* 311–352.

State of Victoria. (2009). Northern region sustainable water strategy. Department of Sustainability and Environment.

State of Victoria. (2016). Water for Victoria. Department of Environment, Land, Water and Planning.

Steffen, W. (2015). Thirsty country: Climate change and drought in Australia. Climate Council of Australia.

Strelein, L. (2009). *Compromised jurisprudence: Native title cases since mabo,* 2nd edn. Aboriginal Studies Press.

Stewart, P. (2013). Indigenous water reserve policy tap turned off. ABC News (Online). 10 October. http://www.abc.net.au/news/2013-10-09/ntindigenous-water-reserve-policy-dropped/5012152.

Tan, P. L. (2002). Legal issues relating to water use in land and water in Australia. In *Property rights and responsibilities, current Australian thinking.* AGPS.

Tan, P.-L., & Quiggin, J. (2004). Sustainable management of the great artesian basin: An analysis based on law and environmental economics. *Australasian Journal of Natural Resources Law and Policy., 9,* 255–303.

Tehan, M. (2003). A hope disillusioned, an opportunity lost: reflections on common law native title and ten years of the native title act. *Melbourne University Law Review, 27,* 523–571.

Turville, A. C., Cullen, S., & Tan, P.-L. (2015). Planning for the future: integrated water management in the ord river catchment. *Water, 41,* 80–86.

US Army Corps of Engineers and The Nature Conservancy. (2011). Sustainable rivers project: improving the health and life of rivers, enhancing economies, benefiting rivers, communities and the nation. US Army Corps of Engineers, The Nature Conservancy.

Western Australian Government. (2006). Parliamentary debates. Legislative Assembly, 21 November. 8443c–8443c. Alan Carpenter, Premier.

Wooding, R. (2009). Populate, parch and panic: Two centuries of dreaming about nation building in Inland Australia. In A. W. Gardner et al (Eds.), *Water resources law*, LexisNexis Butterworths 57.

Part IV
Future Governance Challenges—Cumulative Impacts, Resource Industries and Climate Change

Regulating Cumulative Impacts in Groundwater Systems: Global Lessons from the Australian Experience

Rebecca Louise Nelson

Abstract The regulation of groundwater extraction has shifted, almost seismically, through Australia's water reform era, which culminated in the National Water Initiative and the federal *Water Act 2007* (Cth). Its starting point was minimal regulation, targeted at controlling selected extractions and pursuing sustainability for consumptive purposes. Since then, groundwater laws have imported regulatory concepts from the surface water sphere; among them, regulating extraction through statutory plans and in a broad-based way, and protecting dependent ecosystems. One issue emerges at the heart of both the history and future of groundwater reform: dealing with the cumulative impacts of extraction. This chapter reviews how groundwater reforms in Australia have approached cumulative impacts—albeit often implicitly rather than explicitly—measured against concepts of cumulative impact assessment derived from the scientific environmental assessment literature. It finds that both regulatory and non-regulatory, incentive-based measures have been used. As water reforms have progressed, arrangements for assessing and managing the cumulative impacts of groundwater extraction have taken a dramatically broader view of potentially affected natural systems, as well as consumptive values. However, they have not focused significantly on the implications of connections between natural systems. Reforms have also resulted in a much broader view of the impacting activities that are relevant to an assessment of cumulative impacts, though often with an unclear temporal scope. Remaining weaknesses in approaches to managing cumulative impacts and a lack of empirical evidence as to

---

This is a revised version of a journal article that was first published by Thomson Reuters as Rebecca Nelson, "Broadening Regulatory Concepts and Responses to Cumulative Impacts: Considering the Trajectory and Future of Groundwater Law and Policy" (2016) 33(4) Environmental and Planning Law Journal 356–371. For all subscription inquiries please phone, from Australia: 1300 304 195, from Overseas: +61 2 8587 7980 or online at legal. thomsonreuters.com.au/search. The official PDF version of this article can also be purchased separately from Thomson Reuters at http://sites.thomsonreuters.com.au/journals/subscribe-or-purchase.

R. L. Nelson (✉)
Melbourne Law School, University of Melbourne, Melbourne, VIC, Australia
e-mail: rebecca.nelson@unimelb.edu.au

the effectiveness of existing measures warrant attention. The chapter concludes by charting key challenges that remain for resolution, and setting out an accompanying research and reform agenda.

1 Introduction

Australia's era of water reform, extending over the past three decades and celebrated globally (Pigram 2006, p. 81), has been driven largely by concerns about surface water. Groundwater laws generally have been a forgotten passenger—at least in the public mind. Yet the regulation of groundwater extraction has also shifted dramatically. Whereas groundwater law traditionally focused on controlling selected extractions and pursuing sustainability for consumptive purposes, reform-era regulation controls extraction in a broad-based way, through statutory plans that aim to protect dependent ecosystems as well as human consumptive users. Despite the intensity of this reform period, a critical issue remains largely unaddressed: regulating the cumulative impacts of groundwater extraction on water entitlement holders and dependent ecosystems (groundwater cumulative impacts (GCIs)).

This chapter reviews the trajectory of groundwater law in Australia through the lens of how it has approached—albeit often implicitly rather than explicitly—the regulation of cumulative impacts, and seeks to draw out lessons applicable to future developments in Australia as well as further afield. Section 2 of the chapter introduces the concept of cumulative impacts, seeking to expand recent scholarly discussion of cumulative impacts in the water context by reference to broader and better-established concepts of cumulative impact assessment (CIA) from the environmental assessment literature, which also recognises that cumulative impacts are 'a persistent analytical challenge' (Canter et al. 2010, p. 259) where the legal framework is often found wanting (Lawrence 2013, p. 223). Discussions about groundwater have focused conceptually on ensuring a comprehensive assessment of impacts, for example, regulating 'interception activities', such as those deriving from mining and plantation forestry, but have neglected other broader concepts associated with effective CIA. I argue that effectively addressing GCIs requires a much broader focus, informed by CIA. Section 3 of the chapter reviews the regulation of GCIs through groundwater law and related environmental laws, with a focus on mandatory regulatory controls rather than information-producing requirements, and assesses these vehicles against the key concepts distilled in Sect. 2. Again, scholarly discussions about groundwater have tended to focus on statutory water plans. I argue that a wider and more diverse set of regulatory mechanisms for dealing with these impacts has emerged across national, state and interstate water law and policy, as well as special-purpose environmental law provisions dealing with water, which are heavily influenced by water law reforms. I identify three relevant mechanisms for regulating GCIs: statutory water plans, state-based rules for licensing individual groundwater extractions and other state-level water law controls, and project-level environmental approvals and associated plans (noting that federal strategic assessments relevant to GCIs are nascent). Section 4 synthesises how Australian groundwater law has developed to address GCIs, highlights key remaining challenges, and draws out comparative reflections and lessons of relevance to water governance in Australia and further afield.

2 Cumulative Impacts and Groundwater: Expanding Our Horizons

2.1 The Scientific Perspective

'Cumulative impacts' have been defined in a variety of ways in varying contexts. Their definition has been considered 'evolving, complex and often muddy' (Contant and Wiggins 1991, p. 298). In the environmental impact assessment literature, CIA:

> *systematically [a]nalyses and assesses cumulative environmental change ... focus[ing] on the receiving environment, and on whether individually minor effects will be collectively significant ... [It] simultaneously assesses the positive and negative effects (additive, interactive, synergistic, irregular) on given receptor(s) from existing, planned, proposed, and potential human activities* (Lawrence 2013, p. 222)

It considers not just past, present and expected human activities, but also present and expected environmental conditions (Contant and Wiggins 1991, p. 299). That is, CIA 'place[s] project impacts within the context of all potential impacts on receptors ... and [relative to regular impact assessment] broadens the temporal and spatial scope of analysis to encompass other past, present, and likely future actions' (Lawrence 2013, p. 224). At the strategic level, CIA 'focuses on the resource and activity totality rather than only on assessing a proposed plan or program' (Lawrence 2013, p. 224).

The cumulative impacts of a project have been conceptualised in two major categories: 'effects resulting from a project's relationship to other development activities, and effects produced by an activity's presence within a set of many natural systems' (Contant and Wiggins 1991, p. 298) that act as receptors of those impacts. In an influential paper, Contant and Wiggins further developed the elements of these categories (Contant and Wiggins 1991, pp. 301–3). The first category, the context of other development activities, includes 'incremental effects' resulting from unmonitored past developments (taking into account delayed responses) and current and future developments, which 'may be similar or different, connected or unconnected, to the proposed action' (Contant and Wiggins 1991, p. 302). This category also includes indirectly causally connected effects—growth inducement, where a project changes the rate of development of new activities, or where it causes broader changes to regional structural development. The second major category of complex natural systems captures effects that extend beyond simply adding impacts over space and time, recognising complex effect processes. These include unanticipated or non-accumulative responses, that is, non-linear cause-effect response relationships between a project and its 'receptor' natural systems: 'amplifying or exponential effects, the result of an incremental addition that produces a larger effect than earlier additions ... discontinuous effects, the impact resulting from exceeding a threshold or the crossing of a stability boundary' (Contant and Wiggins 1991, p. 301). This category also includes crowding in time, where a system cannot 'recover from a perturbation before a new one is present' or crowding (overlapping effects) in space (Contant and Wiggins 1991, p. 301). The natural systems context can also display systemic changes such as synergism ('total effects that are qualitatively or quantitatively different from the sum of the effects of the individual disturbances') (Contant and Wiggins 1991, p. 301), structural change (significant effects to the structure of a system later in time, as in exposure to cancer-causing agent), and cycling (involving cycles of impact and recovery, as in clear-cutting forest patches), as well as complex interactions between connected systems.

2.2 When Science Meets Groundwater Law

In contrast with the broad scientific view of cumulative impacts outlined above, scholarly and policy discussions of GCIs in the regulatory context focus on the cumulative impacts of individual minor, unregulated water extractions, and multiple regulated water extractions that tend to be considered in isolation (Bubna-Litic 2015; National Water Commission 2011; Neville 2003) but have not yet analysed the underlying concepts at play in the notion of cumulative impacts. These existing discussions stem from the significant forms of groundwater 'take' that have traditionally either been entirely unregulated (such as plantation forests), or have not required a regular water licence (such as stock and domestic bores). These discussions are also driven by the rise of clusters of groundwater withdrawal activity, notably coal seam gas (CSG) extraction. Yet these issues constitute only part of Contant and Wiggins' (1991) broader view of cumulative impacts (Table 1, cell 1A: incremental effects, and to an extent, cell 2D: interactions across connected natural systems). Analysing groundwater characteristics and problems against Contant and Wiggins' (1991) categories of cumulative impact suggests that a much broader set of issues and concepts are relevant to controlling GCIs. Table 1 sets out a current example of the manifestation of each of these concepts in the groundwater context.

Some of these sub-categories of cumulative impact concepts relate most strongly to the scientific modelling context and ensuring that this context best approximates the complexity of natural systems, for example, cycling and crowding in time. Requirements in relation to gathering data and producing information about impacts and receptors are vital to CIA (Contant and Wiggins 1991) and an area of recognised weakness in CIA arrangements (Lawrence 2013). However, they lie beyond the scope of this chapter to consider in detail. Rather, the focus here is regulatory vehicles for controlling impacts more directly. The categories above usefully point to key concepts that science recommends to a regulatory response to assessing and

Table 1 Examples of a broad view of the cumulative impacts of groundwater extraction, applying Contant and Wiggins (1991) categories to groundwater examples

1 Context of other development activities—groundwater examples	2 Context of complex natural systems—groundwater examples
1A Incremental effects: cumulative effects of stock and domestic wells that are exempt from licensing requirements (Merz et al. 2010)	2A Unanticipated or non-accumulative responses: many free-flowing artesian bores causing water pressure to fall, so that the source becomes sub-artesian, introducing pumping costs for other users and causing spring discharge to cease (Parsons et al. 2011)
1B Future development actions: development of new groundwater-dependent irrigation in the West Kimberley, (Australian Government 2015) alongside mining	2B Crowding in time or space: drilling new irrigation wells as withdrawals from existing wells increase due to drought or as recharge decreases due to climate change (Bates et al. 2008; McCallum et al. 2010); high spatial density of CSG wells
1C Growth inducement: urban growth based on accelerating pumping of groundwater (National Water Commission 2012)	2C Systemic changes: • Synergism: pumping that cumulatively causes saltwater intrusion into a freshwater aquifer (Werner 2010) • Structural changes: cumulative groundwater pumping changing the direction of groundwater flow (Tularam and Krishna 2009) • Cycling: changes in groundwater pumping (and therefore groundwater levels) in response to recurrent drought (National Water Commission 2012)
1D Regional structural development changes: development of a large regional CSG industry with associated industries relying on produced water (RPS Australia East Pty Ltd 2011)	2D Interactions across connected natural systems: hydraulic fracturing that causes inter-aquifer leakage between poor and good quality aquifers (National Water Commission 2012)

managing GCIs. Synthesising and abbreviating for the purposes of further discussion, these are:

- a *broad scope of impacting activities*, encompassing effects caused by other connected or unconnected developments, including, at its broadest, effects on groundwater systems caused by the cumulative effects of pumping and non-pumping activities that indirectly affect recharge by, for example, altering the land surface or contributing to climate change.
- a *broad scope of receptors*, encompassing a comprehensive view of natural systems, and including connected natural systems, such as aquifer-aquifer and aquifer-surface water connections.
- a *broad temporal scope*, encompassing developments in the present, future, and past, the effects of which may be time-lagged (delayed responses are a particular issue in the groundwater context, since impacts can manifest years, decades or even centuries after pumping commences), interacting with changes to natural systems that also occur over longer time scales, such as climate change.
- *clear thresholds*, reflecting a potentially complex natural, and by extension, social response (non-linear, uncertain) to a development.

- *structural changes to development*, caused by inducing growth or causing a change in regional structural development.
- *structural changes to natural systems*.

Scientists and others commonly urge that better law, policy, and associated guidance are vital to effectively managing cumulative impacts (Badr et al. 2004; Bashour 2016; Cooper and Sheate 2002; Folkeson et al. 2013; Marshall et al. 2005; Sinclair et al. 2017). Yet there is often little formal guidance about how to go about assessing cumulative impacts – a situation true in Australia (Howe 2011), as well as elsewhere around the world. Although the drivers for regulatory consideration of GCIs in Australia have tended to emphasise relatively narrow contexts of concern, a wider, scientifically informed conception of GCIs as set out above, should inform the evaluation, formulation and implementation of law related to GCIs.

3 Groundwater Cumulative Impacts in Australian Water Reforms

As a prelude to discussing Australian groundwater law in detail, it is necessary to sketch the key characteristics of Australian groundwater law systems, and how they differ from their major global counterparts. In brief, as is common in most jurisdictions of the western U.S. and the EU, the government has a non-proprietary, regulatory interest in groundwater (Nelson and Quevauviller 2016). Groundwater licensing (known as permitting in some other jurisdictions) is delivered through state law, with an overlay of federal quantitative caps on extraction in the Murray-Darling Basin through an overarching and a series of subordinate, federally accredited statutory water plans—broadly analogous to the EU's Water Framework Directive structure, but largely alien to the western United States (Nelson and Quevauviller 2016).

Although most scholarly and policy attention on GCIs has focused on current incremental effects (as per Table 1, cell 1A), an analysis of the trajectory of Australian groundwater reforms reveals that they have, in fact, adopted a broader view of GCIs and have used a range of regulatory measures to address them. This section uses three key examples briefly to demonstrate the hitherto unrecognised breadth of these measures, while also arguing that there is much scope for further improvement.

3.1 Groundwater Cumulative Impacts and Statutory Water Planning

3.1.1 National Water Initiative

Until the 1990s, policy concerns in Australia about groundwater over-extraction were muted, to put it generously. Key policy documents like the 1994 Council of

Australian Governments' (COAG) National Water Reform Framework did not deal with groundwater in a sustained way. Groundwater-specific policy documents dealt with basic management issues like concepts of sustainable yield and trading (Agriculture Resource Management Council of Australia and New Zealand and Standing Committee on Agriculture and Resource Management ARMCANZ 1996). They did not mention or allude to GCIs. Importantly, however, they supported protecting the ecological values of groundwater (ARMCANZ 1996), broadening the scope of relevant receptors beyond the resource itself, and its consumptive values to human users, to dependent ecological values.

These policy concepts began to solidify when the federal government used funding to states to grease the wheels of state legislative water reforms, under the banner of the Intergovernmental Agreement on a National Water Initiative (NWI), signed between 2004 and 2006 by all Australian governments. These reforms have increasingly recognised and addressed groundwater. Though the NWI did not include the term cumulative impacts, it included critical related concepts and control mechanisms, which apply equally to surface water and groundwater.

Most fundamentally, the NWI required statutory water planning, which scholars had earlier proposed as a key vehicle for precautionary 'caps' to control cumulative effects (Maher et al. 2001, p. 59; Neville 2003, pp. 88–9). Statutory water plans provide a strategic, proactive way to control impacts by reference to overall levels of planned stress, chiefly controlled by a volumetric cap—rather than seeking to rely on licence-by-licence determinations. In CIA terms, plans and caps set clear, acceptable thresholds of impact on a resource, at least on a relatively large geographic scale.

The NWI sought to broaden and develop concepts in water law that are used to address GCIs. The NWI explicitly recognised that legally uncontrolled withdrawals threatened a broad scope of receptors—both water entitlements, in relation to their security, and also environmental and public benefit outcomes (cl 25(i)–(ii)).

The NWI adopted a broader scope of impacting activities than had traditionally been controlled through water law, including 'interception activities' undertaken without needing a water access entitlement (cl 25(i)–(ii)). It specifically pointed to groundwater withdrawals by bores and large-scale plantation forestry (cl 55), but explicitly excluded potentially large and unregulated withdrawals of groundwater associated with minerals and petroleum, which were agreed to be subject to 'special circumstances' and require tailored management (cl 34). The NWI's suggested control mechanisms for responding to interception activities varied from implementing monitoring and licensing requirements to only planning these measures, depending on water allocation levels and the economic and environmental significance of existing and proposed withdrawals (cls 56, 57). States agreed to implement the measures relating to interception activities in accordance with water plans by 'no later than 2011' (Schedule A). However, in 2014, the National Water Commission assessed that 'no state or territory has fully implemented interception arrangements that meet the requirements of … the NWI' (National Water Commission 2014, p. 33).

The NWI marked an important conceptual and policy development in relation to key ideas about controlling GCIs. It formally linked a broad scope of receptors (including groundwater-dependent ecosystems, GDEs), a relatively broad scope for impacting activities (excluding minerals and petroleum), and the need to establish thresholds of impact (though without indicating precisely how to do so), with a control mechanism (requiring entitlements for otherwise uncontrolled extractions, to be subject to a cap in a statutory water plan). However, the NWI did not explicitly contemplate structural changes in receptors or impacts (though it mentioned climate change in a minor way) (cl 82(iii)(c), sch E cl 1(iii)), nor a broad temporal scale appropriate to groundwater. Nonetheless, the NWI paved the way for the introduction of regulatory mechanisms to control GCIs through the federal *Water Act*.

3.1.2 Federal Water Act and Basin Plan

Based on the NWI, the *Water Act 2007* (Cth) (Water Act) further developed a legal framework for dealing with GCIs, albeit a controversial, and arguably incomplete one. Its central element and vehicle for addressing GCIs is the Basin Plan 2012 (Cth), an overarching, cap-based statutory water plan for the Murray-Darling Basin (MDB), with which regional state-prepared water resource plans (WRPs) must be consistent (*Water Act*, s 55(2)). The Basin Plan is prepared by the Murray-Darling Basin Authority (MDBA) and adopted by the federal Water Minister (*Water Act*, ss 41, 44).

The *Water Act* and *Basin Plan* generally adopt a very broad scope of receptors. The *Basin Plan*'s main geographic scope of receptors is its precisely identified WRP areas (*Basin Plan* regs 3.03, 3.06, 3.07). The MDBA and the Minister must also 'have regard to' how the *Basin Plan* may affect the use and management of water resources outside the MDB and, vice versa (*Water Act* s 21(4)(c)(vii)–(viii)). However, the federal *Water Act* does not explicitly permit the *Basin Plan* to include strategies that relate to the extra-Basin causes of risks to the MDB (*Water Act* item 5 of table in s 22(1)), perhaps for constitutional reasons (*Water Act*, s 18B(4)). This indicates an undesirable mismatch between the wide geographic scope of receptors in the *Basin Plan*, and the narrower geographic scope of the impacting activities that it may contemplate.

The *Water Act*'s key controversial element is the degree to which it protects ecological receptors, vis-à-vis economic and social values, which the legislation also mentions (*Water Act* ss 3(c), 4(2), 11(2)(g), 20(d), 21(4)(a), 21(4)(c)(v), item 1 in table in s22(1), 86A, 86C). The key method of protecting ecological receptors is legally binding caps on extraction called 'long-term average sustainable diversion limits' (SDLs). These must reflect an 'environmentally sustainable level of take' (ESLT), defined as 'the level at which water can be taken from that water resource which, if exceeded, would compromise' any of four enumerated matters, being 'key environmental assets', 'key ecosystem functions', 'the productive base' or 'key environmental outcomes' (*Water Act* ss 4(1) 'environmentally sustainable level of take', 23 (1)). The legislation clearly contemplates a broad range of groundwater-dependent ecological receptors, due to the broad definitions of these four elements (noting, however, that in most cases only 'key' elements are protected).

In addition, the legislation adopts broad definitions of 'water resource', 'ground water', and 'water-dependent ecosystem'. 'Water resource' is defined to include 'ground water', 'an aquifer, whether or not it currently has any water in it', and 'water, organisms and other components and ecosystems that contribute to the physical state and environmental value of the water resource' (*Water Act* s 4(1)). Meanwhile, 'water-dependent ecosystem' includes 'a ground water ecosystem, and its natural components and processes … that depends on … significant inputs of water for its ecological integrity and includes an ecosystem associated with' wetlands, streams, lakes, salt marshes, estuaries, karst systems, or ground water systems (*Water Act* s 4(1)). The scope of 'water-dependent ecosystem' is also site-specific, as 'a reference to a water-dependent ecosystem includes a reference to the biodiversity of the ecosystem' (*Water Act* s 4(1)).

Unfortunately, thresholds in relation to these receptors are unclear, since the qualifier 'key' and the term 'compromise' are not further defined. Impacts on non-'key' elements are accepted, as are impacts that do not amount to 'compromising' the four ESLT matters. These crucial thresholds are left to be determined in the absence of a detailed regulatory process. The dangers of this omission appear to have been realised, based on a recent review of groundwater SDLs finding that the modelling criteria used to determine certain SDLs bear 'no obvious' connection to the statutory ESLT elements or that this relationship is 'not immediately clear' (Forbes et al. 2014, p 5; Wade 2014, p 34).

Looking beyond the basic scope of receptors protected under the Basin Plan, its conception of interactions between systems also appears relatively limited. In particular, it adopts a high threshold for recognising groundwater-surface water connections, by global standards (Nelson 2013), and runs against recommendations in favour of assuming that groundwater and surface water are connected until proven otherwise (Cullen 2006; National Water Commission 2011; Fullagar 2004). This effectively reduces the number of groundwater SDLs which must be set conservatively at the baseline level of diversions—a MDBA requirement in relation to connected systems (Murray-Darling Basin Authority 2012)—and allows for greater impacts of GCIs on connected surface waters.

The temporal scope of the Basin Plan and WRPs is not explicitly stated. While these documents clearly have a 10-year life (*Water Act* ss 50, 64(1)), the timespan over which they must consider risks and impacts to water resources is not explicit. This is despite the Act requiring the MDBA and Minister to 'take into account the principles of ecologically sustainable development' (*Water Act* s 21(4)(a)) that relate to decision-making for the long term (*Water Act* s 4(2)); and the explicitly long-term nature of the salinity arrangements that operate outside, but in tandem with the Basin Plan, and which contemplate impacts to the year 2100 (*Water Act* sch 1 cl 6(2)).

The scope of impacting activities theoretically contemplated by the federal *Water Act* is extremely broad. The *Basin Plan* controls activities that 'take' water, defined as 'to remove water from, or to reduce the flow of water in or into, the water resource' including by pumping, permitting water to flow from a well, or impeding the flow of water in the resource (*Water Act* s 4(1)). Importantly, this encompasses withdrawals from bores that are not required to be licensed under state water laws, as well as interception activities that do not involve bores, for example plantation forestry. Interception activities that are individually *or cumulatively* 'significant' must be regulated (*Water Act* ss 22(3)(d), 22(7)), mirroring the NWI, though such activities need not necessarily be subject to a licensing requirement in all cases (s 22(7)(b)). The Act considers indirect and structural changes through requiring consideration of 'risks to the condition, or continued availability, of the Basin water resources' posed by climate change and land use change, but does not regulate the relevant activities (*Water Act*, table to s 22(1)).

The federal Water Act and Basin Plan represent important regulatory development of NWI policy—but in some respects, unrealised promise—in relation to GCIs. The Basin Plan might have taken a truly comprehensive approach to impacts using the broad definition of 'take' that could include all incremental effects; an extremely broad scope of receptors through expansive definitions of groundwater and the four ESLT matters; and explicitly a long temporal scope. The present incarnation of the Basin Plan is an important first attempt to implement these provisions of the Water Act. Ultimately, though, it adopts a narrower view of receptors by narrowly interpreting environmental terms, arguably a weaker regulatory response to interception activities than originally suggested by the NWI, and it appears not to adopt (at least not explicitly) a long temporal scope that is appropriate for groundwater systems. It remains to be seen whether WRPs will merely comply with, or go beyond, this regulatory treatment of GCIs.

3.2 Cumulative Impacts in State Groundwater Regimes: Licensing and Other Approaches

In addition to adopting statutory water plans that are the cornerstone of the water reform era, state-level groundwater laws use separate mechanisms to control GCIs, including through licensing processes. As alluded to in Sect. 3.1.1, uncontrolled GCIs have traditionally been caused, at least to some extent, by gaps in state licensing laws. Traditionally, state water laws have applied exemptions to licensing requirements based on the purpose of use (typically those for stock and domestic purposes), and contemplated only withdrawals effected by a bore, overlooking interception activities like plantation forestry and mine voids below the water table. At the extreme, extractive industry laws may provide for the right to withdraw unlimited amounts of groundwater associated with those activities (*Petroleum and Gas (Production and Safety) Act* 2004 (Qld) s 185(3)), clearly frustrating attempts to regulate GCIs using a broad scope of impacting activities. These 'loopholes' mirror the licence exemptions of other water law systems in the western U.S. (Nelson and Perrone 2016; Nelson and Quevauviller 2016).

The Water Management Act (2000) (NSW) (WMA) and accompanying Aquifer Interference Policy (AIP) fill some important gaps in this traditional approach, or will do, when the relevant WMA provisions are proclaimed to have effect (as of March 2018, the provisions are not yet in effect anywhere in New South Wales). They require traditionally licence-exempt extractive activities like CSG extraction and mine dewatering to hold regular water licences (NSW Department of Primary Industries Office of Water 2012), including separate licences if the activity induces flow between connected water sources, including groundwater-surface water connections (NSW Department of Primary Industries Office of Water 2012). These activities also require special 'aquifer interference approval' for 'aquifer interference activities'—chiefly large-scale extractive industries that use groundwater incidentally (WMA s 91). The AIP quantitatively defines the thresholds of 'minimal harm' to water sources or water-dependent ecosystems, beyond which neither water access licences nor aquifer interference approvals may be granted (WMA, ss 63(2)(b), 97(6)). These thresholds relate to groundwater levels and pressure drawdown and changes to groundwater and surface water quality, which differ according to the type of water source and require proponents to supply increasing assessment information with increasing uncertainty or likely performance of the project closer to the thresholds (NSW Department of Primary Industries Office of Water 2012). Thresholds are expressed in terms of cumulative impacts, for example, 'cumulative variations' from particular groundwater levels at certain distances from high priority GDEs or high priority culturally significant sites, and cumulative pressure head declines at water supply works. Those thresholds take into account the combined impacts of all activities taking place *after* the commencement of the relevant water sharing plan (NSW Department of Primary Industries Office of Water 2012). While this formulation appears to ignore the time-lagged impacts of previous actions, the AIP explicitly recognises that the impacts of current aquifer interference activities can potentially extend centuries after the activities cease; it accordingly requires consideration of how to account for water taken beyond the life of a project (such as in the case of an open cut mine intercepting the groundwater table), for example, by surrendering a licence to cover the project's ongoing water use (NSW Department of Primary Industries Office of Water 2012). In CIA terms, the AIP displays a laudably broad view of incremental effects; a broad natural scope, including interactions between connected water systems (subject to the determination of what constitutes a 'high priority' asset); an explicitly long temporal scope; and clear thresholds.

In addition to altering traditional licensing exemptions, as does the AIP (a well-developed outlier in state practice, joined by some other more minor examples (e.g. Adelaide and Mount Lofty Ranges Natural Resources Management Board 2013; National Water Commission 2014)), other states have adopted a motley collection of additional approaches to addressing GCIs:

- statutory natural resource management plans outside water plan areas may control activities that do not require a water licence (*Natural Resources Management Act* 2004 (SA) s 127(2)).

- a Victorian Water Minister may temporarily or permanently qualify a water right (including, by implication, licence-exempt groundwater withdrawals) in response to a water shortage or to broader economic, social or environmental matters (*Water Act 1989* (Vic) ss 33AAA, 33AAB).
- geographically-defined works limits may prohibit the drilling of licence-exempt stock and domestic bores close to sensitive and highly valued GDEs in NSW (e.g. Water Sharing Plan for the Greater Metropolitan Region Groundwater Sources 2011 (NSW) reg 41).
- a special Queensland regime provides for monitoring and mitigating the impacts of CSG activities in 'cumulative management areas' using 'underground water impact reports' and requirements to 'make good' impacts on consumptive—though not ecological—receptors (*Water Act 2000* (Qld) ss 361–454).

Regular state-level water licensing procedures may also take account of GCIs. Some states' water legislation and policy require a decision-maker in relation to a water licence to consider cumulative impacts and matters relevant to CIA, or other water extractions, in considering whether to grant a licence. As an example, the *Water Act 1989* (Vic) requires that the Minister have regard to adverse effects on various environmental matters, existing and future ('projected') water quality and availability, 'the needs of other potential applicants' (i.e. future users), and 'any other matter that the Minister thinks fit to have regard to' (ss 40, 53). The formal policy detailing how delegates exercise the Minister's powers in considering and granting groundwater licences requires the delegate to impose conditions on a licence relating to the use of water on land, 'that are required to protect the environment and other water users, as the delegate considers appropriate', including 'minimising cumulative impacts [which are not defined] of water use' (Minister for Water, Victoria 2014). These provisions allow for considering a broad scope of impacting activities and receptors, though without clearly specifying a temporal scope or thresholds (other than a cap), nor the need to consider structural changes to development or natural systems. After licences to withdraw groundwater are granted, cumulative impacts can be adaptively (in effect, retrospectively) managed under allocation and restriction policies that influence how groundwater licences can be exercised (Minister for Water, Victoria 2014), for example, reducing allocations where GCIs become evident.

Although each of these state-level mechanisms for controlling water licence-exempt and also licensable uses of groundwater is available in theory, there appears to be no scholarly or other work analysing the degree to which, and manner in which, they have been used in practice and on the ground. The global literature suggests that although cumulative impacts are considered in legislation and applied in practice, their evaluation is beset by methodological and experimental challenges (Foley et al. 2017). The practical implementation challenges experienced in relation to the NWI interception activity provisions suggest the value of empirical investigation into the use of these mechanisms. This is particularly so given the potential for these mechanisms to find new life in a new generation of Basin Plan-compliant WRPs.

3.3 Groundwater Cumulative Impacts at the Intersection of Water and Environmental Law

In contrast to the first approaches to EIA laws (notably the *National Environmental Policy Act* in the US), Australia's EIA laws usually impose substantive controls on environmental impacts, rather than serving a function restricted to providing information to decision-makers and the public. Australia's federal *Environment Protection and Biodiversity Conservation Act* 1999 (Cth) (*EPBC Act*) requires assessment and approval of actions that are likely to have a significant impact on a 'matter of national environmental significance' (MNES). Until 2013, these MNES provisions were silent on cumulative impacts. In June 2013, the EPBC Act was amended to introduce a new statutory 'water trigger'—a new category of controlled actions. This filled a key gap in the implementation of cumulative impact controls in water law. The new water trigger prohibits a person from taking an action that involves CSG development or 'large coal mining development' if the action will have, or is likely to have, a 'significant' impact on a water resource (*EPBC Act* s 24D). The amendments also established in law the Independent Expert Scientific Committee on CSG and Large Coal Mining Development (IESC) (*EPBC Act* s 505C), previously established under a National Partnership Agreement as the source of non-binding advice to state project decision-makers (Commonwealth of Australia et al. 2012). The IESC advises the Minister on CSG and large coal mining developments, matters relating to technical bioregional assessments ('scientific analys[e]s of the ecology, hydrology and geology of the area for the purpose of assessing the potential direct and indirect impacts of coal seam gas development or large coal mining development on water resources') (*EPBC Act* s 528), research projects, and scientific information (EPBC Act s 505D).

Cumulative impacts are clearly within the scope of the definition of 'significant impact' for the purposes of the water trigger, since they are explicitly included in the definition of the relevant actions. 'CSG development' is defined as:

> *any activity involving coal seam gas extraction that has, or is likely to have, a significant impact on water resources (including any impacts of associated salt production and/or salinity):*
>
> (a) *in its own right; or*
> (b) *when considered with other developments, whether past, present or reasonably foreseeable developments* (EPBC Act, s 528).

'Large coal mining development' is defined analogously (*EPBC Act* s 528). The explicit inclusion of incremental effects of other projects is particularly important: recent Federal Court case law suggests, in relation to the other MNES, that the Minister need not consider the cumulative impacts of a project with other causally unrelated projects but rather, only the direct and indirect consequences of the action at hand (Tarkine National Coalition v Minister for the Environment (2015) FCAFC 89). In other words, the narrower understanding of 'significant impact' applicable to other MNES requires consideration of causally connected structural changes to development (such as growth inducement) but not other unrelated incremental effects. By contrast, water law does cover these impacts using statutory caps on 'take'.

Policy guidelines set out criteria to assist in determining whether an action is likely to have a significant impact (Department of the Environment, Australian Government 2013). The guidelines adopt an apparently broad scope of receptors, and also acknowledge the potential for structural changes to natural systems. They encompass effects on water quality or quantity or hydrological or hydrogeological characteristics that reduce 'the current or future utility of the water resource for third party users, including environmental and other public benefit outcomes' (Department of the Environment, Australian Government 2013, pp. 16–17). The guidelines include a wide geographic scope for considering impacts, covering 'the local, aquifer or catchment, and regional scale (including in connected or potentially connected hydrological systems)' (Department of the Environment, Australian Government 2013, p. 21). Unlike the Basin Plan, they include a rebuttable presumption of connections between aquifers and with surface water (Department of the Environment, Australian Government 2013). The guidelines also encourage adopting a long temporal scope, potentially extending beyond the life of the action, '[g]iven the length of time that impacts may take to become observable in groundwater systems' (Department of the Environment, Australian Government 2013, p. 20).

The water trigger relates to federal water reforms in a number of ways. Firstly, the EPBC Act uses the federal *Water Act*'s very widely scoped definition of 'water resource' (*EPBC Act* s 528) as the key receptor. Even a scope restricted to 'significant' impacts on this receptor would appear to be wider than the narrower *Water Act* definitions of the ESLT elements, which focus on a subset of 'key' assets, functions and outcomes.

Secondly, the concept of 'utility' (Department of Environment 2013, pp. 13–21) for the purposes of assessing the significance of the action's impact derives from an ecosystem services perspective (Department of the Environment, Australian Government 2013). The concept of ecosystem services also features in the federal *Water Act*'s objectives (s 3(d)(ii)) and its definition of 'environmental assets' that must not be compromised by 'take' of water (s 4(1); Reid-Piko et al. 2010). The current *Basin Plan* does not discuss ecosystem services in any detail, and it is not apparent that a wide definition was used in deriving SDLs for groundwater. Further developing the concept of ecosystem services in either the federal *Water Act* or EPBC Act context would usefully inform the scoping of relevant receptors in both contexts.

A third connection between the water trigger and water reforms arises through reliance of the former on the latter. As a matter of policy, the Department of the Environment will consider that a referral of a project under the *EPBC Act* is less likely to be necessary if all of the water used in an action is authorised under water entitlements that are consistent with a NWI-consistent statutory water plan (Department of the Environment, Australian Government 2013). This apparent reliance on state arrangements could be inappropriate in some circumstances.

Some state agencies may do little to assess an individual extraction proposal that falls within a water plan cap, and may not consider impacts on local-scale GDEs on this basis (Nelson 2013). If applied with gusto, the current *EPBC* policy could 'lock in' inadequate assessments of local-scale impacts, despite this being a recognised concern of 'leading practice principles' on CSG regulation (Standing Council on Energy and Resources 2013).

Overall, the *EPBC Act* water trigger compares favourably to the *Water Act* in terms of its scope of impacting activities and receptors, temporal scope, and recognition of structural changes to development and natural systems. The *EPBC Act* threshold of 'significance' is accompanied by clearer policy guidance (though comparatively brief by global standards) (Council on Environmental Quality 1997) than the *Water Act*'s threshold of compromising ESLT characteristics, though the relevant administrative guidelines are not legally binding (McGrath 2005). Future use of other *EPBC Act* regulatory mechanisms to control GCIs, notably strategic environmental assessments (for which the first relevant to groundwater was recently approved at the time of writing) (BHP Billiton 2016) may further develop these concepts.

4 Synthesis, Comparative Reflections, and Lessons

This chapter began by considering the concepts of cumulative impacts that groundwater law should consider in assessing and managing or mitigating impacts, with an analysis inspired by CIA in the environmental assessment literature. It has identified three key vehicles for assessing and managing GCIs:

- statutory water plans and associated caps on extraction;
- rules for licensing individual groundwater extractions and a miscellany of other rules of use in managing GCIs outside statutory water plans and licensing rules, for example, ministerial orders and qualifications of water rights; and
- approvals under the water trigger of the *EPBC Act*.

Australia's trajectory of water reforms, which has led to direct interactions between water law and environmental law, has generally seen a continuous increase in the degree to which laws for addressing GCIs reflects a broader understanding of both the development context and the natural systems context of groundwater withdrawal. This section synthesises the progress evidenced in the water reform era by reference to two broad contexts, development (i.e. impacting activities) and natural systems, elaborated in Part II.

As water reforms have progressed, arrangements for assessing and managing GCIs have taken a dramatically broader view of potentially affected natural systems. The initial focus solely on consumptive use values has expanded to encompass a potentially very broad range of GDEs and resource characteristics, as typified by the definitions of water resource, and derivatively, groundwater, used in the federal *Water Act* and *EPBC Act*. This potential may, however, remain unrealised due to the wide discretion embodied in statutory qualifiers like 'key' or 'high priority' assets and 'significant' impacts. Compared to this broader view of the valuable components of groundwater systems, there has been little explicit acknowledgement of the potential for non-accumulative responses or systemic or structural changes to natural systems. Law recognises the possibility of interactions between systems—between connected aquifers or between groundwater and surface water. However, managing impacts that arise because of these connections is constrained in federal water law by constitutional limitations and relatively high thresholds for recognising interactions between groundwater and surface water, with which the rebuttable presumption of connectedness under the *EPBC Act* water trigger guidelines contrasts favourably. Australia's water plan-based approach to recognising this aspect of GCIs provides an interesting contrast with better-developed licensing mechanisms for controlling the impacts of groundwater pumping on surface water in other jurisdictions (Nelson 2013). Yet these developments appear to contrast favourably with the general lack of consideration given to GDEs in water allocation processes in the western US, with European law also advanced on this front (Nelson and Quevauviller 2016).

Water reforms have also resulted in a much broader view of the development context. The NWI and federal *Water Act* focus on interception activities, and by using an expansive definition of 'taking' water, contemplate a broad range of potentially impactful activities (though requirements to manage these impacts only arise if they are 'significant'). However, they cannot regulate indirect impacts of taking water, such as land use change. By contrast, policy guidelines accompanying the *EPBC Act* water trigger address a narrow scope of impacting activities (CSG and large coal mine developments), particularly relative to the *EPBC Act*'s global counterpart environmental impact assessment laws. Positively, the water trigger does explicitly recognise structural changes, and unlike the situation for other MNES, the legislation includes unrelated developments as relevant to assessing the cumulative impacts of a project.

Despite the significant progress made in recognising and establishing legal vehicles to address GCIs in Australian law, it is clear that remaining weaknesses as well as nascent opportunities raise research issues warranting attention. In many cases, the same key issues for research have also arisen in the global environmental assessment literature (Lawrence 2013). Thus, these reflections on the Australian experience serve to bolster existing calls for similar investigations in the global literature. Notable issues for research include:

1. Empirically comparing the effectiveness and efficiency of different law and policy vehicles for managing GCIs (or combinations of vehicles), keeping in mind calls for 'smart regulation' in the water sphere (National Water Commission 2011, pp. 118–20). This work would review, for example, individual environmental impact assessments and IESC advice relating to groundwater withdrawals and state groundwater licensing decisions, and analyse the adequacy of the concepts and approaches used to deal with cumulative impacts in practice.

2. Exploring new or lesser-developed law and policy vehicles for managing GCIs, particularly economic tools (ARMCANZ 1996), and the workability of offsets and 'make good' obligations, as used in Queensland's cumulative management area arrangements (Randall 2012). New vehicles could build on existing tools in a variety of ways, for example, building on the nascent use of ecosystem services concepts in the federal *Water Act* and *EPBC Act*, for example, through 'location-dependent pumping fees that reflect the monetised adverse impacts of groundwater pumping' (Nelson 2013, p. 560).
3. Undertaking comparative analysis of the treatment of GCIs under the U.S. *National Environmental Protection Act*, which has been subject to extensive judicial consideration in relation to such impacts (Contant and Wiggins 1991), and the relatively recent European Strategic Environmental Assessment Directive, which specifically addresses 'secondary, cumulative, synergistic, short, medium and long-term permanent and temporary, positive and negative effects' on the environment (Commission of European Communities 2001, Annex 1).
4. Examining the justifications for, consistency of, and implications of, the apparently varying thresholds used in different law and policy vehicles for assessing and managing GCIs, including:

    - 'significance' for the purposes of managing interception activities under the *Water Act*, 'significance' for the purposes of the *EPBC Act* water trigger, as well as potential models for better defining these thresholds; and
    - the selection of GDE receptors for protection, using qualifiers such as 'key' and 'high priority'.

5. Analysing laws that deal with the gathering and sharing of groundwater data between public and private parties, the production and sharing of groundwater models, and how they deal with the inevitable remaining uncertainty about impacts using concepts such as the precautionary principle and risk-based management.

5 Conclusion

As Australia braces itself for looming water developments, including the expansion of unconventional gas mining and the agricultural development of the north, the issue of cumulative impacts will rise to the fore. More generally, increasing pressures on natural resources across the globe suggest the critical importance of developing robust regulatory approaches to cumulative impacts that correspond with scientific understanding of the issues. Examining cumulative impacts in the context of groundwater extraction—complicated as it is by potentially long time lags between actions and effects, profound uncertainty, and difficult data collection

—can offer valuable paths for dealing with cumulative impacts in natural resources regulation more broadly.

Australia's trajectory of groundwater law development shows that there is further work to do to ensure that regulatory conceptions of cumulative impacts in the groundwater context match broad scientific concepts of cumulative impacts. In a global context, Australia's laws appear comparatively underdeveloped in some respects (as with recognising the cumulative impacts of groundwater pumping on surface waters); and in others, comparatively advanced (as with broad conceptions of GDEs, at least on paper). It also suggests that we should broaden our horizons of the regulatory tools that could address these impacts. Current experience in this regard includes using caps under statutory water plans, state water law licensing rules and a motley collection of other state water law tools, and the recent awakening of federal environmental regulation in this sphere. Notable outstanding issues for further research include empirically assessing the effectiveness of current approaches, developing alternatives to these approaches, investigating comparative experience, and addressing particularly difficult issues surrounding thresholds, data and uncertainty.

References

Adelaide and Mount Lofty Ranges Natural Resources Management Board. (2013). *Water Allocation Plan: Western Mt Lofty Ranges*. Adelaide: Government of South Australia.
Agriculture and Resource Management Council of Australia and New Zealand and Standing Committee on Agriculture and Resource Management. (1996). *Allocation and use of groundwater: A national framework for improved groundwater management in Australia—Policy position paper for advice to States and Territories*. Canberra: Commonwealth of Australia.
Australian Government. (2015). *Our north, our future: White paper on developing Northern Australia*.
Badr, E.-S. A., Cashmore, M., & Cobb, D. (2004). The consideration of impacts upon the aquatic environment in environmental impact statements in England and Wales. *Journal of Environmental Assessment Policy & Management, 6*(1), 19–49.
Bashour, L. (2016). Comparative analysis of enabling legislation for EIA follow-up in Lebanon, Palestine and Jordan (and Aqaba). *Impact Assessment and Project Appraisal, 34*(1), 72–78.
Basin Plan. (2012). (Cth).
Bates, B., et al. (2008). *Climate change and water*. Geneva: Intergovernmental Panel on Climate Change.
BHP Billiton. (2016). Draft MNES Program, Pilbara Strategic Assessment. http://www.bhpbilliton.com/-/media/bhp/regulatory-information-media/iron-ore/western-australia-iron-ore/0000/mnes-program/160316_ironore_waio_pilbarastrategicassessment_commonwealth_draftmnesprogram.pdf. Accessed 30 Jan 2017.
Bubna-Litic, K. (2015). Fracking in Australia: The future in South Australia? *Environmental and Planning Law Journal, 32*(5), 437–454.
Canter, L., Atkinson, S. F., & Sadler, B. (2010). Introduction to a special issue on cumulative effects assessment and management. *Impact Assessment and Project Appraisal, 28*(4), 259–260.

Commission of European Communities. (2001). Directive 2001/42/EC of the European Parliament and the Council of 27 June on the assessment of the effects of certain plans and programmes on the environment. *Official Journal of the European Communities* 197(30), Annex I.

Commonwealth of Australia. (2004). *Intergovernmental Agreement on a National Water Initiative*.

Commonwealth of Australia et al. (2012). National Partnership Agreement on Coal Seam Gas and Large Coal Mining Development. https://www.ehp.qld.gov.au/management/impact-assessment/pdf/partnership-agreement.pdf . Accessed 30 Jan 2017.

Contant, C. K., & Wiggins, L. L. (1991). Defining and analyzing cumulative environmental impacts. *Environmental Impact Assessment Review, 11*(4), 297–309.

Cooper, L. M., & Sheate, W. R. (2002). Cumulative effects assessment: a review of UK environmental impact statements. *Environmental Impact Assessment Review, 22*(4), 415–439.

Council on Environmental Quality. (1997). *Considering cumulative effects under the National Environmental Policy Act*. United States: Executive Office of the President.

Cullen, P. (2006). *Flying blind: The disconnect between groundwater and policy*. Canberra: National Water Commission.

Department of the Environment, Australian Government. (2013). *Significant impact guidelines 1.3: Coal seam gas and large coal mining developments—impacts on water resources*. Canberra: Australian Government.

Environment Protection and Biodiversity Conservation Act. (1999). (Cth).

Foley, M. M., et al. (2017). The challenges and opportunities in cumulative effects assessment. *Environmental Impact Assessment Review, 62*, 122–134.

Folkeson, L., Antonson, H., & Helldin, J. O. (2013). Planners' views on cumulative effects: A focus-group study concerning transport infrastructure planning in Sweden. *Land Use Policy, 30*(1), 243–253.

Forbes, M., et al. (2014). Report to the MDBA by the review panel for the Goulburn-Murray sedimentary plain SDL area in Victoria. http://www.mdba.gov.au/sites/default/files/pubs/GM-sed-plain-groundwater-review-panel.pdf . Accessed 30 Jan 2017.

Fullagar, I. (2004). *Rivers and aquifers: Towards conjunctive water management* (*workshop proceedings*). Canberra: Bureau of Rural Sciences.

Howe, P. (2011). *Framework for assessing potential local and cumulative effects of mining on groundwater resources—project summary report*. Canberra: National Water Commission.

Lawrence, D. P. (2013). *Impact assessment: Practical solutions to recurrent problems and contemporary challenges*. Hoboken: Wiley.

Maher, M., Neville, J., & Nichols, P. (2001). *Improving the legislative basis for river management in Australia (stage 2 report: final report)*. Canberra: Land and Water Australia.

Marshall, R., Arts, J., & Morrison-Saunders, A. (2005). International principles for best practice EIA follow-up. *Impact Assessment and Project Appraisal, 23*(3), 175–181.

McCallum, J. L., et al. (2010). Impacts of climate change on groundwater in Australia: A sensitivity analysis of recharge. *Hydrogeology Journal, 18*(7), 1625–1638.

McGrath, C. (2005). Avoiding the legal pitfalls in the EPBC Act by understanding its key concepts. *National Environmental Law Review, 32*, 37.

Merz, S. K., CSIRO, & Bureau of Rural Sciences. (2010). *Surface and/or groundwater interception activities: Initial estimates*. Canberra: National Water Commission Waterlines Report Series No 30.

Minister for Water, Victoria. (2014). *Policies for managing take and use licences 2014*.

Murray-Darling Basin Authority. (2012). *The proposed groundwater baseline and sustainable diversion limits: Methods report*. Canberra: Murray-Darling Basin Authority.

National Water Commission. (2011). *The National Water Initiative—securing Australia's water future: 2011 assessment*. Canberra: National Water Commission.

National Water Commission. (2012). *Groundwater essentials*. Canberra: National Water Commission.

National Water Commission. (2014). *Australia's water blueprint: National reform assessment 2014*. Canberra: National Water Commission.

Natural Resources Management Act. (2004). (SA).

Nelson, R. (2013). Groundwater, rivers and ecosystems: Comparative insights into law and policy for making the links. *Australian Environment Review, 28,* 558–566.

Nelson, R., & Perrone, D. (2016). Local groundwater withdrawal permitting laws in the south-western US: California in comparative context. *Groundwater, 54*(5), 747–753.

Nelson, R., & Quevauviller, P. (2016). Groundwater law. In Tony Jakeman et al (Eds.), *Integrated groundwater management: Concepts, approaches and challenges* (pp. 173–196). Online: Springer.

Neville, J. (2003). Managing the cumulative effects of incremental development in freshwater resources. *Environmental Planning and Law Journal, 20*(2), 85–94.

NSW Department of Primary Industries Office of Water. (2012). *NSW aquifer interference policy: NSW government policy for the licensing and assessment of aquifer interference activities.* New South Wales: NSW Department of Primary Industries.

Parsons, S., et al. (2011). *Evolving issues and practices in groundwater-dependent ecosystem management.* Canberra: National Water Commission Waterlines Report Series No 46.

Petroleum and Gas (Production and Safety) Act. (2004) (Qld).

Pigram, J. (2006). *Australia's water resources: From use to management.* Collingwood: CSIRO.

Randall, A. (2012). Coal seam gas—toward a risk management framework for a novel intervention. *Environmental and Planning Law Journal, 29,* 152–162.

Reid-Piko, C., et al. (2010). *Ecosystem services and productive base for the Basin Plan: final report.* Canberra: Murray-Darling Basin Authority.

RPS Australia East Pty Ltd. (2011). *Onshore co-produced water: Extent and management.* Canberra: National Water Commission.

Sinclair, J. A., Doelleb, M., & Duinker, P. N. (2017). Looking up, down, and sideways: Reconceiving cumulative effects assessment as a mindset. *Environmental Impact Assessment Review, 62,* 183–194.

Standing Council on Energy and Resources. (2013). National harmonised regulatory framework for natural gas from coal seams. http://www.coagenergycouncil.gov.au/sites/prod.energycouncil/files/publications/documents/National-Harmonised-Regulatory-Framework-for-Natural-Gas-from-Coal-Seams_1.pdf. Accessed 30 Jan 2017.

Tarkine National Coalition v Minister for the Environment. (2015). FCAFC 89.

Tularam, G. A., & Krishna, M. (2009). Long term consequences of groundwater pumping in Australia: A review of impacts around the globe. *Journal of Applied Sciences in Environmental Sanitation, 4*(2), 151–166.

Wade, A. (2014). Goulburn-Murray Sedimentary Plain Groundwater Sustainable Diversion Limit Review Synthesis Report. Richmond North: Aquade Pty Ltd. http://www.mdba.gov.au/sites/default/files/pubs/GM-sed-plain-synthesis-report.pdf. Accessed 30 Jan 2017.

Water Act. (1989). (Vic).

Water Act. (2000). (Qld).

Water Act. (2007). (Cth).

Water Management Act. (2000). (NSW).

Water Sharing Plan for the Greater Metropolitan Region Groundwater Sources. (2011). (NSW).

Werner, A. D. (2010). A review of seawater intrusion and its management in Australia. *Hydrogeology Journal, 18*(1), 281–285.

Compromising Confidence? Water, Coal Seam Gas and Mining Governance Reform in Queensland and Wyoming

Poh-Ling Tan and Jacqui Robertson

Abstract This chapter considers water reform in the context of resources industries particularly coal seam gas (CSG) and mining in Queensland, Australia and Wyoming, USA. We start with federal level guidance relating to governance of water in the CSG and mining resources industries then focus on reform measures taken in Queensland comparing these with measures in Wyoming. Contrary to national water policy, Queensland's petroleum and gas activities were designated as exceptions to the water permitting process and given a statutory right to water. As a safeguard, environmental impacts are addressed mainly through obligations to monitor, report and manage impacts including using 'make good' obligations to neighbouring landholders. Recent litigation illustrates how a bifurcated legislative framework, where issues of quality and quantity are separately assessed, proved a barrier to substantive consideration of the impact of development, in this case mineral extraction, on underground water. The statutory right to take water has now been extended to the mining industry. Although the regulatory intent is for adaptive management of the water resource, regulatory change in Queensland appears to be reactionary to immediate concerns. In both Queensland and Wyoming, it has proved difficult to introduce and implement a water governance model that promotes sustainability.

P.-L. Tan (✉) · J. Robertson
Griffith Law School, Griffith University, Nathan, QLD, Australia
e-mail: p.tan@griffith.edu.au

J. Robertson
e-mail: jacqui.robertson@griffithuni.edu.au

© Springer Nature Singapore Pte Ltd. 2018
C. Holley and D. Sinclair (eds.), *Reforming Water Law and Governance*,
https://doi.org/10.1007/978-981-10-8977-0_12

1 Introduction

Large-scale mining and unconventional gas projects evoke polarised opinions. CSG,[1] in particular, is perceived to be in direct competition with agriculture and pastoralists for water resources.[2] As CSG extraction requires that water is first pumped out to produce the gas, significant amounts of water are co-produced along with the gas.[3] During its infancy, the industry's impact on water was overlooked in the 1994 COAG Agreement (COAG 1994). By 2004, despite rising public unrest, the National Water Initiative (NWI) allowed mining and petroleum industries to be subject to special policies and measures outside of the COAG agreement (NWI 2004, cl 34).

One of the most significant environmental issues relating to CSG is the volume of groundwater that is co-produced with the gas (Bryner 2004). Some criticise inadequate regulation of the vast quantity of water produced by the CSG industry (Tan et al. 2015; Nelson 2012). While water extraction from conventional gas production has decreased to about 1 gigalitre (GL) per year in the Surat Cumulative Management Area of Queensland, water extracted by the CSG industry in the same area has increased to about 65 gigalitres (GL) per year (OGIA 2016, p. xii). This amount surpasses the 53 GL of water that is taken on average from the same groundwater source, the Great Artesian Basin, for other uses in the same area (OGIA 2016, p. xii).[4] Others point to inadequate assessment of the short and long term risks of environmental impact by the resources industries[5] (Jensen 2010); cumulative impact of multiple CSG wells on water tables, potential damage or pollution of the aquifer (Williams et al. 2012; Hancock et al. 2012; Davies et al. 2015; Tan et al. 2015) and have criticized the fragmented nature of assessment of the above by the federal and Queensland regulatory regimes (NWC 2014b, p. 2).

---

[1] In Australia, the methane that is produced from coal seams is referred to as 'coal seam gas' and in the USA as 'coal bed methane'. As this chapter is primarily focussed on the Australian position, the term 'coal seam gas', or 'CSG' is used. This chapter discusses the law and commentary as at 1 May 2017.

[2] The gas is held in place by hydrostatic pressure in seams found over very large areas such as across the Surat Basin which covers 300,000 square kilometres (km^2) of Queensland and parts of NSW in Australia, (see Geoscience, Undated).

[3] The production of water is greatest in the initial stages of mining. The most recent assessment in the Surat Cumulative Management Area in Queensland sets the figure at 65 GL/y increasing to about 110 GL/Y in the next three years with the average water extraction predicted to be 70 GL/y, (OGIA 2016, at p. 89). The Surat Cumulative Management Area is the geographic location of most CSG projects and an area managed collectively by the Queensland government.

[4] Total extraction in the geographic area from other users is reported as 203GL which includes 144GL taken from overlying alluvial and volcanic aquifers and 6GL from the deeper Bowen Basin formations (OGIA 2016, p. xii).

[5] In contrast to the former National Water Commission (NWC), we use the term 'resources industries' to refer exclusively to petroleum, gas and mining industries as distinct from the extractive industries being quarrying for sand, gravel and rock which are regulated differently to the petroleum, gas and mining activities in Queensland.

We consider these recurring issues in Queensland where most of Australia's CSG production is located (see Fig. 12.1 below) and compare how they are identified and managed in Wyoming, a major producer of unconventional gas in the USA. Starting with a brief description of water governance with respect to the resources industry, and the approvals and assessment framework for resources projects, we consider recent court decisions and changes that have taken place in the governance of the mining sector. Finally, using nationally accepted principles established by the NWI, we assess whether reforms have been effective, efficient, equitable and resilient.

We find Queensland's governance of water and CSG has tended towards prioritising energy projects and short-term economic interests over longer-term environmental protection. Tensions between economic, ecological and socio-political agendas have detrimentally affected sustainable use and integrated water management. Inherent uncertainties relating to scientific understanding and impacts on water by the CSG sector have not been managed well, with reforms in 2004 conferring a statutory right to water on CSG developers. Recent amendments in 2014–2016 that extended the statutory right to water to the mining sector are viewed as unwarranted. In resolving a key problem regarding the bifurcated nature of assessment which segregated quantity from quality issues, decision-makers have exacerbated issues regarding competition to water.

2 Federal and State Regulatory Approaches

2.1 The NWI's Provisions for the Resources Industry

Clause 34, NWI provides for special measures for resources industries:

> … specific project proposals will be assessed according to environmental, economic and social considerations, and that factors specific to resource development projects, such as isolation, relatively short project duration, water quality issues, and obligations to remediate and offset impacts, may require specific management arrangements outside the scope of this Agreement.

In five years of auditing implementation of the NWI, the National Water Commission (NWC)[6] found disarray amongst the jurisdictions as to appropriate measures and policies for resource industries (NWC 2010). Thus, in 2010, the NWC issued a Position Statement recommending CSG activities should be incorporated into NWI-consistent water frameworks, stressing that the exception in clause 34 should rarely be used.

---

[6]Reports were prepared in 2007, 2009 and 2011.

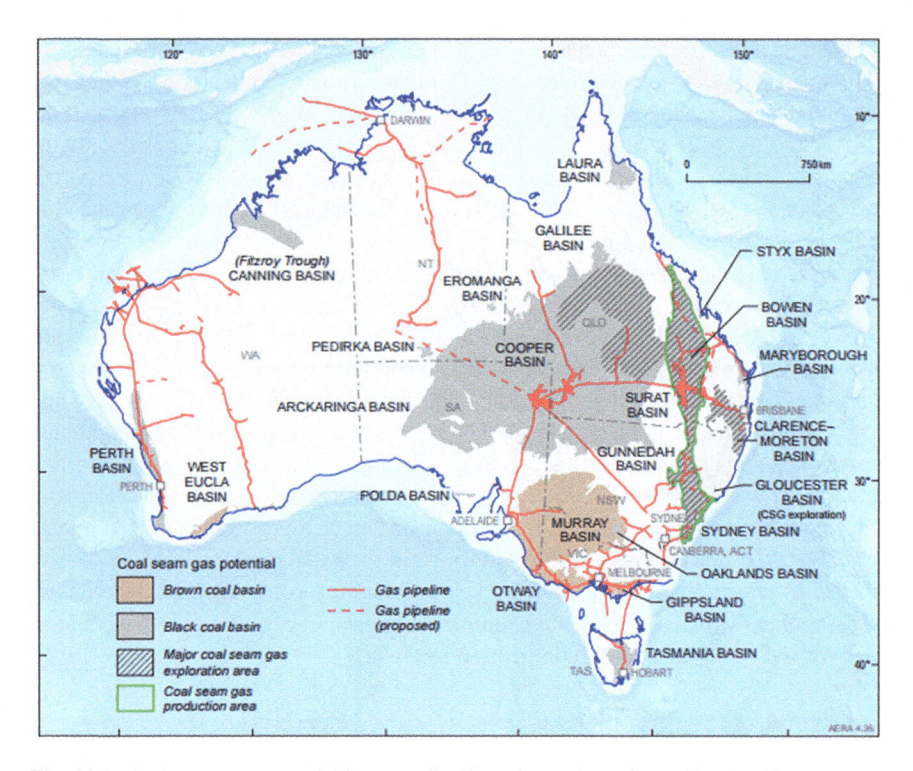

Fig. 12.1 Coal seam gas potential in Australia (Geoscience Australia and BREE 2014, p. 118)

Eleven principles were proposed by the NWC including:

- The interception of water by CSG extraction should be licensed to ensure it is integrated into water sharing processes from their inception.
- Jurisdictions should work to achieve consistent approaches to managing the cumulative impacts of CSG extraction. Such arrangements should consider and account for the water impacts of CSG activities in water budgets and manage those impacts under regulatory arrangements that are part of, or consistent with, statutory water plans and the NWI.
- Jurisdictions should undertake water and land-use change planning and management processes in an integrated way to ensure that water planning implications of projects are addressed prior to final development approval.
- Water produced as a by-product of CSG extraction, that is made fit for purpose for use by other industries or the environment, should be included in NWI-compliant water planning and management processes (NWC 2010).

The NWC's fourth audit report recommended that the NWI should guide the way all water is allocated and managed for all users including the resources industry (NWC 2014a, pp. 10–11). The NWC made the point more strongly in a targeted 2014 report, 'Water for Mining and Unconventional Gas under the National Water

Initiative', that alternative measures and policies instituted for the resources sector had led to preferential arrangements over other users thus reducing confidence in the water planning (NWC 2014b, p. 1).

2.2 Water Rights, Planning and Allocation in Queensland

Historically, the English common law and riparian doctrine applied (Stoeckel et al. 2012, p. 17 citing *Mabo v Queensland (No.2)* (1992) 175 CLR 1 at 38 (per Brennan J, (Mason CJ and McHugh J agreeing), and pp. 79 and 86 per Deane and Gaudron JJ and p. 138 per Dawson J; and *Gartner v Kidman* (1962) 108 CLR 12 at 23). In 1910, the Crown (in right of the state) adopted the power in respect of rights to 'the use, flow and to the control of water' including artesian groundwater (*Rights in Water and Water Conservation and Utilization Act of* 1910 Sect. 5). The *Water Act 2000* (Qld) installs statutory planning as the main allocation system for the sustainable development of water resources (*Water Act 2000* (Qld) Chap. 2, part 2). Water plans[7] for 23 catchments in Queensland set out (among other things) how water is shared between consumptive and environmental use, the volume of unallocated water reserved, plan objectives, strategies and outcomes (*Water Act* Sect. 43). The plans were developed through extensive community, technical and scientific consultation to establish triple-bottom line objectives (Tan 2006).

Licences or allocations provide rights to access water (*Water Act* Chap. 2, part 3). Each access right has a nominal volume, adjusted by an announcement by the regulator through the year; 'shortage powers' may limit extraction in times of scarcity (DNRM, undated). Rules under the relevant water plan specify and manage environmental flows while consumptive access rights are a share of the pool of water for a given catchment and equally share in limitations of supply. Any exceptions to this framework (such as for resource industries or stock and domestic uses), means these uses are not accounted for, thereby undermining water planning.

2.3 Approvals Required for Resources Projects

Mineral and natural resources fall within state jurisdiction, thus the Commonwealth[8] government has limited controls over most of these concerns. The *Environment Protection and Biodiversity Conservation Act* (1999) (EPBC Act) (Cth) provides for ecologically sustainable use of natural resources, and is the main

---

[7]These replaced the earlier 'Water Resource Plans' by the *Water Regulation* 2016 which also repealed the *Water Regulation* 2002.

[8]'Commonwealth' refers to the 'Commonwealth of Australia' which is the federal government in Australia.

Commonwealth regulatory mechanism relating to resource industries and the environment. Should actions impact on matters of national environmental significance,[9] the EPBC Act requires an environmental approval for the specific project.

Assessment under the EPBC Act requires an Environmental Impact Statement (EIS) and the Commonwealth Department of Environment takes advice from the Independent Expert Scientific Committee (IESC 2016) on impacts on water by proposed CSG and large coal mining development (EPBC Act Sect. 131AB). The IESC, established in 2012 in response to rising concerns over resources industry activities, often considers the adequacy of the draft EIS submitted by an applicant (IESC). In 2013, after a Senate enquiry into the CSG industry, a 'water trigger' was inserted into the EPBC Act requiring a federal approval for coal and CSG mining actions that could impact on water resources (EPBC Act ss 24D and 24E). Prior to that date, these actions only required approval on other grounds such as impact on listed species, but did not require analysis in respect of the water impacts and any subsequent Commonwealth conditions imposed on the project would not directly relate to water impacts.

At the state level, in Queensland, the *Mineral Resources Act* 1989 (Qld) (MR Act) governs rights to access mineral resources while CSG production is primarily regulated by the *Petroleum and Gas (Production and Safety) Act* 2004 (Qld) (P&G Act). Corresponding to the resource tenure, operators must also obtain an environmental authority for the specific activity under the *Environmental Protection Act* (1994) (Qld) (EP Act). If the project also requires federal approval under the EPBC Act, the one assessment is conducted covering matters to be considered under both Acts.[10] The initial consideration and assessment of a project's impacts may be coordinated under the provisions of the *State Development and Public Works Organisation Act* (1971) (Qld) (SDPWO Act). The EIS is not an approval itself but information is fed-back into other federal or state approval processes.[11]

In Queensland, different legislative instruments exercise regulatory control of water quality and quantity. Controls regarding quality of water, as well as environmental qualities generally, are governed primarily through the EP Act, and impacts of a proposed project that relate to water quality, are considered by environmental assessments as discussed above. However, issues relating to the actual right to extract water, that is, issues specifically of quantity, are governed by the *Water Act*. Where a specific project involves the taking of water and falls within the ambit of the *Water Act*, a separate application would be required under that Act

---

[9]These include listed threatened species and communities, listed migratory species, wetlands of international importance (Ramsar wetlands), nuclear actions, Commonwealth marine areas, world heritage properties and national heritage places and the abovementioned 'water trigger'.

[10]This is because the Commonwealth and Queensland governments have entered into an agreement in order to coordinate approval processes (known as the 'assessment bilateral agreement') (under Chap. 3 of the EPBC Act); (Cth and Qld 2009).

[11]This would be the federal EPBC Act, and respective state legislation being the MR Act or P&G Act and EP Act. For an excellent flow chart of the EIS process as at 5 April 2016 see McGrath 2016.

separate from the environmental assessment approval process described above. For users that are exceptions to the allocation provisions of the *Water Act*, no separate additional approval under that Act is required.

2.4 P&G Industry's Statutory Right to Take Groundwater

Queensland introduced the P&G Act to 'boost the coal seam gas sector and allow many developments to proceed with certainty' (Qld Hansard 2004, p. 881). Specifically, the legislation facilitated the new industry to provide a regulatory framework where coal and gas resources exist in the same deposit, an issue not considered by the existing *Petroleum Act* (1923). The P&G Act came into force in September 2004, just months after the NWI was signed by the Queensland Premier.

The P&G Act gave petroleum tenure holders a powerful statutory right to take underground water produced during (or resulting from) the carrying out of authorised activities (P&G Act s 185 (reprint 1)).[12] This statutory right extended to extraction of water by resource companies for incidental uses such as providing water for workers' camps, dust suppression and even construction (P&G Act s 185 (1)(c) (reprint 1)). As a safeguard, the statutory right to take water was coupled with obligations to monitor, report and manage impacts including using 'make good obligations' to neighbouring landholders (s. 250 P&G Act reprint 1). Only a cursory explanation of the operations of the new provisions rather than their justification was given in the explanatory notes to the Bill.

As extraction of the gas is dependent on the extraction of the water that keeps the gas in place, limits on the extraction of this water would indirectly limit the extraction of the gas. The regulator for the *Water Act*, the Department of Natural Resources and Mines has also opined that licensing prior to pumping is difficult because the quantity of water required to be extracted is unknown prior to CSG operations (AREPC 2014). Another reason for the exception is that the water commonly associated with gas extraction is of such poor quality it is not seen as a 'resource'.[13]

With CSG activities being exceptions to the water permitting process, no volumetric controls apply.[14] All other states (except for the Northern Territory) in

---

[12]Previously, the petroleum tenure holder would be required to obtain permission under the *Petroleum Act* (1923). Prior to the granting of an application under the *Petroleum Act* (1923), the application would be referred to the Water Act regulator and investigations and recommendations would be made under that Act to the decision-maker under the *Petroleum Act* (1923) (see s. 86 *Petroleum Act* (1923) reprint 6).

[13]Whilst this may be true in some circumstances it doesn't account for the fact that many agricultural water licences are in fact given for the Walloon Coal Measures, which is the main source of the gas and its associated water.

[14]Other amendments have been made to the regime that do not affect the statutory right to extract enjoyed by the petroleum and gas industry with are largely attempts to manage the impacts of the statutory right.

Australia require water entitlements for mining and CSG activities (NWC 2014b p. 3). Whilst extensive environmental assessments are undertaken under Queensland's EP Act and the federal EPBC Act, no further decision in respect of the right to a huge quantity of water under the *Water Act* is considered. The governance framework addresses impacts on other users by imposing underground water obligations including the preparation of underground water impact reports and 'make good' obligations (Comino et al. 2014). All these activities fall outside of water resource planning. Consideration of impacts, after their occurrence, is postponed to subsequent planning, and this is considered by scholars to be an unacceptable risk (Tan et al. 2015).

Unlike CSG, mining activities were, until December 2016, subject to the water legislation. Separate applications under the *Water Act* were necessary for any taking of water during operations (MR Act s 235(3), reprint dated 27 September 2016). Water seeping into mine pits from intersected aquifers as well as capture of overland flow is a common problem. Where mine pits require dewatering, projects could not start before obtaining water licences despite having environmental approval for the mining activity. We turn next to consider examples of this bifurcated assessment of water quality and quantity.

3 Key Issues with Regulatory Framework and Respective Reforms

This framework distinguishing between water quality and water quantity impacts, on the one hand, and also distinguishing between water users on the other, has created two key problems for the sustainable governance of water resources.

3.1 Bifurcated Assessment of Water Quality and Quantity Impact

As it was only mining activities and not CSG operations that were subject to separate assessment processes, the two key cases that sparked reform of the bifurcated assessment process were mining applications. These developments to mining in Queensland reflect upon CSG's statutory right to take water because that right has now been extended to miners.

The recent developments illustrate how a bifurcated legislative framework (issues of quality and quantity separately assessed) proves a barrier for substantive consideration of the impact of development, in this case mineral extraction, on underground water. However, the reform that was put in place, partly to address this issue, has exacerbated the second problem in the regulatory framework – the exception of certain users from allocation frameworks.

It is important to note that the two particular mines' applications pre-dated the insertion in 2013 of the 'water trigger' under the federal EPBC Act. Since 2013, large coal mines and CSG development must have the environmental impacts on water resources included in their environmental assessment at a Commonwealth level. Nonetheless, until recently, the mining projects would not have had these impacts assessed at the state level as part of the environmental permit.

3.1.1 Wandoan and Alpha Coal Mines

In *Xstrata Coal Qld Pty Ltd v Friends of the Earth* (2012) QLC 13 (the *Wandoan* case), objectors' concerns included the extraction and diversion of water at a proposed mining project at Wandoan in the Surat Basin. However, MacDonald P ruled that she could not make recommendations with respect to the extraction and diversion of water at the proposed mine because they were outside the scope of an environmental authority under the EP Act. MacDonald P highlighted that these issues would need to be fully canvassed in a separate hearing dealing with proposed water licences. What the judge ruled on were the impacts on water *directly resulting from mining activities*, and allowed the mining lease with additional recommendations for conditions relating to groundwater impacts.[15]

MacDonald P stressed the unsatisfactory nature of the current regulatory position where:

> the impacts of water diversions and extractions associated with the project [which] are highly relevant to any consideration of whether the project should be approved or refused... [are not] properly assessed under the Water Act until after the project has been approved under the MR Act and the EP Act (para [608], p. 140).

Two years after the *Wandoan* case, the *Alpha* case in the Galilee Basin came before the Land Court: *Hancock Coal Pty Ltd v Kelly & Ors and Department of Environment and Heritage Protection (No. 4)* (2014) QLC 12. Member P A Smith followed the reasoning of MacDonald P in making a distinction between taking or using underground water (which was a matter under the *Water Act*, and not before the court) and that of direct impacts of water from mining activities. Smith acknowledged that while the extraction or taking of this water from the pit would require a licence under the *Water Act*, the extension of the pit below the aquifer and therefore interfering with the flow of the underground water could occur without a mandate under the *Water Act*. Smith argued:

> It would appear nonsensical for this Court, as part of these proceedings, to be permitted to consider the question of, and consequences which flow from the interference with

---

[15]These related to further monitoring and the requirement that make good agreements be a prerequisite to the grant of the environmental authority. The proposed Wandoan project which would have included an open-cut coal mine, a coal handling and preparation plant, and support facilities, has now been put on hold.

groundwater, but not any aspect or consequence which arises from the taking or diversion of groundwater (para [121], page 47).

Accordingly, Smith recommended that either the mining lease and draft environmental authority be refused or that the mining lease be granted subject to Hancock first obtaining water licences, that the draft environmental approval be conditional on obtaining water licences, and further monitoring and 'make good' agreements.

The *Alpha* case was subject to judicial review by the Supreme Court of Queensland (*Coast and Country Association of Queensland Inc v Smith and Ors* 2015 QSC 260, Douglas J) and a later appeal to the Court of Appeal (*Coast and Country Association of Queensland Inc v Smith and Ors* 2016 QCA 242, McMurdo P, Fraser and Morrison JJA). Both the application and later appeal were argued in respect of largely unrelated issues to the water aspects of the case and each were dismissed upholding the Land Court decision. An application for special leave to appeal to the High Court by the Coast and Country Association of Queensland Inc. was dismissed with costs on 7 April 2017.

Both the *Wandoan* and *Alpha* cases reveal the inherent difficulties in drawing the line between environmental impacts of a project generally and water extraction and highlight that they can be very much related issues. While the *Alpha* decision extended the assessment of the project to include the later water extraction that would be necessary, a second application would still be required before the project could proceed.

3.1.2 Controversial Queensland Reform in 2014 and 2016: WROLA 2014 and EPOLA 2016

Soon after the Land Court *Alpha* case, the government introduced *Water Reform and Other Legislation Amendment Bill* (2014) (WROLA) which included a key reform measure: the extension of the statutory right to take enjoyed by the petroleum and gas industry to the mining industry.[16] Extraction of water that is inextricably connected to the extraction of the mining resource would no longer require *Water Act* approval. At the same time, a restriction on the petroleum and gas industry's statutory right to extract would also be implemented; only water taken as part of the extraction of the gas would be covered. The explanation given for these changes by the Government at the time was to provide consistency for managing the underground water impacts of both the mining and petroleum and gas sectors (Explanatory Notes WROLA 2014; Queensland Hansard 2014, pp. 3258–3262).

---

[16]WROLA included many other reforms that are not considered in detail in this chapter such as establishing a new purpose for the Water Act, introducing a watercourse identification map, streamlining the water planning and allocation process, enacting health and safety reforms for overlapping tenures among other amendments. Some of the reforms were ultimately reversed or amended.

The WROLA 2014 was passed but the changes to the statutory right to take water and corresponding underground water obligations were deferred as further consultation was deemed necessary.

A change of government months after the passing of the WROLA 2014 saw a modification in some aspects of government policy with respect to the resources industries and water. The new government made election commitments to repeal some of the WROLA 2014.

Stakeholder consultation occurred for two years before the *Environmental Protection (Underground Water Management) and Other Legislation Amendment Bill* 2016 (EPOLA) was introduced. Yet the government persevered with restricting the P&G industry's broad statutory right to take water and extending the statutory right to take 'associated water' by the mining industry. Proposed miners' rights would however be tweaked by transitional arrangements for projects part way through the assessment process by the introduction of 'associated water licences' (EPOLA Bill 2016 cl10). The EPOLA Act 2016 was narrowly passed in November 2016.

Future mining projects no longer require licences to take water extracted as part of activities, and impacts of this statutory right to water will be assessed as part of the environmental assessment process under the EP Act. As noted above, since the introduction of the 'water trigger', water impacts of mining and gas projects have been assessed at the federal level and would necessarily be included in the environmental impact statement. Since December 2016, the impacts for mining on water would also be considered in the environmental assessment at the state level. They will be firmly within the jurisdiction of the Land Court on any objections hearing in that Court, thus addressing the issues raised in the Land Court decisions discussed above.

The recent amendment of the *Water Act*, EP Act, MR Act and P&G Act, took two years, two separate government administrations and was only recently passed by one vote in Parliament, reflecting just how difficult it is to separate issues of quality and quantity when activities impact both.

3.2 The Problem of Exceptions in Water Planning

In resolving the bifurcated nature of assessments of impacts, reform measures now have implications for the allocation of water. In contrast with the NWC's recommendations since 2010, now all resources industries have a consistent statutory right to water in Queensland. Even though the impacts of the quantity of extraction will be considered as part of the EP Act process, no active decision needs to be made with respect to the quantity extracted as part of both mining and CSG activities. To address the issue of two separate assessments, one with respect to quality and the other to quantity, the state essentially relinquished control over quantity. Since 2016, no resources activities will share in limitations of the quantity of the resource leaving likely impacts on other users. It will only be after extraction of the water that the quantities will be able to be included in the planning process.

The implications of the changes in WROLA 2014 and EPOLA (2016) on water planning and allocation of water was not lost on stakeholders. The Queensland Resources Council (QRC),[17] argued that while the underground water rights and obligations framework applicable to the petroleum and gas industry had been seen as working well by the Queensland Competition Authority (QCA 2014), it had been developed over many years specifically in respect of the unique nature of the CSG industry (QRC 2014; QRC 2015). Doubt was raised whether the wholesale application of the framework to the mining industry would be as appropriate given the mineral industry's unique and different operations and needs (QRC 2014; QRC 2015). Furthermore, QRC noted that there was a strong preference from its members that changes would be better introduced through the water framework established with respect to catchment areas (QRC 2015, p. 4). Opposition was given by environmental groups on the basis that removing the need for mines to undertake the publicly notified water assessment processes under the *Water Act* would lead to inequality between that sector and other users (namely agriculture) (Lock the Gate Alliance 2014; EDO Qld 2014 p. 3). It was also suggested that reform would also relax controls on the environmental impacts of those projects (EDO Qld 2014), and that the amendments were contrary to the obligations in the NWI (EDO Qld 2014, p. 16; WWF 2014). The existing exception of petroleum and gas projects was criticized (Wilderness Society 2014, p. 6). The Agriculture, Resources and Environment Parliamentary Committee tasked with making recommendations to Parliament in respect of the WROLA 2014 acknowledged:

> that there are divergent views between stakeholders on the statutory right to take and interfere with associated water for the resources sector…[but] notes that the amendments proposed in the Bill are consistent with government's policy direction to reduce red tape for the mining industry (AREPC 2016, p. 48).

In our opinion, the recent changes that have extended the statutory right to extract water enjoyed by the petroleum and gas industry to the mining industry, have now exacerbated the impacts on water planning and governance in Queensland.

3.3 Wyoming's Approach to Water Quality Protection, Water Allocation and CSG Activities

Wyoming became a state in 1890 and its Constitution adopted the doctrine of Prior Appropriation in respect of surface water, and became the first state to require a permit for the use of water (Wyo Const. Art 1 §31 and 8§§1 and 3; MacDonnell

---

[17]QRC is the peak representative organization of the minerals and energy sector in Queensland.

2014). This position was extended to groundwater in 1945 (1945 Wyo. Sess. Laws, chap. 139§3; MacDonnell 2014).

The doctrine includes the following principles:

- The principle of priority to address water shortages, 'First in Time, First in Right'.
- The appropriation (or water use) must be for a beneficial purpose[18] (MacDonnell 2015).

As early as 1890, Wyoming pursued active supervision of water resources through adjudication, which considers the various rights of the different water users from a particular source (Potter CJ in *Farm Investment Co. v Carpenter* 9 Wyo 110, 61 P. 258 at 260 (Wyo 1900). Providing an initial data base of water use, adjudication involves inspectors checking what water is actually used rather than what is claimed to be used. It prevents speculation by tying water rights to actual use, and provides planning, indirectly avoiding over-allocation by assessing competing rights at the time of granting the right.

The nature of a water right, applying equally to groundwater, is set out in Wyoming Statute § 41-3-101 as:

> ... a right to use the water of the state, when such use has been acquired by the beneficial application of water ..... Beneficial use shall be the basis, the measure and limit of the right to use water at all times.

Wyoming similarly has separate assessment of water quality and water quantity with respect to its CSG industry. Impacts on water quality are regulated by the Wyoming Department of Environmental Quality (WDEQ) under the Wyoming *Environmental Quality Act* (Wyoming Statutes Title 35 Chap. 11) and the federal *Safe Drinking Water Act* (42 USC §300f et seq) and *Clean Water Act* (also known as the *Federal Water Pollution Control Act Amendments* 1972 33 USC §1251 et seq), while allocations of the resource are governed by the Wyoming State Engineer's Office (WSEO) (Wyoming Statute Title 41).

The WDEQ considers that it lacks power to regulate water quantity issues (MacKinnon and Fox 2006), a view supported by Wyoming's Attorney General (Crank 2006), contrary to scholarly opinion that release of large amounts of water impact on its qualities (MacKinnon and Fox 2006 p. 394). WDEQ's position means that quantity issues lie with the WSEO.

As the WSEO decides what is a 'beneficial use' (MacKinnon and Fox 2006 p. 376, FN22), the State Engineer has designated the extraction of water as part of CSG production as fulfilling this criterion (WSEO 2004). While each CSG well requires a permit to appropriate the groundwater, rather than the Wyoming Oil and Gas Conservation Commission being solely responsible for the permitting of these wells (and relying on its powers with respect to waste production found in Wyoming Statutes § 30-5-104), the State Engineer has assumed jurisdiction for the

---

[18]The notion 'beneficial use' in this context relates to the primary extraction and compares to a requirement to beneficially re-use the water.

initial diversion of water from the ground. This is presumed to be due to the large quantities of water that were produced from the early wells for significant periods of time without producing any gas (Darin 2002, p. 324).

Several exceptions apply to CSG wells. Generally, a well must be adjudicated once it is completed,[19] but an exception has been applied to CSG permits because of the presumably shorter time frame for the dewatering stage (MacKinnon and Fox 2006). Additionally, Wyoming allows the pumping of water during the extraction process without a preliminary inquiry into the harm this may cause existing water rights holders (Mudd 2012).

During the peak producing years of the CSG industry in Wyoming, many scholars criticised the WSEO's administration of the CSG permits and the legislative framework (Darin 2002; MacKinnon and Fox 2006; MacKinnon 2006; Kwasniak 2007) arguing that the by-product water going to waste was against the beneficial use principle (MacKinnon and Fox 2006; Mudd 2012).

A public interest test applies to permits (Wyo. Const. Art. 8, § 3; Wyo. Stat. § 41-3-931 and § 41-4-503 (2015)). How the State Engineer applies this 'public interest' test has been questioned (Herlihy 2009), with scholars recommending a positive duty for the WSEO to conduct a public interest test during the CSG permitting process; that is, to consider whether it is in the public interest to extract the quantities of water along with the gas (MacKinnon and Fox 2006; Herlihy 2009). However, the 'public interest' is not defined in the Wyoming legislation (Herlihy 2009).

This issue came to a head in the 2009 case of *William F Ranch, LLC v Tyrrell* 206 P3d 722 (Wyo 2009), in the Supreme Court of Wyoming, where ranchers challenged the state's application of the public interest test in Wyoming's Constitution and the Wyoming statutes and the lack of notice given to landholders prior to the issuance of CSG water permits. The Supreme Court dismissed the petition for relief on the grounds of lack of justiciability (at para [22] at p. 730) thus did not rule on the substantive issues, saying:

By ruling that the Court does not have jurisdiction over this case, we do not want to leave the impression that we approve of the State's administration of CBM [CSG] water. West and Turner raise serious allegations of damages to their property from CBM [CSG] water and failures on the part of the State to properly regulate CBM [CSG] water statewide (Para [48] at p. 737.)

The WSEO has adopted administrative measures to address these complaints including:

- In 2007, it cancelled permits of wells producing water with very little gas for substantial periods (Herlihy 2009).
- Since 2010, it has required CSG wells that were three years old to meet a water/gas ratio of 10bbl/mcf to maintain compliance with their permits (WSEO 2011).

---

[19] At that time the WSEO conducts a field inspection of the works to ensure, among other things, the accuracy of mapped location details and the yield of the well.

- In 2012, it amended the conditions attached to each CSG permit to include a self-reporting requirement in respect of the water/gas ratios (WSEO 2012).

Effecting legislative changes regarding CSG and water impact is also difficult in Wyoming as 'between 1997 and 2007, 39 bills were proposed, 12 were passed, none of which focussed on beneficial use or re-use of produced water' (Valorz 2010). Perhaps due to the widespread community concern and warning from the Wyoming Supreme Court, two Bills were introduced in 2011 (SF0126) and 2012 (SF0076) that sought to statutorily define 'beneficial use'. Presumably, this would then take the matter outside of any discretion of the WSEO, although it is unclear as to why the amendments were introduced. Both attempts at amending the legislation failed. In January 2017, an amendment was introduced (SF0122) to require the WSEO to maintain a publicly available list of beneficial uses as provided by Statute or the WSEO. This further attempt at legislative change was postponed indefinitely.

4 Discussion and Concluding Analysis

We assess how effective, efficient, equitable and resilient is Queensland's approach to designing and implementing CSG and water governance using the frame of the NWI, which is based on a hybrid of collaborative planning, regulatory and market theory (see Chap. 1).

As early as 2009 the NWC had reservations as to how water used and produced by the resources industry was not integrated with the broader water planning framework. The NWC cautioned that:

given the high level of uncertainty around water impacts on the temporal nature of CSG development, this ... requires a precautionary approach that demands innovation from water managements and planners ... (NWC 2010).

Besides WROLA 2014 and EPOLA 2016, Queensland has responded to this call by attempting to protect regionally significant water sources such as the Condamine Alluvium under the *Regional Planning Interests Act* 2014 (Qld), its Coal Seam Gas Water Management Policy and the requirement for water management plans as part of the environmental approval process under the EP Act 1994, as well as the introduction of 'make good' obligations in the *Water Act* and other green-tape reduction and water supply safety measures all of which have been outside the scope of this chapter. It also set up a new agency, the Office of Groundwater Impact Assessment to attempt to protect against cumulative impacts (Comino et al. 2014).

Although there has been reform, Queensland has not complied with many of NWC's principles for managing CSG activities and water, for economic and political reasons. There has been innovation in the sense that a statutory right to water has been conferred on CSG developers and now miners, a right that is not found elsewhere in Australia, bar the Northern Territory. Although considered through an EIS process, water impacts for projects are not part of a rigorous water planning process and, as an exception from the process, is seen by water users to be a threat (Tan et al. 2012, 2015). Therefore, confidence in and the effectiveness of water planning frameworks are jeopardised.

Efficiency is usually implemented by 'user-pays' provisions. Potential third party impacts from resource industries are dealt with through 'make good' arrangements and collection of monitoring data by the developers. In this sense, there are attempts at transparency as data is publicly available. However, given the nature of groundwater, and the risk of irreversible impacts on groundwater levels and water quality, externalities may only be realised decades after the projects have terminated. In addition, the make good arrangements are targeted at addressing supply to other users rather than remediation of the impacted aquifers. With good reason, scholars question the efficiency of these arrangements.

The exceptions created from Queensland's water allocation frameworks for CSG, now also the mining sector, have attracted much criticism as being inequitable as between users and departing from the established expectation of integrated water governance. Perceived inequity in how a resource is shared influences the social acceptance of water reform (Gross and Dumaresq 2014). Therefore, social acceptance of these measures will require adequate communication of the rationale for the changed water sharing paradigm. As noted by Gross and Dumaresq, it will also necessitate contemplating and addressing the unintended consequences of the reforms (Gross and Dumaresq 2014, p. 2491).

Reform measures transferring the water extraction issues to the EP Act framework and out of the *Water Act* framework, result in streamlined review of administrative decisions relating to the impact on water by resource development but have resulted in oversights being watered down (EDO 2016). Objections on draft administrative decisions for environmental permits and mining leases (the EP Act avenue) are heard by the Land Court, which makes a recommendation to the administrative decision-makers. That decision is open to judicial review of the administrative process, rather than an appeal on substantive grounds. Besides the issues of distributional justice, environmentalists consider this a backward step in equity – prior to the reform, an appeal of an administrative decision on a water licence (which must take place after first an internal review of the decision, see *Water Act* 2000 Chap. 6, part 2) was an appeal to the Land Court which had broad powers to set aside, or amend the administrative decision (*Water Act 2000* s 882).

Resilience is defined as:

the capacity of a system to absorb disturbance and reorganize while undergoing change so as to still retain essentially the same function, structure, identity, and feedbacks (Walker et al. 2004; Folke et al. 2010).

General resilience refers to whether social-ecological systems are able to withstand all kinds of shocks, including novel ones (Folke et al. 2010). Applying this terminology to the question at hand, we find economic and political agendas favour employment, development and royalties over environmental concerns. The influence of these factors has continued to provide a barrier to integrated water resource planning. Several successive Queensland governments of both political persuasions have endorsed and extended the exception for resource industries from integrated water resource planning. In that sense, the CSG industry has proved resilient.

However, when viewed from the perspective of adaptive management which requires that environmental management actions need to change in response to changes in the social-environmental system, then we observe that threats to groundwater systems are monitored yet feedback loops will be post-facto, therefore any damage and management responses are likely to be impotent. Thus, general resilience is compromised by socio-ecological systems being subject to thresholds of change which could result in an unstable landscape, particularly with respect to the water planning and allocation framework.

The issue of timing is a barrier to the integration of CSG water into the water planning process. Each water plan takes over three years from the initial scoping papers to the final plan. At the beginning of the process, CSG mining may not even have been contemplated in that catchment or groundwater basin. However, the industry has been in existence for many years now. It is entirely within the authority of the water regulator to reopen planning now the impacts of the CSG industry are becoming clearer. This is the case for the Surat and Bowen Basins where CSG development is well established. However, social and environmental factors, although strongly articulated within those basins' water plans, have resulted in piece meal reform, not integration.

In this space as well as overall water reform, the Commonwealth government and its agencies have provided much needed policy analysis and evaluation. Two Senate Inquiries (Senate Rural Affairs and Transport References Committee 2011; Senate Rural and Regional Affairs and Transport References Committee 2013) and political currents at federal level gave weight to the call for a 'water trigger' in the EPBC Act. In its independent Biennial assessments and policy papers, the NWC played an invaluable role. However, political support for water reform came to an end in 2014. In May 2014, with a change of Commonwealth government it was announced that the NWC would be abolished and its audit and monitoring functions would be transferred to existing government agencies in order to save $20.9 million over four years (Australian Government 2014). Papas notes that the abolition of the NWC, along with other proposed reforms at the time, would reduce the role of the Commonwealth Government in water reform progress (Papas 2015, p. 90).

The historical roots of Queensland's approach to water governance could not be more different to Wyoming's Prior Appropriation rule. Evolution of Wyoming's Prior Appropriation doctrine requires a public interest test before issuing a water permit for CSG use *and* the beneficial use of water and adjudication. As opposed to a statutory right to unlimited water in Queensland for resource industries, MacKinnon and Fox explain that the concept of 'beneficial use' limits rights of extraction to that which is reasonably necessary for the beneficial purpose to discourage waste (MacKinnon and Fox 2006; Mudd 2012).

However deeper analysis of the administration of Wyoming's governance framework, leads to a finding of congruence of both regulatory frameworks for the CSG industry. First, we find a struggle to provide a system that adequately governs both water quality and quantity aspects as well as providing an equitable water sharing framework. Secondly, we find that although permitting is required for CSG water use in Wyoming, in practice CSG permits are not taken to adjudication, thus missing the opportunity of ensuring a catchment is not over-allocated. Thirdly we find that the 'public interest' test in Wyoming was, in the opinion of some scholars, not justifiably implemented.

It is noteworthy that since the landmark Tyrrell case was filed, the Wyoming administration has used its powers to provide some limits on water taken during the CSG process. Where legislative efforts in both jurisdictions have tended towards exacerbating issues, administrative action appears to have provided some positive steps.

In conclusion, this chapter reveals the difficulties faced by successive Queensland governments in reforming water and resources legislation for multiple purposes: promoting development of the CSG industry; accommodating competing demands on water by consumers: aligning and integrating processes for environmental assessment of impact for the CSG and mining industries; reducing green-tape to make regulation more efficient; and lastly, ensuring their particular government's own political survival. These issues are broadly similar to Wyoming where the public interest was not vigorously pursued in regulating CSG development. The exceptions for the resources industry in Queensland to the water planning regime undermine fundamental aspects of the NWI. While Queensland has placed numerous conditions on project approvals to attempt to achieve environmental and public benefit outcomes, these will operate post-facto. Make good provisions are again post-impact, and relate to consumptive use, not the environment. Unforeseen post-tenure impacts, for example depressurisation, ongoing monitoring, and legacy infrastructure should be but are not addressed by regulation. Despite onerous monitoring and data collection obligations placed on resource companies that exercise their statutory right to underground water, security of access for other users and the environment is placed at risk. It is useful to compare this set of factors with the history of irrigation development, particularly in the Murray Darling Basin. As numerous scholars have noted, a developmental regime from the 1880s to 1970s led to Basin communities adopting a view that irrigation water underpinned their way of life (Smith 1998, Powell 1989, Pigram 2006, Connell 2007, Gross and Dumaresq 2014, Guest 2017). Even when many irrigators could see the benefit of water reform of the 1990s and 2000s, it is undeniable that their interests met with severe challenges. Remediation of unintended consequences of past regulatory arrangements imposes immense social, political and financial costs. It is this history that we keep in mind.

References

Agriculture, Resources and Environment Parliamentary Committee. (2014). Water Reform and Other Legislation Amendment Bill 2014, Report No. 52. http://www.parliament.qld.gov.au/documents/tableOffice/TabledPapers/2014/5414T6459.pdf. Retrieved April 10 2017.

Australian Government. (2014). Budget Paper No. 2 - Budget Measures, Budget 2014–2015. http://budget.gov.au/2014-15/. Retrieved April 10 2017.

Bryner, G. (2004). Coalbed methane development in the Intermountain West: Producing energy and protecting water. *Wyoming Law Review* 4(2), 541–557. 541-*Wyoming Law Review*.

Comino, M., Tan, P.-L., & George, D. (2014). Between the cracks: Water governance in Queensland, Australia and potential cumulative impacts from mining coal seam gas. *Journal of Water Law*, 23(6), 219–228.

Commonwealth of Australia and the State of Queensland. (2009). A Bilateral Agreement Between the Commonwealth and the State of Queensland Under section 45 of the Environment Protection and Biodiversity Conservation Act 1999 Relating to Environmental Assessment. https://www.environment.gov.au/system/files/pages/b44206bc-d8e5-450b-a05e-4d7c26d8afa1/files/qld-bliateral-agreement-assessment-amended-2014.pdf. Retrieved April 10 2017.

Connell, D. (2007). *Water politics in the Murray-Darling Basin*. Leichhardt: Federation Press.

Council of Australian Governments. (1994). Water Reform Framework. Extracts from Council of Australian Governments: Hobart, 25 February 1994 Communique. https://www.environment.gov.au/system/files/resources/6caa5879-8ebc-46ab-8f97-4219b8ffdd98/files/policyframework.pdf. Retrieved April 10 2017.

Crank, P. (2006). Formal Opinion No. 2006-001. (Attorney General of Wyoming, April 12, 2006) http://www.powderriverbasin.org/assets/Uploads/files/testimonies/opinionattorneygeneralapril122006.pdf. Retrieved April 10 2017.

Darin, T. F. (2002). Waste or wasted? Rethinking the regulation of coalbed methane byproduct water in the Rocky Mountains: A comparative analysis of approaches to coalbed methane produced water quantity. Legal issues in Utah, New Mexico, Colorado, Montana and Wyoming. *Journal of Environmental Law and Litigation* 17(2): 281–341.

Davies, P. J., Gore, D. B., & Khan, S. J. (2015). Managing produced water from coal seam gas projects: implications for an emerging industry in Australia. *Environmental Science and Pollution Research*, 22(14), 10981–11000.

Department of Natural Resources and Mines. Undated. Announced Entitlements and Announced Allocations'. https://www.business.qld.gov.au/industry/water/managing-accessing/accessing-water/authorisations/announced-entitlements. Retrieved April 10 2017.

Environmental Defenders Office Queensland. (2014). Submission No. 40 on the WROLA. (2014). https://www.parliament.qld.gov.au/documents/committees/AREC/2014/26-WaterReformOLA14/submissions/040EDOQ.pdf. Retrieved April 10 2017.

Environmental Defenders Office Queensland. (2016). Submission No. 28 on the EPOLA. (2016). https://www.parliament.qld.gov.au/documents/committees/AEC/2016/12EPUWMOLAB2016/submissions/028.pdf. Retrieved April 10 2017.

Explanatory Notes to the Water Reform and Other Legislation Amendment Bill. (2014). https://www.legislation.qld.gov.au/Bills/54PDF/2014/WaterReformOLAB14E.pdf. Retrieved February 2 2017.

Folke, C., Carpenter, S. R., Walker, B., Scheffer, M., Chapin, T., Rockstrom, J. (2010). Resilience thinking: integrating resilience, adaptability and transformability. *Ecology and Society*, 15(4), 20. http://www.ecologyandsociety.org/vol15/iss4/art20/. Retrieved April 10 2017.

Gross, C., & Dumaresq, D. (2014). Taking the longer view: Timescales, fairness and a forgotten story of irrigation in Australia. *Journal of Hydrology*, 519, 2483–2492.

Geoscience Australia. Undated. Surat Basin. http://www.ga.gov.au/scientific-topics/energy/province-sedimentary-basin-geology/petroleum/onshore-australia/surat-basin. Retrieved April 10 2017.

Geoscience Australia and BREE. (2014). Australian energy resources assessment. 2nd edn. Canberra, Australia. Geoscience Australia. ISBN 978-1-925124-22-4 (web). https://d28rz98at9flks.cloudfront.net/79675/79675_AERA.pdf. Retrieved April 10 2017.

Guest, C. (2017). Sharing the water – one hundred years of River Murray politic. Canberra: Murray Darling Basin Commission. ISBN 978 1 925599 04 6.

Hancock, S., & Wolkersdorfer, C. (2012). Renewed demands for mine water management. *Mine Water Environment*, 31(2), 147–158.

Herlihy, C. S. (2009). Trading water for gas: Application of the public interest review to coalbed methane produced water discharge in Wyoming. *Wyoming Law Review*, 9(2), 455–482.

Independent Expert Scientific Committee on Coal Seam Gas and Large Coal Mining Development. (2016). The IESC. http://iesc.environment.gov.au/iesc. Retrieved April 10 2017.

Jensen, T. (2010). Water issues in jurisdictional planning for mining: an overview of current practice- Waterlines Report. National Water Commission. ISBN 978-1-921107-99-3. http://www.mdba.gov.au/kid/files/1016%20-%20update_Water_issues_in_jurisdictional_planning_for_mining.pdf. Retrieved April 10 2017.

Kwasniak, A. J. (2007). Waste not want not: A comparative analysis and critique of legal rights to use and re-use produced water - Lessons from Alberta. *University of Denver Law Review, 10*(2), 357–390.

Lock the Gate Alliance. (2014). Submission No. 42 on the WROLA 2014. https://www.parliament.qld.gov.au/documents/committees/AREC/2014/26-WaterReformOLA14/submissions/042LTG.pdf. Retrieved April 10 2017.

MacDonnell, L. J. (2014). The development of Wyoming water law. *Wyoming Law Review, 14*(2), 327–378.

MacDonnell, L. J. (2015). Prior appropriation: A reassessment. *Denver Water Law Review, 18*(2), 228–311.

MacKinnon, A. (2006). Historic and future challenges in western water law: The case of Wyoming. *Wyoming Law Review, 6*(2), 291–330.

MacKinnon, A., & Fox, K. (2006). Demanding beneficial use: Opportunities and obligations for Wyoming regulators in coalbed methane. *Wyoming Law Review, 6*(2), 369–400.

McGrath, Christopher James. (2016). Major Environmental Impact Assessment (EIA) Processes in Queensland. http://envlaw.com.au/wp-content/uploads/Handout-for-lecture-5-EIA-processes-1.pdf. Retrieved April 10 2017.

Mudd, M. B. (2012). Montana v Wyoming: An opportunity to right the course for coalbed methane development and prior appropriation. *Golden Gate University Environmental Law Journal, 5,* 297–341.

National Water Commission. (2010). Position Statement Coal Seam Gas Water. Australian Government. http://webarchive.nla.gov.au/gov/20160615065541/, http://archive.nwc.gov.au/library/position/coal-seam-gas. Retrieved April 10 2017.

National Water Commission. (2014a). National Water Commission Assessment of the 2004 National Water Initiative Part 1. https://web.archive.org/web/20150310211643/, http://nwc.gov.au/__data/assets/pdf_file/0008/37673/Part-1-accessible-PDF-for-web-NWC-Australias-water-blueprint_national-reform-assessment-2014.pdf. Retrieved April 10 2017.

National Water Commission. (2014b). Water for Mining and Unconventional Gas under the National Water Initiative. https://web.archive.org/web/20160302162506/, http://nwc.gov.au/__data/assets/pdf_file/0008/37691/Water-for-mining-and-unconventional-gas-under-the-National-Water-Initiative.pdf. Retrieved April 10 2017.

National Water Initiative, Commonwealth of Australia and the Governments of New South Wales, Victoria, Queensland, South Australia, the Australian Capital Territory and the Northern Territory. (2004). Intergovernmental Agreement on a National Water Initiative. http://webarchive.nla.gov.au/gov/20130904113546/, http://www.environment.gov.au/water/australia/nwi/index.html. Retrieved April 10 2017.

Nelson, R. L. (2012) Unconventional gas and produced water. In *Australia's unconventional energy options,* ed. Committee for Economic Development in Australia (CEDA). ISBN: 0 85801 282 0. http://adminpanel.ceda.com.au/folders/Service/Files/Documents/15347~cedaunconventionalenergyfinal.pdf. Retrieved April 10 2017.

Office of Groundwater Impact Assessment (OGIA). (2016). Underground Water Impact Report for the Surat Cumulative Basin Management Area. https://www.dnrm.qld.gov.au/__data/assets/pdf_file/0007/345616/uwir-surat-basin-2016.pdf. Retrieved April 10 2017.

Papas, M. (2015). The way forward: Are further changes to Australian water governance inevitable. *Environmental and Planning Law Journal, 32*(1), 75–90.

Pigram, J. J. (2006). Australia's water resources: from use to management. Collingwood: CSIRO Publishing. ISBN 9780643094420 (pbk.).

Powell, J. M. (1989). Water the garden state: water, land and community in Victoria 1834–1988. Sydney: Allen & Unwin. ISBN 0043640249.

Queensland Competition Authority. (2014). Coal Seam Gas Review- Final Report. http://www.qca.org.au/getattachment/aaaeab4b-519f-4a95-8a65-911bc46cc1d3/CSG-investigation.aspx. Retrieved April 10 2017.

Queensland Hansard. (2004). Ministerial Statement on the Coal Seam Gas Industry by the Hon S Robertson on 12 May 2004, at pp. 881–882 https://www.parliament.qld.gov.au/documents/hansard/2004/2004_05_12_WEEKLY.pdf. Retrieved February 2 2017.

Queensland Hansard. (2014). Ministerial Introduction of the Water Reform and Other Legislation Amendment Bill by the Hon A P Cripps, 11 September 2014, pp. 3258–3262. https://www.parliament.qld.gov.au/work-of-assembly/sitting-dates/dates/2014/2014-09-11 Retrieved April 10 2017.

Queensland Resources Council. (2014). Submission No. 41 on the WROLA 2014. https://www.parliament.qld.gov.au/documents/committees/AREC/2014/26-WaterReformOLA14/submissions/041QRC.pdf. Retrieved April 10 2017.

Queensland Resources Council. (2015). Submission No. 65 on the WLAB 2015. https://www.parliament.qld.gov.au/documents/committees/IPNRC/2015/WLAB2015/submissions/065.pdf. Retrieved April 10 2017.

Senate Rural Affairs and Transport References Committee (2011). Management of the Murray Darling Basin, Interim report: the impact of mining coal seam gas on the management of the Murray Darling Basin. ISBN 978-1-74229-559-6. http://www.aph.gov.au/Parliamentary_Business/Committees/Senate/Rural_and_Regional_Affairs_and_Transport/Completed%20inquiries/2012-13/mdb/interimreport/index. Retrieved April 10 2017.

Senate Rural and Regional Affairs and Transport References Committee. (2013). The management of the Murray-Darling Basin. ISBN 978-1-74229-781-1.
http://www.aph.gov.au/Parliamentary_Business/Committees/Senate/Rural_and_Regional_Affairs_and_Transport/Completed%20inquiries/2012-13/mdb/report/index. Retrieved April 10 2017.

Smith, D. I. (1998). *Water in Australia: Resources and Management*. Melbourne: Oxford University Press Australia.

Stoeckel, K., Webb, R., Woodward, L., & Hankinson, A. (2012). Australian water law. Pyrmont Australia: Thomson Reuters (Professional) Australia Limited. ISBN 9780455228679 (pbk.).

Tan, P.-L. (2006). Legislating for adequate participation in allocation of water in Australia. *Water International, 31*, 21–22.

Tan, P.-L., Baldwin, C., White, I., & Burry, K. (2012). Water planning in the Condamine Alluvium, Queensland: Sharing information and eliciting views in a context of overallocation. *Journal of Hydrology, 474*(12), 38–46.

Tan, P.-L., George, D., & Comino, M. (2015). Cumulative risk management, coal seam gas, sustainable water, and agriculture in Australia. *International Journal of Water Resources Development, 31*(4), 682–700.

Valorz, N. J. (2010). The need for codification of Wyoming's Coal Bed Methane produced groundwater laws. *Wyoming Law Review, 10*(1), 115–140.

Walker, B., C. S. Holling, S. R. Carpenter and A Kinzig. (2004). Resilience, adaptability and transformability in social-ecological systems. *Ecology and Society 9*(2), 5. http://www.ecologyandsociety.org/vol9/iss2/art5/. Accessed 10 April 2017.

Wilderness Society (2014). Submission No. 25 on WROLA 2014. https://www.parliament.qld.gov.au/documents/committees/AREC/2014/26-WaterReformOLA14/submissions/025WSQ.pdf. Retrieved April 10 2017.

Williams, J., Stubbs, T., Milligan, A. (2012). An analysis of coal seam gas production and natural resource management in Australia Issues and ways forward. https://www.aie.org.au/AIE/Documents/Oil_Gas_121114.pdf. Retrieved April 10 2017.

World Wildlife Fund- Australia. (2014). Submission No. 38 on WROLA 2014. https://www.parliament.qld.gov.au/documents/committees/AREC/2014/26-WaterReformOLA14/submissions/038WWF.pdf. Retrieved April 10 2017.

Wyoming State Engineer's Office. (2004). Guidance: CBM/Ground Water Permits. https://2ce3bd20-a-84cef9ff-s-sites.googlegroups.com/a/wyo.gov/seo/seo-files/CBM%20Guidance%2004.pdf?attachauth=ANoY7coLZ-J0NhqgrPhjtsf2kOI8nxXAXPUQFYVRkOkhfwrIcEZrXgfn1S58fiVCLsDOvAhFcmeLPhwzf2luRp-xED95_73R2aKNyRuiAe79A7ujiUwRof1FfMzjMp7kplw3Y0Bg1YYOtznJaUWCLNjNvIh-kV8SZC3g9pfYZtUPIm14cq5r89sUsa0oTr1P_NpNDFn-FMhIa3Dub0g4m5gmak4Yam-Ptw%3D%3D&attredirects=0>. Accessed 10 April 2017.

Wyoming State Engineer's Office. (2011). 2011 Annual Report. (October 1, 2010 through September 30, 2011). https://2ce3bd20-a-84cef9ff-s-sites.googlegroups.com/a/wyo.gov/seo/seo-files/2011AnnualReport.pdf?attachauth=ANoY7crAkRq0-GVEBrwh90LAtGqbeGd2qtuU8VFvAIo2TW6sVD2GFxHkcYfXlGit3zrcCb0rynOO_a6dXraVclPtpo1hCYNZnSOgm_nxgS4chSbRaXndxKSIs4dIdRCm-9uZZjdan4HkcwNfwT9g5BzBM8Defzu-z2eumkwokIpnWlYtjzpiN1UGlW1OvsFDk8CcsGbuWDkaQXf4z_MJ8PuEoB5Acl8R5g%3D%3D&attredirects=0. Retrieved April 10 2017.

Wyoming State Engineer's Office (2012). 2012 Annual Report. (October 1, 2011 – September 31, 2012). https://2ce3bd20-a-84cef9ff-s-sites.googlegroups.com/a/wyo.gov/seo/seo-files/2012AnnualReport.pdf?attachauth=ANoY7cr7PoXeHpcg3xYdZ_6DyozdzZ9OndX1YjLNcdtURlcK9boPL-aA0ttCz4TmoQX4BcrL1E2KtfWU7S_xKoUkD6o5N-Zmjb9PNZbFsqwULdwFgVsC6S4EeEhuQ8JW7h3MdFRt1hl1atsBnONHXJKSYroRWiN0n7NvTpv8fl-juB_acWlszMEj4a6nS2UadQPOIG50Fd5tTw2gVsWoDpn4pruUF4g2fQ%3D%3D&attredirects=0. Retrieved April 10 2017.

Cases:
Australian:
Coast and Country Association of Queensland Inc v Smith & Ors (2015) QSC 260.
Coast and Country Association of Queensland Inc v Smith & Ors (2016) QCA 242.
Gartner v Kidman (1962) 108 CLR 12 at 23.
Hancock Coal Pty Ltd v Kelly & Ors and Department of Environment and Heritage Protection (No. 4) (2014) QLC 12.
Mabo v Queensland (No.2) (1992) 175 CLR 1.
Xstrata Coal Qld Pty Ltd v Friends of the Earth (2012) QLC 13.
USA:
Farm Investment Co. v Carpenter 9 Wyo 110, 61 P. 258 at 260 (Wyo. 1900).
William F West Ranch v Tyrrell 206 P.3d 722 (2009), Supreme Court of Wyoming.
Legislation:
Commonwealth:
Environment Protection and Biodiversity Conservation Act 1999 (Cth).
Queensland:
Bills:
Water Reform and Other Legislation Amendment Bill 2014. https://www.legislation.qld.gov.au/Bills/54PDF/2014/WaterReformOLAB14.pdf. Retrieved April 10 2017.
Environmental Protection (Underground Water Management) and Other Legislation Amendment Bill 2016. https://www.legislation.qld.gov.au/Bills/55PDF/2016/B16_0114_EnvironmentalProUWMOLAB16.pdf. Retrieved April 10 2017.
Acts and Regulations:
Environmental Protection Act 1994 (Qld). https://www.legislation.qld.gov.au/LEGISLTN/CURRENT/E/EnvProtA94.pdf. Retrieved April 10 2017.
Environmental Protection (Underground Water Management) and Other Legislation Amendment Act 2016 (Qld). https://www.legislation.qld.gov.au/LEGISLTN/ACTS/2016/16AC061.pdf. Retrieved April 10 2017.
Mineral Resources Act 1989 (Qld), *Reprint dated 27 September 2016*. https://www.legislation.qld.gov.au/LEGISLTN/SUPERSED/M/MineralReA89_160927.pdf. Accessed 10 April 2017.
Mineral Resources Act 1989 (Qld). https://www.legislation.qld.gov.au/LEGISLTN/CURRENT/M/MineralReA89.pdf. Retrieved April 10 2017.

Petroleum and Gas (Production and Safety) Act 2004 (Qld). https://www.legislation.qld.gov.au/LEGISLTN/CURRENT/P/PetrolmGasA04.pdf. Retrieved April 10 2017.
Petroleum and Gas (Production and Safety) Act 2004 (Qld) Reprint 1. https://www.legislation.qld.gov.au/LEGISLTN/SUPERSED/P/PetrolmGasA04_001_041231.pdf. Retrieved April 10 2017.
Petroleum Act 1923 (Qld) reprint 6. https://www.legislation.qld.gov.au/LEGISLTN/SUPERSED/P/PetrolmA23_006_041231.pdf. Retrieved April 10 2017.
Regional Planning Interests Act 2014 (Qld). https://www.legislation.qld.gov.au/LEGISLTN/CURRENT/R/RegionPlanIntA14.pdf. Retrieved April 10 2017.
Rights in Water and Water Conservation and Utilization Act of 1910 (1 GEO, No.25). http://www.austlii.edu.au/au/legis/qld/hist_act/riwawcauao19101gvn25652/. Retrieved April 10 2017.
State Development and Public Works Organisation Act 1971 (Qld). https://www.legislation.qld.gov.au/LEGISLTN/CURRENT/S/StateDevA71.pdf. Retrieved April 10 2017.
Water Act 2000 (Qld). https://www.legislation.qld.gov.au/LEGISLTN/CURRENT/W/WaterA00.pdf. Retrieved April 10 2017.
Water Reform and Other Legislation Amendment Act 2014 (Qld). https://www.legislation.qld.gov.au/LEGISLTN/ACTS/2014/14AC064.pdf. Retrieved April 10 2017.
Water Regulation 2002 (Qld). https://www.legislation.qld.gov.au/LEGISLTN/REPEALED/W/WaterR02_160701.pdf. Retrieved April 10 2017.
Water Regulation 2016 (Qld). https://www.legislation.qld.gov.au/LEGISLTN/CURRENT/W/WaterR16.pdf. Retrieved April 10 2017.

USA:
Safe Drinking Water Act (also known as Title XIV of the Public Health Service Act: Safety of Public Water Systems; 42 USC §300f et seq.) https://www.epa.gov/sdwa/title-xiv-public-health-service-act-safety-public-water-systems-safe-drinking-water-act. Retrieved April 10 2017.
Clean Water Act (also known as the *Federal Water Pollution Control Act Amendments* 1972) 33 USC §1251 et seq). https://www.epa.gov/laws-regulations/summary-clean-water-act. Retrieved April 10 2017.

Wyoming:
Bills:
'Water- beneficial use,' SF0126 (2011) 11LSO-0544. http://legisweb.state.wy.us/2011/Introduced/SF0126.pdf. Retrieved April 10 2017.
'Water- beneficial use,' SF0076 (2012) 12LSO-0277. http://legisweb.state.wy.us/2012/Introduced/SF0076.pdf. Retrieved April 10 2017.
'Water development amendments,' SF0122 (2017) 17LSO-0475. http://legisweb.state.wy.us/2017/Engross/SF0122.pdf. Retrieved April 10 2017.

Acts:
Wyoming Constitution. http://legisweb.state.wy.us/LSOWEB/StatutesDownload.aspx. Retrieved April 10 2017.
Wyoming Statute Title 30 chapter 5 (Oil and Gas). http://legisweb.state.wy.us/LSOWEB/StatutesDownload.aspx. Retrieved April 10 2017.
Wyoming Statute Title 41 (Water). http://legisweb.state.wy.us/LSOWEB/StatutesDownload.aspx. Retrieved April 10 2017.
Environmental Quality Act (Wyoming Statutes Title 35 Chap. 11). http://legisweb.state.wy.us/LSOWEB/StatutesDownload.aspx. Retrieved April 10 2017.

Governing the Freshwater Commons: Lessons from Application of the Trilogy of Governance Tools in Australia and the Western United States

Barbara Cosens

Abstract Common pool resources, by definition, are difficult to govern. The governance of water as a common pool resource is even more complex because of its unique physical characteristics, our special relation to it, and its necessity for life. Western societies have developed three approaches to governance of common pool resources: (1) government regulation; (2) the division of the resource into private property; and (3) self-organisation involving both governmental and nongovernmental actors. This chapter asserts that the unique nature of water requires the availability of all three approaches for use in combination if society is to rise to the challenge presented by the impact of growing population and climate change on water supply and demand. Long-term drought provides a window on potential future climate scenarios. In the face of long-term drought, Australia and the western United States have responded through innovation in water governance. This provides an opportunity for comparative analysis of reform to begin to understand the combination of governance approaches that lead to greater adaptive capacity.

1 Introduction

Governments have long wrestled with the appropriate means to govern common pool resources including water (Ostrom 1990). Today, with the intersection of climate change and population growth, society faces what appears to be the biggest challenge to equitable and sustainable governance of water resources in history. The combination of change in supply from historic average, change in timing due to alteration in snowpack, and increase in demand corresponding to the response of vegetation to higher temperatures (Jiménez Cisneros 2014) alone would challenge governance approaches adopted under the assumption of stationarity (Milly et al. 2008; Craig 2010a). Add to this the prediction that global population may grow

---

B. Cosens (✉)
College of Law, University of Idaho, Moscow, ID, USA
e-mail: bcosens@uidaho.edu

from its current level of 7.4 billion to 11.2 billion in this century (United Nations 2017), and the demand side of the water sufficiency equation becomes even worse. Compounding these two factors is the likelihood that the increase in energy in the atmosphere that is precipitating these changes will also result in greater extremes, including flood and drought beyond the historic record (Field et al. 2014). This unprecedented change and rate of change requires a degree of flexibility and adaptability in governance that has yet to be attempted. If these changes were to unfold slowly, society might fail to act. Yet with increased variability, flood and long-term drought presents sufficient crisis to open a window of opportunity for experimentation (Olsson et al. 2006). The responses to the prolonged Millennium Drought in Australia and the five-year drought in California and parts of the Inland Northwest of the United States are just such experiments. Comparison of the approaches to water law reform in Australia and the western United States in response to extended drought provides an opportunity to draw lessons from these experiments.

This chapter begins with a general discussion of how the challenge of governing the commons manifests in water law. This background sets the stage to explore specific aspects of water law reform in Australia and the western United States in the face of drought. The chapter will conclude with a discussion of whether these measures place their societies in a better position to respond to change and increase in variability, and what the western United States and Australia might learn from each other.

2 Governing the Water Commons

A common pool resource is a 'resource system that is sufficiently large as to make it costly (but not impossible) to exclude potential beneficiaries from obtaining benefits from its use' (Ostrom 1990, p. 30). Water is a common pool resource with unique characteristics that include: society uses water both in and out of its place of origin; water flows; in many cultures, water is sacred; and water is essential to human life, thus to deny access is the functional equivalent of genocide.

Garrett Hardin identified two mechanisms to avoid over-exploitation of a common pool resource: government regulation, and private property rights to the resource (Hardin 1968). While the division of a common pool resource into private property may have advantages in more directly connecting benefits and costs and in allowing adaptation through the development of markets, property use may also have costs that spill over onto neighbouring property (i.e. externalities). In application to water resources, the fact that water flows and is used both in situ and out of the place of origin means that external consequences of use are the rule rather than the exception. In addition, strict adherence to private property may deny access to some, violating fundamental human rights (United Nations General Assembly 2010). As a result, privatisation of the resource alone is a problematic approach to sustainable water governance (Ostrom 1990).

In contrast, government regulation allows prospective consideration of potential harm, adjustment of allocation as conditions change, and attention to the fact that the resource is connected. However, as the sole means of governance of a resource with the size and complexity of a water basin, government regulation is problematic. It requires perfect knowledge and enforcement to be adaptable in the face of change—an unlikely and costly scenario.

Nobel laureate Elinor Ostrom added a third possibility by documenting the self-organisation of local interests to maintain the sustainability of a resource they rely on (Ostrom 1990, 2009). Self-organisation, as the third prong of management of common pool resources, may have advantages in being more adaptive than government regulation while providing a collective means to address the systemic effects of private property and of market failure. In addition, self-organisation at the scale of the resource tightens feedback loops between the resource and those managing it. Nevertheless, self-organisation alone may have problems with local capacity, corruption and willingness to make difficult tradeoffs (Ostrom 2009).

This chapter begins with the assumption that the most adaptable legal framework for water governance includes a mix of all three approaches (Fig. 1) and the flexibility to move among them to navigate the drought-flood-sustainability challenge to come. Thus, any legal framework for water management must have authority for:

- Clearly defined, marketable, water use rights (the private property prong).
- Adaptive water planning through local participatory processes, with capacity built through governmental assistance (the self-organisation prong).

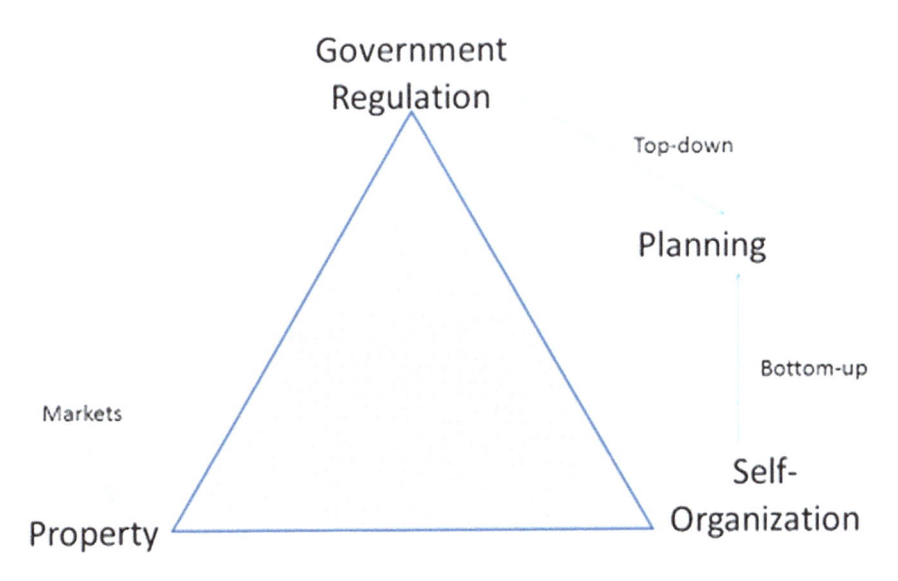

Fig. 1 The trilogy of approaches to governing common pool resources (Reproduced from Cosens 2016)

- Governmental authority to adjust water allocation as circumstances change or new information becomes available, governmental oversight of markets, and governmental facilitation of planning (the government regulation prong).

While the three approaches to governance of the commons—marketable private property, government regulation, and local self-organisation—are generally championed by those with differing political views and presented as requiring a choice of one approach over the others, it is entirely possible for all three to exist in concert, providing an expanded toolbox for the water manager. This chapter asserts that the water laws that provide the highest level of adaptive capacity in the face of change will facilitate aspects of all three with the authority to emphasise one over the other depending on the circumstances. Before discussing how water law reform in Australia and the western United States fits within this framework, it is necessary to understand the particular problem each is suited to resolve and thus the appropriate context for its emphasis.

2.1 Marketable Property Rights

Division of a common pool resource into private property provides incentive for stewardship and reduces the likelihood of over-exploitation by directly linking benefits and burdens (Hardin 1968). Studies also show that clear definition of rights enhance enforcement and compliance with use allocation (Ostrom 1990, 2009). Markets in any commodity facilitate rapid adaptation (coined 'The Invisible Hand' of the market by Adam Smith 1904). Thus, the severance of water rights from the land, and implementation of water markets, is one approach to increasing adaptive capacity in the face of variable water supply. Property rights nevertheless have increasing issues with externalities when applied to a fugitive resource like water (Ostrom 1990). For example, the extraction of water from its natural watercourse may harm both ecological (e.g. aquatic species) and human use dependent on its use in situ (e.g. navigation). The marketing of water outside the system of origin can harm irrigators who are interdependent on shared delivery systems. Transfer out of the system may also harm the ecosystem downstream that depends on return flow. Thus, both use and transferability require an overlay of regulation (Dellapena 2000).

2.2 Government Regulation

Governmental regulation of the use of common pool resources has the advantage over systems of property of allowing prospective consideration of the consequences of allocation. The prospective approach informs investment and allows a systemic approach to consideration of the secondary impacts of management actions. Yet, the inability to contextualise regulation, a problem caused by lack of resources and restraints on agencies put in place to assure uniform implementation of environmental laws, has led to dissatisfaction with top-down governmental regulation in the United States (for a concise summary of the criticisms of command and control environmental regulation in the United States, see Orts 1995). Perfect knowledge is essential to perfect implementation of a regulatory approach. Regulation of a resource the size and complexity of a water basin will encounter substantial difficulty without major expenditures in monitoring, forecasting, and enforcement. Even with substantial resources, the uncertainty associated with climate change makes this degree of predictability unlikely. Thus, government regulation without local capacity to act and individual capacity to transfer water and adjust water use is unlikely to prove robust in the face of variable supply.

2.3 Self-organisation

Self-organisation of combinations of private, economic and governmental actors, is observed in many collaborative governance processes referred to with a variety of terms reflecting both context and discipline of the observer. These terms include new governance, collaborative governance/co-management and adaptive governance, and are thought to be an emergent (i.e. self-organising) phenomenon (Karkkainen 2004; Dietz et al. 2003; Brunner et al. 2005; Folke et al. 2005; Gunderson and Light 2006; Chaffin et al. 2014a). These approaches enhance the use of local knowledge and locally imposed restrictions tend to carry greater legitimacy. Planning, particularly if locally driven, may enhance adaptive capacity by being itself adaptive (Arnold 2010).

Collaborative processes are not timely in the face of crisis such as flood, unless used prospectively to develop scenarios and capacity for emergency response. Collaborative processes lacking external pressure, such as the threat of government regulation or the failure of the market, may also encounter difficulty in making tradeoffs in times of scarcity, and may carry a greater risk of corruption (Ostrom 2009). Local collaboration for management of a resource as complex as a water basin facing climate change requires substantial capacity building through access to external governmental resources (Cosens et al. 2017). Recent studies have also shown that law may facilitate as well as hinder the emergence of these processes (Cosens et al. 2017).

2.4 Combining Governance Approaches

In the face of a variable water supply, the availability of all three approaches to management of a common pool resource such as water provides the greatest adaptive capacity. In crisis situations (e.g. flood), governmental intervention and strong control may be essential, but citizens within the flood zone must also rely on strong local capacity for initial response and recovery. The use of government regulation and planning, in concert with local self-organisation, provides, on the one hand, local buy-in (thus reduced resources needed for enforcement and enhanced legitimacy), and use of local knowledge allowing tailoring; and, on the other hand, a check on local corruption.

The timeframe of extended drought represents an intermediate point between a slowly changing climate and the immediate crisis of a flood. It is the assertion of this chapter that all three approaches are necessary to navigate drought and that their development in that context will build capacity to respond to the more long-term effects of climate change.

3 Water Law Reform in Australia and the Western United States

Water allocation law of both the United States and Australia has its roots in the common law of England which recognised the right to use water as an incident of riparian land ownership (*Gartner v Kidman* (1962) 108 CLR 12; *Tyler v Wilkinson* 24 Fed Cas 472 (Cir Ct DRI, 1827)). Yet both Australia and the western United States altered their approach to account for the arid nature of the landscape and the corresponding fact that the location of the desired use may not be riparian to any water source (*Gartner v Kidman* (1962) 108 CLR 12, 23; Avey and Harvey 2014). Thus, even prior to recent effects of climate change, Australia and the United States have tailored their water allocation laws to the reality of aridity and variability.

In recent decades, both Australia and the western United States have experienced extreme drought. Global climate change is likely to result in future droughts of similar severity to these recent droughts which occurred in Australia from 1997–2011 (the Millennium Drought) (Van Dijk et al. 2013), and in California from 2011–2016 (Williams et al. 2015). A shared understanding of how both regions have responded to this challenge provides an opportunity for both to learn. In keeping with the chronological order of the Millennium and California droughts, the following paragraphs will describe the selected aspects of water law reform related to the three approaches to governance of common pool resources, first in Australia and then in the United States.

3.1 The Millennium Drought and Water Reform in Australia

Water allocation is a matter of state law in Australia and, despite its origin in English common law, all Australian states and territories have adopted statutory schemes for the management and allocation of water that allow diversion away from land riparian to the water source (*Australian Constitution* Sect. 51; Stoeckel et al. 2012). As an example of state water allocation prior to the water reform during the Millennium Drought, South Australia's initial statutory scheme, reflected in the *Water Resources Act 1990* (SA), allocated water under licences to landowners. Water shortage is shared among those holding a licence. A government-issued licence is an interest for a specific use that may be cancelled by the government, thus the state is viewed as the owner of the water itself (*ICM Agriculture Pty Ltd v Commonwealth* (2009) 240 CLR 140; 84 ALJR 87; 170 LGERA 373; [2009] HCA 51, 55). The water licence attached to the land and the licence expired on transfer of the land (Avey and Harvey 2014).

As the Millennium Drought unfolded, the Commonwealth began to play an increasing role in water through its power over interstate commerce (*Australian Constitution* Sect. 51), and spending (*Australian Constitution* Sect. 96; Stoeckel et al. 2012). In addition, the Council of Australian Governments (COAG) (with membership including the Prime Minister, state and territory Premiers and Chief Ministers, and the President of the Australian Local Government Association), entered into the *Intergovernmental Agreement on a National Water Initiative* (NWI). The governments of New South Wales, Victoria, South Australia, Queensland, the Northern Territory, and the Australian Capital Territory signed in 2004, Tasmania signed in 2005, and Western Australia signed in 2006 (Stoeckel et al. 2012).

The NWI recognises water management as a national issue (NWI clause 3), while leaving implementation to the states and territories (NWI clause 20). The NWI calls for planning to identify a consumptive pool and a non-consumptive level of environmental flows (NWI clause 23), and an accounting of the consumptive use portion of individual water entitlements to assure they remain within that consumptive pool (NWI clause 28). Rights to water are recorded as shares in a particular consumptive use pool. However, the water allocation available to a given water entitlement is quantified in proportion to the share of water available in a specific season.

In addition, the NWI sought to remove barriers to water trading (NWI clause 23). Essential to facilitation of water transfers is the clear definition of the water right and the public availability of that information. The NWI calls for a registration of all water entitlements and trades (NWI clause 59), development of uniform pricing, and for states and territories to reduce transaction costs (NWI clauses 58 and 60). The property nature of a water entitlement allows it to be mortgaged independently of land and to be traded separate from land (NWI clause 31). The NWI contemplates both permanent trades of water entitlements/licences, and temporary trades of the seasonal water allocation (NWI Schedule G). The NWI sets forth the principles the state parties should use to establish consistent approaches to water transfers, maintain consistency with water plans and avoid impact on environmental and cultural values when considering transfers (NWI Schedule G).

The Water Registry for South Australia provides an example of state approaches. It is available to the public online and information on registered water licences include water source, quantity, type of use, and duration (South Australia, nd). South Australian law provides for the temporary and permanent transfer of all or part of a water entitlement (*Natural Resource Management Act* of 2004, NRM Act, part 3, Sect. 150), and of a seasonal water allocation (NRM Act, part 3, Sect. 157). Approval of a transfer by the Minister requires consistency with the water plan and that the transfer is in the public interest, and allows the Minister to alter the water entitlement or allocation accordingly (NRMA part 3, Sects. 150 and 157).

One of the consequences of water transfers is the third party impacts caused by the fact that water, even when diverted and used, is part of a system physically shared with others. Thus, water users who share diversion structures or who are located downstream from the return flow or wastewater outlet of another user are dependent on aspects of the use of other water users. The requirement in South Australia water law that a transfer need only comply with the plan substantially reduces transaction costs by avoiding individualised study of third party impacts. A study done following implementation of marketable water rights in the Murray-Darling Basin suggests that this has substantially increased the ease of water transfers and thus eased the economic impact of drought (Young 2011). However, Australia may have gone too far in reducing transaction costs at the expense of local economies and ecosystems. Much of the commentary on Australia's new water markets has focused on the concern with transfers that remove water from a local area, because in doing so the transfer may harm environmental and cultural values as well as local economies (McKay 2011). Third parties and communities harmed by a transfer that is nevertheless consistent with the water plan have no recourse (Young 2011).

Australia's response to drought may also have decreased local self-organisation. Again, South Australia provides an example of how this played out. With the *Water Resources Act 1997* (SA), South Australia implemented a planning approach, with catchment boards that allowed local representation to lead the effort (Avey and Harvey 2014). This approach can be characterised as government assisted self-organisation. In 2004, South Australia revised its water resources management with passage of the NRM Act, reflecting a goal of landscape-scale integrated natural resources management (Avey and Harvey 2014). Development of the NRM Act occurred in parallel with the multi-state/federal effort to develop the National Water Initiative, and thus the NRM Act reflects the goals of the NWI (Avey and Harvey 2014). The perception of implementation of the NRM Act is that it replaced the prior locally driven planning reflected in the catchment boards with a top-down approach (Mitchell 2014). Studies of the perceived effectiveness of the change in South Australia to integrated management suggest that some legitimacy was lost in the process by eliminating longstanding community relations with board representatives from the prior governance entities (Mitchell 2014). Over time, it is possible that the top-down approach will decrease local capacity to respond to a water crisis. In the short term, the loss of legitimacy may be a factor in some of the backlash to regulation since the end of the Millennium Drought.

Groundwater law reform in Australia is a secondary consequence of drought. Groundwater is the storage that irrigators turn to in the face of drought. The need for reform in groundwater law stems from the fact that, if groundwater is not managed appropriately, over pumping and, in some cases, irreversible loss of storage due to subsidence may occur. In addition, the historic tendency of the law to ignore connection between ground and surface water may result in over estimation of supplies and harm to ecosystems and people using groundwater-fed surface water. Over-drafting of aquifers for agriculture in response to drought has caught water managers by surprise in both Australia and the United States (Brodie et al. 2007; Dimick 2014). Lack of integrated management of surface water and groundwater has resulted in double counting of available water in both countries (Brodie et al. 2007; *American Falls Reservoir District No 2 v Idaho Department of Water Resources* 154 P3d 433 (2007)), with consequences for both surface water users (*American Falls Reservoir District No 2 v Idaho Department of Water Resources* 154 P3d 433 (2007)), and ecological features (Harrington and Cook 2014).

Groundwater management in Australia began with the common law of England that gave landowners the right to exploit the groundwater beneath their land (*Dunn v Collins* (1867) 1 SALR 126; McKay 2007). In the era of hand-dug wells, this was probably an adequate approach. With the development of modern drilling technology, and the adaptation of turbine pumps used in the oil industry for use in irrigation wells beginning in the 1940s (Ganzel n.d.), the right to capture whatever groundwater you could beneath your land has created significant third party impacts. Australia has seen a substantial increase in groundwater extraction since the 1980s (Harrington and Cook 2014). Between 1983 and 1997, groundwater extraction increased by 58% due to the combination of drought and caps on water use that applied to surface water but not groundwater (Brodie et al. 2007). The NWI sought to change this by bringing groundwater within the system for surface water regulation.

The NWI calls for: the treatment of connected surface water and groundwater as a single source (NWI clause 23(x)); integrated accounting where there is close interaction between streams and aquifers (NWI clause 82(iii)(b)); and recognition of the contribution of groundwater to environmental benefits (NWI clauses 5, 25, 78–79, Holley et al. 2016). The NWI also calls for the inclusion of groundwater in water plans (NWI clause 36) and, in particular, those plans intended to address over-allocated systems of both surface water and groundwater (NWI clause 26(i)). What is most notable about the NWI is not that groundwater is mentioned, but that its mention is pervasive. Every policy or action articulated in relation to surface water also applies to groundwater.

Individual states have discretion in how to implement the NWI, and the South Australian NRM Act once again provides an example. The NRM Act allows designation of prescribed areas in which plans are required to address the impact of groundwater extraction on water quantity and quality as well as ecosystems (NRMA Sect. 76). The Act authorises the Minister to impose reductions in both surface and groundwater diversion during drought if taking water from a surface source will impact groundwater quality or groundwater extraction will cause subsidence/aquifer collapse (NRM Sect. 132).

3.2 Drought and Water Law Reform in the Western United States

Similar to Australia, water allocation is a matter of state law in the United States. Historically, a water right was established simply by diversion and application to a beneficial use. Most western states have developed either an administrative or judicial system for the issuance of permits for the right to use water. Unlike Australia, most western states allocate water in times of scarcity in order of priority, with priority determined in order of the first development of the right or the issuance of a permit (see, e.g., *Idaho Statutes*, n.d., *California Water Code*, n.d.). As full allocation was reached on many western water systems, states recognized the need to record rights developed prior to the institution of permit systems. This has been accomplished in recent decades through massive adjudications in which all those claiming rights on a system must participate (see, e.g., Idaho Water Adjudications n.d., Montana Water Court n.d., Wyoming Bighorn Adjudication n. d., Arizona General Stream Adjudication n.d.). The adjudication of a water right is analogous to the determination of an entitlement in South Australia in that it determines a maximum right to water that will nevertheless vary according to water availability and beneficial use in any given year (see, e.g., *American Falls Reservoir District No 2 v Idaho Department of Water Resources,* 154 P3d 433 (2007), 446–449).

Despite the recognition of governmental authority to establish permit systems and adjudicate water rights, there is reluctance to recognise the authority of government to adjust allocation as circumstances change. Many states recognise state ownership of water (Trelease 1957, as an example of embedding the concept in the state constitution, see *Idaho Constitution* article XV), and clearly view the water right owner to hold a use right (referred to as usufructry right) (Tarlock 2014 [3.10] 'Usufructory nature of water rights'). The common law view of the nature of the private property interest is captured in an Idaho statute declaring that water in its natural state is 'the property of the state' and that the right to use water 'shall not be considered as being a property right in itself' but only as a 'complement' to 'the land or other thing' to which it is applied (*Idaho Water Use Act* Title 42, Sect. 101).

This would suggest that states have at least as much ability to regulate water allocation to prevent harm as they do with respect to land. At present, this is not the case. The U.S. Federal Circuit Court held that any governmental reduction in water use by a water right holder is a physical taking requiring compensation (*Casitas Municipal Water District v United States* 543 F3d 1276 (2008)). This interpretation of government authority over water limits the ability of government to use regulation to respond to variable water supply or changing needs. Interestingly, while recognising the same dichotomy between state ownership and private use rights as the United States, Australia interprets the government to have the authority to reduce water use rights in times of scarcity (*ICM Agriculture Pty Ltd v Commonwealth* (2009) 240 CLR 140; 84 ALJR 87; 170 LGERA 373; [2009] HCA 51 [55]; *Arnold v Minister Administering the Water Management Act 2000* (2010)

240 CLR 242; 84 ALJR 203; 172 LGERA 82; [2010] HCA 3. Note that both Australian cases involve licences to groundwater.).

Confusion between interpreting the rights of water users vis á vis the government that owns the water, and the relative rights between water users is one possible source of the strained interpretation of government authority over water in the United States (MacDonnell 2015). The courts also seem to struggle with the difficult problem of conceptualising a 'use' right. The case referenced above found a government 'taking' for any reduction in use (*Casitas Municipal Water District v United States* 543 F3d 1276 (2008)), and then awarded no compensation since no harm had been shown to beneficial use (*Casitas Municipal Water District v United States* 708 F3d 1340 (2013)). However, beneficial use is one of the elements defining a water right. Thus, the court confused the definition of the right with the finding of harm. Although not yet before the U.S. Supreme Court, this approach currently has a chilling effect on the assertion of government regulation.

Nevertheless, the exercise of emergency powers in the United States is exempt from the Constitutional requirement of compensation for a taking of a property right (Craig 2010b). Thus, California was able to exercise greater regulatory authority over water use by declaration of a drought emergency (Brown 2014). While emergency powers provide an avenue for drought response, it is not clear that such powers would be available during a long-term climate change scenario.

In the context of more slowly developing scarcity as a consequence of overdevelopment and relatively mild drought, states have just begun testing the common law concept that the 'beneficial use' of the water not only determines the type of use allowed, but the amount. What is beneficial must be reasonable under the circumstances, with scarcity, efficiency, and new technology being factors, among others (see, e.g., *American Falls Reservoir District No 2 v Idaho Department of Water Resources*, 154 P3d 433 (2007) 446-449, Benson 2012). While the concept of beneficial use may not upset the priority system, it does provide a tool for government to ensure that no water user faces reduced supplies while a more senior water user wastes water.

Despite the higher value placed on the 'property' nature of the right to water in the western United States, water transfers are heavily regulated, leading to much less robust water markets than developed in Australia's Murray-Darling Basin (see e.g. Holley and Sinclair's chapter in this volume). Western states in the United States allow the transfer of water rights generally through state-administered processes. The process used by Idaho is illustrative. A water user seeking to change the point of diversion, place, type or period of use, including in the process of marketing the water right, must apply to the Idaho Department of Water Resources (*Idaho Water Use Act* Title 42, Sect. 222). The department must conduct an inquiry into, among other things, impacts on other rights, the public interest and the conservation of water resources, and any person with an interest may file an objection (*Idaho Water Use Act* Title 42, Sect. 222(1)).

Although the individualised process for determining change in use avoids some of the secondary impacts of concern in Australia's process, it is also criticised for greatly increasing transaction costs and the timeframe required for a water transfer

(Colby 1990; MacDonnell 1990). The interdependency of irrigation regions due to shared conveyance facilities, return flow, and ecosystems that have grown dependent on the inefficiency of agricultural water application (Spagat 2015), have stood as barriers to numerous attempted transfers (Colby 1990; MacDonnell 1990). A number of states, including Idaho, have turned to water banks to reduce transaction costs (see, e.g., Idaho Statutes Title 42, Sect. 1761; see, generally, Clifford et al. 2004). Water banking reduces transaction costs by eliminating the need for a perfect match between buyer and seller. In essence, it provides a central paper repository for banked water that others may then withdraw from, similar to a financial bank. Water 'banked' is not subject to the rules requiring forfeiture of unused rights (see, e.g., *Idaho Water Use Act*, n 64, s 42-222(2) (regarding forfeiture), Sect. 42-223(5) (regarding exemption from forfeiture while banked)), thus water may be banked before the need to withdraw is identified. Additional means of reducing transaction costs while avoiding secondary impacts that the western United States should consider include redefining irrigation water rights to identify (unbundle) the portion of the right diverted that is actually consumed (Johnson et al. 1981).

Water planning in the western United States is much less extensive than in Australia and is generally a state-based rather than a watershed-based process. Watershed focused efforts develop on an ad hoc basis when conflict arises (Chaffin et al. 2014b). In the face of recent drought, California has taken a major step forward in the development and implementation of a State water action plan that seeks both integration and flexibility in water management. The California Water Action Plan has been developed to meet three broad objectives: more reliable water supplies; the restoration of important species and habitat; and a more resilient, sustainably managed water resources system (water supply, water quality, flood protection, and environment) that can better withstand inevitable and unforeseen pressures in the coming decades (California Natural Resources Agency et al. 2016). What is lacking in this state-led approach is the step toward devolution of specific watershed-based planning to the local level. The top-down approach adopted foregoes the benefits of local capacity building, buy-in, and the opportunity to tailor approaches to specific watersheds.

Similar to Australia, groundwater regulation in the United States began with adoption of the common law of England concept that a land owner had the right to exploit the water beneath their land, but has evolved in the western United States into three different doctrines. The common law approach continues to be followed by Texas, but now with an overlay of regional regulation (*Houston & Texas Central Railroad Co v East,* 98 Tex 146 (1904); Texas Water Code, Sect. 35.001). A number of states apply the doctrine of prior appropriation to groundwater, including Idaho, Montana, Nevada, New Mexico, Utah, Washington and Wyoming (Bryner and Purcell 2003), thus managing shortage in order of the date on which the particular groundwater right was first developed. Even before the recent drought, several states had experienced conflict over aquifer use and begun to experiment with governance approaches to allow more intensive regulation in areas of concern. Thus, Arizona allows designation of an active groundwater management area (AMA), generally in a region in which severe overdraft is already occurring. Within an AMA, substantial restrictions in use, and in particular on new use, may be imposed (*Arizona Revised Statutes*, Title 45, article 2 'Arizona Groundwater Code'). Montana allows designation of a controlled groundwater area in which limits may be placed on new development and pumping may be regulated without adherence to the doctrine of prior appropriation (*Montana Code*, Sect. 85-2-506). Montana's controlled groundwater designation has been used adjacent to Yellowstone National Park to prevent harm to the park's spectacular hydrothermal system, including through loss of recharge (*Montana Code*, Sect. 85-20-401 article IV 'Montana-National Park Service Compact').

Until 2014, California followed the doctrine of correlative rights under which a landowner has the right of access to groundwater beneath their land, but that right is tempered by the rule that the use must be reasonable and shortage is shared in proportion to ownership of land overlying a shared aquifer (*Katz v Walkinshaw*, 141 Cal 116 (1903)). Unfortunately, mere decline in the water table is not considered "shortage" and, in the recent extended drought, irrigators in California have chased falling aquifers resulting in levels of subsidence so substantial as to be measurable using remote sensing (Famiglietti et al. 2014). In response, California's legislature passed the *Sustainable Groundwater Management Act* (SGMA) in 2014 (California Assembly Bill 1739, SB 1168, 1319, 2014, codified in *California Water Code*, Sect. 10720 et seq.; see Kiparsky et al. 2016). The SGMA requires development of local sustainable groundwater management plans in areas of overdraft, leaving the formation of the appropriate management entity to local control. Nevertheless, the SGMA allows the state to step in if the plan is not developed, is inadequate, or if impacts to senior surface water rights occur. The approach of the SGMA can be characterised as government assisted self-organisation. It remains to be seen if this approach will facilitate local decisions on tradeoffs.

4 Discussion and Conclusion

Long-term drought led to water supply crises in both Australia and the Western United States, providing a theatre for experimentation on responses to climate change. Viewing their experiments in water reforms as occurring within a framework of three approaches to governance of the water commons—government regulation, marketable property rights, and self-organisation—allows the following critique of the adaptive capacity of these regions to climate change.

In response to the Millennium Drought, Australia increased the role of government regulation, defined limited property rights to water and 'unbundled' water from the land allowing water transfer independent of land sale. The recognition of the authority of government as owner of the water to adjust water allocations increased the capacity to use government regulation to adapt quickly and to respond to greater variability in water supply. In addition, Australia devoted substantial resources to planning. These measures substantially increased Australia's ability to cope with long-term drought. Nevertheless, minor adjustments to the approach would substantially increase Australia's adaptive capacity.

By limiting restrictions on water transfers to compliance with the water plan, Australia substantially reduced the transaction costs of water transfers. However, in avoiding even an expedited inquiry into the third party and ecological impacts of the transfer, the approach opens the door to substantial harm to irrigation communities and water dependent ecosystems. In addition to the potential impact on individual irrigators who share delivery systems, water transfers are frequently pursued through implementation of efficiency improvements and transfer of 'saved' water. The problem with this approach is that not all efficiency improvements achieve savings. Up to a certain point, increased efficiency increases crop yield. Increased yield comes at the expense of increased water consumption and corresponding reductions in return flow (Young 2011; MacDonnell 2012).

Water planning moved from a locally driven catchment process to a top-down approach under the National Water Initiative. What is lost in this more efficient and controlled approach is the opportunity to build local capacity, integrate local knowledge in the plan, and tighten feedback loops between the water system and those who play a role in planning.

The western United States, with California in the lead, has also increased its adaptive capacity. Use of emergency powers and a tightening in the interpretation of 'beneficial use' provide avenues for government intervention. Both California and Idaho are experimenting with a combination of government regulation and local self-organisation to address the over-development of aquifers and the historic failure to treat connected surface and ground water as a single water source. The success of California's experimentation with government assisted local development of groundwater management plans is yet to be determined. Planning is not an adequate intervention in the midst of an emergency. However, the local capacity building that must occur to develop plans, and the plans themselves, will place California in a better position when the next long-term drought occurs. Nevertheless, the western United States has substantial room for improvement.

The lack of clarity and principled development of the application of the U.S. Constitution's 'Takings' clause to regulation of water chills intervention through government regulation. The use of emergency powers may not be available in the slow unfolding of climate change. The high transaction costs of the transfer-by-transfer inquiry undertaken by western states places them at the other extreme from Australia. Development of an expedited process for use during drought and greater use of water banks is needed. Finally, except for the California experiment in locally driven groundwater planning, water planning in western states remains at the state level. As in Australia, this misses the opportunity to build local capacity, integrate local knowledge in the plan, and tighten feedback loops between the water system and those who play a role in planning.

Developing the authority and capacity to use all three approaches to governance of the water commons is the first step. The second step requires careful consideration of when each tool is appropriate. Australia's response to the Millennium Drought illustrates the need for governmental intervention during crisis. It also illustrates that if that intervention is not based on local capacity building and planning that is already in place when crisis occurs, errors may occur in the approach to intervention and there may be backlash when the crisis ends (see e.g. Carmody's chapter in this volume). The massive subsidence due to groundwater withdrawal in parts of California also illustrates that if planning and an approach to regulation is not in place when crisis occurs, it will be too late. The experience of both countries illustrates the substantial adaptive capacity of water markets. Yet the huge transaction costs in the United States dampen that adaptive capacity, and the avoidance of inquiry into impacts in Australia could lead to irreversible harm. These experiments illustrate that we must build into our efforts at water reform: (1) an adaptive law approach by allowing government regulation while balancing local versus State control; (2) a degree of regulation on markets; and (3) investment in planning to build local capacity to adjust with the water cycle.

Finally, it would be incomplete to end this analysis of water reform without acknowledging that legal reform alone is insufficient. The massive investment in infrastructure undertaken in Australia during the Millennium Drought has no parallel in the United States (with the caveat that 10 more years of drought might have changed that) (Turner et al. 2016). In a non-stationary world in which water supply and demand will not only change but also fluctuate beyond historic extremes, governance reform, investment, and new technology are all essential. Indeed, we are past the point that restoration of historic conditions is possible, and quickly learning that sustaining the status quo is also not an option. We must embrace a new paradigm of reconciliation—a rethinking of the interactions between society and the natural systems. Reconciling our needs with the planet's limits will require investment in new technology and the development of governance that is itself adaptive to our evolving water social-ecological systems.

References

Arizona general stream adjudication. (n.d.). https://www.superiorcourt.maricopa.gov/SuperiorCourt/GeneralStreamAdjudication/Index.asp. Accessed 19 Mar 2017.

Arnold, C. A. (Tony). (2010). Adaptive watershed planning and climate change. *Environmental and Energy Law and Policy Journal, 5*, 417–488.

Avey, S., & Harvey, D. (2014). How water scientists and lawyers can work together: A "Down Under" solution to a water resource management problem. *Journal of Water Law, 24*, 45.

Benson, R. (2012). Alive but irrelevant: The prior appropriation doctrine in today's western water law. *Colorado Law Review, 83*(3), 674–714.

Brodie, R., Sundaram, B., Tottenham, R., Hostetler, S., & Ransley, T. (2007). An adaptive management framework for connected groundwater-surface water resources in Australia. Bureau of Rural Sciences, Canberra. http://www.southwestnrm.org.au/sites/default/files/uploads/ihub/brodie-r-et-al-2007-adaptive-management-framework-connected-groundwater.pdf. Accessed 20 Mar 2017.

Brown, E. G. Jr. (Governor of California). (2014). A proclamation of a state of emergency. https://www.gov.ca.gov/news.php?id=18368. Accessed 25 Mar 2017.

Brunner, R. D., Colburn, C. H., Cromley, C. M., Klein, R. A., & Olson, E. A. et al. (Eds.). (2005). *Adaptive governance: Integrating science, policy, and decision making.* New York: Columbia University Press.

Bryner, G., & Purcell, E. (2003). Groundwater law sourcebook of the Western United States. Natural Resources Law Center, University of Colorado. http://scholar.law.colorado.edu/books_reports_studies/74/. Accessed 18 Mar 2018.

California Natural Resources Agency, California Environmental Protection Agency and California Department of Food and Agriculture. 2016 update of January 2014 Plan. California Water Action Plan 2016 Update http://resources.ca.gov/docs/california_water_action_plan/Final_California_Water_Action_Plan.pdf. Accessed 23 Mar 2017.

California Water Code. (n.d.). http://leginfo.legislature.ca.gov/faces/codesTOCSelected.xhtml?tocCode=wat. Accessed 18 Mar 2018.

Chaffin, B. C., Gosnell, H., & Cosens, B. (2014a). A decade of adaptive governance scholarship: Synthesis and future directions. *Ecology and Society, 19*(3), 56. https://doi.org/10.5751/ES-06824-190356.

Chaffin, B. C., Craig, R. K., & Gosnell, H. (2014b). Resilience, adaptation, and transformation in the Klamath River Basin social-ecological system. *Idaho Law Review, 51,* 157–193.

Clifford, P., Landry, C., & Larsen-Hayden, A. (2004). Analysis of water banks in the western Sstates. (Report for the Washington Department of Ecology, July 2004) https://fortress.wa.gov/ecy/publications/publications/0411011.pdf. Accessed 14 Mar 2017.

Colby, B. G. (1990). Transaction costs and Efficiency in western water allocation. *American Journal of Agricultural Economics,* 1184–1192.

Cosens, B. (2016). Water law reform in the face of climate change: Learning from drought in Australia and the western United States. *Environmental and Planning Law Journal, 33,* 372–387.

Cosens, B. A., Craig, R. K., Hirsch, S., Arnold, C. A., Benson, M. H., DeCaro, D. A., Garmestani, A. S., Gosnell, H., Ruhl, J., & Schlager, E. (2017). The role of law in adaptive governance. Ecology and Society 22(1):30. https://doi.org/10.5751/ES-08731-220130 in Chaffin, B., Gunderson, G., & Cosens, B. (Eds.). Special issue in Ecology and Society: Practicing panarchy: Assessing legal flexibility, ecological resilience and adaptive governance in U.S. regional water systems experiencing climate change. https://www.ecologyandsociety.org/issues/view.php?sf=122 . Accessed 18 Mar 2018.

Craig, R. K. (2010a). Stationarity is dead: Long live transformation: Five principles for climate change adaptation law. *Harvard Environmental Law Review, 34*(1), 9–75.

Craig, R. K. (2010b). Adapting water law to public necessity: Reframing climate change adaptation as emergency response and preparedness. *Vermont Journal of Environmental Law, 11,* 709–756.

Dellapenna, J. W. (2000). The importance of getting names right: The myth of markets for water. *William and Mary Environmental Law and Policy Review, 25*(2), 317–377.

Dietz, T., Ostrom, E., & Stern, P. C. (2003). The struggle to govern the commons. *Science, 302,* 1907–1912. https://doi.org/10.1126/science.1091015.

Dimick, D. (2014). If you think the water crisis can't get worse, wait until the aquifers are drained. National Geographic News. http://news.nationalgeographic.com/news/2014/08/140819-groundwater-california-drought-aquifers-hidden-crisis/. Accessed 20 Mar 2017.

Famiglietti, J. S., Thomas, B. F., Reager, J. T., Castle, S. L., David, C. H., Thomas, A. C., et al. (2014). Satellite observations of epic California drought. Abstracts AGU Fall Meeting, 15–19 December 2014. <https://agu.confex.com/agu/fm14/meetingapp.cgi#Paper/14433.

Field, C. B., Barros, V. R., Dokken, D. J., Mach, K. J., Mastrandrea, M. D., Bilir, T. E., et al. (Eds.). (2014). *Climate change 2014: Impacts, adaptation, and vulnerability. Part A: Global and sectoral aspects. Contribution of working group II to the fifth assessment report of the intergovernmental panel on climate change,* New York: Cambridge University Press.

Folke, C., Hahn, T., Olsson, P., & Norberg, J. (2005). Adaptive governance of social-ecological systems. *Annual Review of Environmental Resources, 30,* 441–473. https://doi.org/10.1146/annurev.energy.30.050504.144511.

Ganzel, B. (n.d.) Irrigation pumps. Farming in the 1940s: Wessels living history farm. http://www.livinghistoryfarm.org/farmlnginthe40s/water_04.html. Accessed 22 Mar 2017.

Gunderson, L. H., & Light, S. S. (2006). Adaptive management and adaptive governance in the Everglades ecosystem. *Policy Sciences, 39,* 323–334. https://doi.org/10.1007/s11077-006-9027-2.

Hardin, G. (1968). The Tragedy of the Commons. *Science, 162,* 1243–1248.

Harrington N., & Cook, P. (2014). Groundwater in Australia. National Centre for Groundwater Research and Training, Australia. http://www.groundwater.com.au/media/W1siZiIsIjIwMTQvMDMvMjUvMDFfNTFfMTNfMTMzX0dyb3VuZHdhdGVyX2luX0F1c3RyYWxpYV9GSU5BTF9mb3Jfd2ViLnBkZg/Groundwater%20in%20Australia_FINAL%20for%20web.pdf. Accessed 14 Mar 2017.

Holley, C., Sinclair, D., Lopez-Gunn, E., & Schlarger, E. (2016). Collective management of groundwater. In T. Jakeman, O. Barreteau, R. Hunt, J.-D. Rinaudo, & A. Ross (Eds.), *Integrated groundwater management: Concepts, approaches and challenges.* Springer.

Idaho Statutes. (n.d.). Idaho Water Use Act and Groundwater Act Title 42. https://legislature.idaho.gov/statutesrules/idstat/Title42/. Accessed 18 Mar 2018.

Idaho Water Adjudications. (n.d.). https://www.idwr.idaho.gov/water-rights/adjudication/. Accessed 18 Mar 2018.

Jiménez Cisneros, B. E., Oki, T., Arnell, N. W., Benito, G., Cogley, J. G., Döll, P., et al. (2014). Freshwater resources. In C. B. Field, V. R. Barros, D. J. Dokken, K. J. Mach, M. D. Mastrandrea, T. E. Bilir, M. Chatterjee, K. L. Ebi, Y. O. Estrada, R. C. Genova, B. Girma, E. S. Kissel, A. N. Levy, S. MacCracken, P. R. Mastrandrea, & L. L. White (Eds.), *Climate change 2014: Impacts, adaptation, and vulnerability. Part A: Global and sectoral aspects. Contribution of working group II to the Fifth Assessment Report of the Intergovernmental Panel on Climate Change* (pp. 229–269). New York: Cambridge University Press.

Johnson, R. N., Gisser, M., & Werner, M. (1981). The definition of a surface water right and transferability. *Journal of Law and Economics, 24*(2), 273–288.

Karkkainen, B. C. (2004). "New governance" in legal thought and in the world: Some splitting as antidote to overzealous lumping. *Minnesota Law Review, 89,* 471–497.

Kiparsky, M., Owen, D., Nylen, N. G., Christian-Smith, J., Cosens, B., Doremus, H., et al. (2016). Designing effective groundwater sustainability agencies: Criteria for evaluation of local options. Report of the Wheeler Water Institute, Center for Law, Energy and the Environment, University of California, Berkeley, School of Law, Berkeley, CA, USA. https://www.law.berkeley.edu/research/clee/research/wheeler/groundwater-governance-criteria/. Accessed 19 Mar 2017.

MacDonnell, L. J. (1990). The water transfer process as a management option for meeting changing water demands. Vol I. USGS Grant Award No 14-08-0001-G1538, University of Colorado.

MacDonnell, L. J. (2012). Montana v. Wyoming: Sprinklers, irrigation water use efficiency and the doctrine of recapture. *Golden Gate University Environmental Law Journal, 5,* 265–295.

MacDonnell, L. (2015). Prior appropriation: A reassessment. *University of Denver Water Law Review, 18,* 228–310.

McKay, J. (2007). Groundwater as the Cinderella of water laws, policies, and institutions in Australia. In G. Bergkamp, & J. McKay (Eds.), *The global importance of groundwater in the 21st Century: Proceedings of the International Symposium on Groundwater Sustainability.* National Groundwater Association International Symposium on Groundwater Sustainability 2007, IUCN-UNESCO, Spain. (pp. 317–331).

McKay, J. (2011). Water markets and yrading. In C. Davis, & B. Swinton (Eds.), *Securing Australia's water future.* Focus Publishing.

Milly, P. C., Betancourt, J., Falkenmark, J., Hirsch, M., Kundzewicz, R. M., Lettenmaier, Dennis P., et al. (2008). Stationarity is dead: Whither water management? *Science, 319,* 573–574.

Mitchell, B. (2014). Evolving regional, integrated and engagement approaches for natural resources management in South Australia. Report as part of the ANZSOG-Goyder Institute Visiting Professor in Public Sector Policy and Management Program.

Montana Water Court. (n.d.). Montana water court. http://courts.mt.gov/courts/water. Accessed 18 Mar 2018.

Olsson, P., Gunderson, L. H., Carpenter, S. R., Ryan, P., Lebel, L., & Folke, C., et al. (2006). Shooting the rapids: Navigating transitions to adaptive governance of social-ecological systems, *Ecology and Society*, *11*(1), 18. http://www.ecologyandsociety.org/vol11/iss1/art18/. Accessed 18 Mar. 2018

Orts, Eric W. (1995). Reflexive environmental law. *Northwestern University Law Review*, *89*, 1227–1340.

Ostrom, E. (1990). *Governing the Commons: The evolution of institutions for collective action*. Cambridge University Press.

Ostrom, E. (2009). A general framework for analyzing sustainability of social-ecological systems. *Science*, *325*, 419–422.

Smith, A. (1904). *An inquiry into the nature and causes of the wealth of nations* (5th ed). Methuen & Co Ltd.

South Australia. (n.d.). Water Permit and Licence Register. Water Connect https://www.waterconnect.sa.gov.au/Systems/WLPR/Pages/default.aspx.

Spagat, E. (2015). Salton Sea, California's largest lake, threatened by urban water transfer. *89.3KPCC*, http://www.scpr.org/news/2015/06/03/52175/salton-sea-california-s-largest-lake-threatened-by/. Accessed 17 Mar 2017.

Stoeckel, K., Webb, R., Woodward, L., & Hankinson, A. (2012). *Australian water law*. Thomson Reuters.

Tarlock, A. D. (2014). *Law of water rights and resources*. Thomson Reuters Publications.

Trelease, F. J. (1957). Government ownership and trusteeship of water. *California Law Review*, *45*, 638–654.

Turner, A., White, S., Chong, J., Dickinson, M.A., Cooley, H., & Donnelly. K. (2016). *Managing drought: Learning from Australia*. prepared by the Alliance for Water Efficiency.

United Nations. (2017). World Population Prospects: The 2017 Revision. https://www.un.org/development/desa/publications/world-population-prospects-the-2017-revision.html

United Nations General Assembly. (2010). Resolution A/RES/64/292 The Human Right to Water and Sanitation.

Van Dijk, A. I. J. M., Beck, H. E., Crosbie, R. S., de Jeu, R. A. M., Liu, Y. Y., Podger, G. M., et al. (2013). The Millennium Drought in southeast Australia (2001–2009): Natural and human causes and implications for water resources, ecosystems, economy, and society. *Water Resources Research*, *49*(2), 1040–1057. http://onlinelibrary.wiley.com/doi/10.1002/wrcr.20123/full.

Williams, A. P., Seager, R., Abatzoglou, J. T., Cook, B. I., Smerdon, J. E., & Cook, E. R. (2015). Contribution of anthropogenic warming to California drought during 2012–2014. *Geophysical Research Letters*, *42*, 6819–6828.

Wyoming Bighorn Adjudication. (n.d.). http://bhrac.courts.state.wy.us/. Accessed 18 Mar 2018.

Young, M. (2011). The role of the unbundling water rights in Australia's Southern Connected Murray Darling Basin, Evaluating Economic Instruments for Sustainable Water Management in Europe. IBE Review Report No D6.1, 19 Dec 2011 http://www.feem-project.net/epiwater/docs/d32-d6-1/CS23_Australia.pdf. Accessed 22 Mar 2017.

Printed by Printforce, the Netherlands